DIALOGUES AMONG CIVILIZATIONS AND CULTURES

HOLOCENE

PERSPECTIVES, ENVIRONMENTAL DYNAMICS AND IMPACT EVENTS

Dialogues among Civilizations and Cultures

Additional books in this series can be found on Nova's website under the Series tab.

Earth Sciences in the 21st Century

Additional E-books in this series can be found on Nova's website under the E-book tab.

CONTENTS

Editor's Note

The Holocene (11.5 ka BP onwards), directly linked to the evolution of modern society, encompasses increase and impacts of the human species world-wide, including by and large striking changeover towards living in the present. The Holocene climate is also responsible for expansion/ migration/decline of the human cultures. Because of this, a period termed as Anthropocene (age of man) is also proposed as the latest part of this epoch since the recent human impacts have been found on the Earth and such impacts are unquestionably of global implication for our future.

Holocene climate has been very uneven, and that there are multiple controls that must have been responsible for this variability. Although a number of the Holocene Impact Working Group scientists from Australia, France, Ireland, Russia and USA and many other countries hypothesize that the meteorite impacts on Earth are more common than were thought before, yet the extreme climatic events and their impacts are not known accurately. Taking this into consideration, a widespread research is now focusing on abrupt Holocene climatic changes at millennial and centennial scales all over the world by using a number of archives and proxies.

The climate during the Holocene was never the same all through the globe. Generally speaking, the Holocene Climate Optimum (HCO) was a warm period during roughly the interval 9,000 to 5,000 yr BP during which the northwest Europe experienced warming, while there was cooling in the south. The signature of the 8.2 ka event, one of the most prominent cold excursions in the Holocene is observed between 8,300 and 6,000 yr BP, shown by the abrupt increase in $\delta^{13}C$ values and low troughs in susceptibility. It is the major rapid climate change event of the Holocene and has been documented by a number of studies, such as marine foraminifera, lacustrine sediments, speleothems, geochemistry, mineral magnetism, pollen and ice-cores etc.

With rising $\delta^{13}C$ values seems to be significant for climatic interpretation, particularly for the 4.2 ka event. Variation in sensitivity of sites and proxies, reliability of dating techniques and most importantly the coarser sampling intervals with very low resolution data in some parts especially in southwest Asia may account for the paucity of records of some of these events, especially those with a very short duration. Speleothems are one of the best continental archives which provide comparable long-spanning, high-resolution and absolutely dated time series of the palaeoenvironmental variables. The dating of stalagmites by U/Th offers a reliable chronology for the period of stalagmite formation. For a sub-sample of about 100mg and ^{238}U concentration of about 1 ppm, current TIMS and MC-ICP-MS capabilities translate to precise ages with small 2σ error such as 500±8 years, 10,000±45 years, 50,000±180 years, 500,000±15,000 years--clearly expressing that the speleothems are the most ideal candidates to reconstruct the past climatic changes within adequate time duration.

Similarly, it is widely believed that the LIA (Little Ice Age) has impacted both the Hemispheres but perhaps not in a globally synchronous way because of the uncertainty about the inception and duration of this event in different parts of the globe. For example, variations in general circulation result in positive anomalies in some regions at times when negative anomalies are recorded elsewhere The proxy data propose drier environment in the northern tropics while wetter conditions in the southern tropics due to the southward shift of the Inter Tropical Convergence Zone. Also, the hydrological conditions during the LIA show variation across different regions.

Undoubtedly, the Holocene events are observed throughout and can be correlated with global events. There may be slight variation in commencement and duration of the events from east to west, which can be attributed to the latitude location, response time, temporal resolution and also non-climate factors (e.g., tectonics). The only aspect remains is to exercise the most dependable archives and proxies with annual to multi-annual resolution to find out the periodicities of the climate change and that will have a straight application to the modern society.

This volume contains seventeen invaluable manuscripts submitted by the authors from diverse countries (e.g., U.K., U.S.A., Spain, Russia, India, France, Portugal and Japan). While most of them are related to the climatic events and environmental dynamics during the Holocene, two each focus on the palaeoseismology and human migration/anatomy.

I am personally obliged to the NOVA publishers (U.S.A.) for giving me this opportunity to edit this volume and bestowing confidence in me. I am exceedingly gratified to Mr. Lalit Mohan Joshi whose untiring help in organizing this volume made me stress-free. My gratitude to the ISRO-GBP, Ahmedabad, Centre of Advanced Study in Geology and Uttarakhand Centre on Climate Change of the Kumaun University for financial support and requisite amenities to create this volume. I am also thankful to all the contributors who without delay submitted their manuscripts and also to the reviewers who replied promptly and punctually. I firmly believe that the volume will be remarkably precious to those working on the Holocene climate and events.

B. S. Kotlia
Editor, NOVA Holocene volume
Co-Coordinator, Uttarakhand Centre on Climate Change
Centre of Advanced Study, Department of Geology
The Durham, Kumaun University, Nainital, India, 263002
E mail: bahadur.kotlia@gmail.com
31st March, 2012

In: Holocene
Editor: Bahadur Singh Kotlia

ISBN: 978-1-62257-722-4

Chapter 1

HOLOCENE: DEFINITION AND CURRENT STRATIGRAPHIC STATUS IN THE GEOLOGICAL TIME SCALE

Devesh K. Sinha*[*1], Ashutosh K. Singh[1], Kirtiranjan Mallick[2] and Vikram Pratap Singh[1]

[1]Department of Geology, Centre of Advanced Studies,
University of Delhi, Delhi-110007 India
[2]Department of Geology, Utkal University, Bhubaneshwar-751004 India

ABSTRACT

Holocene is the youngest Epoch/Series in the recent Geological Time Scale which spans to the present. This is the only formal chronostratigraphic division in the geological time-scale whose boundary (the Pleistocene–Holocene boundary) is defined climatostratigraphically in an ice core in contrast to other Phanerozoic Epoch boundaries which have been defined biostratigraphically in the sedimentary rock record. The base of the Holocene is fixed in the North Greenland Ice Core Project (NGRIP) ice-core record at the horizon showing the clearest signal of climatic warming, an event that marks the end of the last cold episode (Younger Dryas Stadial/Greenland Stadial 1) of the Pleistocene in calendar years, the base of the Holocene would be close to 11,700 yrs BP (Before Present refers to before 1950 CE). Till date, there have been no proposals to formally subdivide the Holocene Epoch/ Series into smaller intervals and as a result , casual usage of "early", "middle" and "late Holocene" has been much prevalent. In this chapter, we propose that in the absence of any significant faunal events in marine as well as the terrestrial records, like its base, the subdivisions of the Holocene be made chronometrically with a climatostratigraphic bias and should roughly correspond to major climate change in the North Atlantic. The North Atlantic climate changes have also been found to affect the strength of the Asian Monsoon. We propose that the Holocene be subdivided into an Early, Middle and Late Holocene based on the major changes in North Atlantic climate, e.g., Bond events. The boundaries may correspond roughly to the Bond Events which mark significant change in the thermohaline circulation and thus would be possible to

* E mail: deveshksinha@yahoo.com.

determine in the marine records worldwide. This will also to a great extent minimize the problem of casual usage of the early, middle and late terms as adjective to Holocene.

Keywords: Geological Time Scale, Holocene, Younger Drays, Bond events, Asian Monsoon

INTRODUCTION

Holocene distinguishes it from other formal intervals of the Geological time scale by its continuance in the modern time. It has also got a unique distinction of perhaps the most extensively studied geological time interval and offers best resolved archives of the interaction between marine terrestrial, cryospheric and atmospheric processes (Lowe and Walker, 1997; Walker et al., 2009). Holocene also offers the scientists a unique opportunity to observe the "process" and "product" relationship in the geological context. The interval provides us a unique scenario where there is scope of establishing a variety of empirical relationship between observed processes and their products, which otherwise becomes much speculative with more and more old age. For example, a ripple mark left recently by river can be related to its "known" direction (Reineck and Singh, 1973), a change in the lacustrine biota can be related to "known" changing climatic conditions (Rosenzweig et al., 2007), the formation of modern sand dunes and their migration can be related to known wind speed and direction (Gary Nichols, 1999), the current movements of the plates and tectonic movements within the plates can be measured using Global Positioning System and can be related to the structural features; and the modern sediment flux in oceans at various depths and locations can be related to physico-chemical parameters of the ocean water masses and changing climate (Joint Global Ocean Flux Studies (JGOFS); Curry et al., 1992). All the above examples strengthen our interpretation of the processes based on our investigations of the "products" applying cautiously the Huttonian Principle of "Present is the key to the past". A stratigraphic record of older ages has to be interpreted in terms of our understanding of the present processes but many a times one does not encounter a typical modern analogue of the old record and thus interpretation often involves subjectivity. The geological literature is full of dual interpretations of the same stratigraphic record. But the interpretation of the Holocene stratigraphic record tends to be substantially objective because the processes involved in producing the record can be observed. The science of Geology dealing with the time of origin of the Earth till present day is largely based on the multitude of "products" resulting from "various geological processes" interplaying independently and many a times interacting with each other. This connectivity is both the beauty of the Geology as a science as well as the problem while interpretations are made. Most of our interpretations do have to follow the Huttonian Uniformitarianism though with caution. The major handicap for the geologists is to infer the processes with the help of product which is nothing but a "geological record'. Holocene is a geological interval where we see the products as well the processes.

The study of Geology is principally limited to the study of the chronology and sequential order of events in the Earth's history. The "geological events" are defined and interpreted based on material and convincing evidences left in the rocks by various kinds of geological activities in different time periods. The stratigraphic record of Holocene is a store house of various geological processes including changes in ocean circulation, sea-level changes, climate change before and after the intensive anthropogenic activities, geomorphological processes, changes in cryosphere, geophysical processes, vegetational changes, faunal relocation and finally the effect of modern civilization. Despite this rich and varied palaeoenvironmental archive, and in spite of the fact that the Holocene has long been recognized as a stratigraphic unit of "Series" status a definition of the base of the Holocene (the Pleistocene– Holocene boundary) has only recently been formally ratified by the International Union of Geological Sciences (Walker et al., 2009).

Holocene is the only geological time interval whose fate also depends on one of its species "*Homo sapiens sapiens*" as we are aware of the pressure put by us on the Holocene environment. It provides us a typical time interval to be studied from time it was not besmirched by human influence to the modern time when much of the environmental changes are being attributed to anthropogenic activities.

Definition of Holocene and Its Relation with Quaternary

The term Holocene was first proposed by Gervais in 1867. It comes from the Greek words holos (meaning "whole") and cene (meaning "recent"). Holocene is the youngest subdivision having the rank of Epoch/Series in the Geological Time Scale (GTS-2009-Walker, Geissman and compilers, 2009). Chronostratigraphically, it is placed above the Calabrian Stage. In the GTS-2009, the Holocene Series/Epoch has been placed as the later/upper part of the Quaternary System/Period, though much debate still prevails on the status of the Quaternary itself (e.g., Van Couvering et al., 2009). Though, two more formal Stages namely Ionian Stage and Tarantian Stage in ascending stratigraphic order, have been proposed above Calabrian and below Holocene, they are still to be ratified by the IUGS. Following suggestions from a large number of workers, the base of the Quaternary has been made coincident with the base of Pleistocene and is now been placed at the base of Gelasian Stage.

The base of the Gelasian Stage is defined as the base of the marly layer overlying Sapropel MPRS 250, located at 62m in Monte San Nicola section, Italy. The astronomical age of Mediterranean Precession Related Sapropels (MPRS)-250 (mid-point), corresponding to precessional cycle 250 from the present, is 2.588 Ma BP, which can be assumed as the age of the boundary (Rio et al., 1998). This definition of the base of Quaternary is within the Marine Isotope Stage 103 and has been accepted by the International Commission on Stratigraphy (ICS).

The base of the Quaternary and Pleistocene can be approximated biostratigraphically by the last occurrence of *Discoaster pentaradiatus* which occurs in most low and mid-latitude areas close to isotopic stage 99, some 80 ka above the boundary (Rio et al., 1998) however there are large number of planktic foraminiferal events which have been recorded close to the Matuyama Gauss boundary like *Globorotalia inflata* FAD; *Globorotalia tosaensis* FAD; *Dentoglobigerina altispira* LAD; *Globorotalia plesiotumida* LAD; *Globorotalia merotumida* LAD (Sinha and Singh, 2008). This new definition of Quaternary and Pleistocene has stretched the lower limit of both which was earlier 1.88 Ma BP. The upper limit of the Quaternary and Pleistocene is defined by the base of Holocene.

Table 1. Details of Global Stratotype Section and Point for Pleistocene and Holocene (Taken from International Commission on Stratigraphy, 2009). http://www.stratigraphy.org/column.php?id=GSSPs

Stage	Age (Ma) CGTS 2008	GSSP Location	Latitude Longitude	Boundary Level	Correlation Events	Status	References
Phanerozoic Eon							
Cenozoic Era							
Quaternary System							
Holocene Series							
Holocene Stage	0.011784	North GRIP Ice Core, Central Greenland	$75^0$10.00' N $42^0$30.00' W	1492.45m depth in Borehole NGRIP2	Climatic -end of the Younger Dryas, Cold Spell which is reflected in a shift in deuterium excess values	Ratified 2008	Episodes 31/2, 2008. J Quaternary Sci. vol. 24.pp.3-17, 2009
Pleistocene Series							
Tarantian Stage	0.126	Amsterdam Terminal Bore Hole Netherlands	$52^0$37.92N $04^0$91.44E	63.5m below surface	Climatic—Base of the Eemian interglacial stage (= base of marine isotope stage 5e) before final glacial episode of Pleistocene. Base of Tyrrhrenian regional stage of Mediterranean	Accepted by ICS in 2008, on hold by IUGS	Episodes 31/2, 2008
Ionian Stage	0.781	Candidate Sections in Italy Montalbano Jorica or Valle de Manche) and Japan (Chiba)			Magnetic—Brunhes-Matuyama magnetic reversal (base of chron 1n)	Anticipated 2009	
Calabrian Stage	1.806	Vrica Italy	$39^0$03.85N $17^0$13.48E	Base of the marine claystone overlying the sapropelic marker bed"e"(Mediterranean precision related sapropel , MPRS 176)	Magnetic—15 kyr after end of Olduvai (C2n) Normal Polarity Chron	Ratified 1985	Episodes 8/2, pages, 116-120, 1985
Gelasian Stage	2.588	Monte San Nicola, Sicily Italy	$37^0$1469N $14^0$20.35W	Base of marly layer overlying sparopel MPRS 250	Magnetic—Matuyama/Gauss boundary (C2r/C2An) is 1m below GSSP. GSSP level is within Marine isotope Stage 103	Ratified 1996	Episodes 21/2, pp.82-87, 1998

Global Stratotype Section and Point (GSSP) for Holocene

The Holocene, our present interglacial period, (i.e., the last 10,000 years), though considered to be a period of relatively unvarying warmth, was punctuated by intervals of abrupt changes such as the 8.2 ka BP cold event (Mayewski et al., 1994; Alley et al., 1997) and Bond events (Bond et al., 1997). The Holocene is the only interval in the geological time-scale whose boundary (Pleistocene–Holocene boundary) is defined climatostratigraphically in contrast to other Phanerozoic Epoch boundaries which have been defined biostratigraphically. The base of the Holocene is fixed in the North Greenland Ice Core Project (NGRIP) ice-core record at the horizon showing the clearest signal of climatic warming, an event that marks the end of the last cold episode (Younger Dryas Stadial/Greenland Stadial 1) of the Pleistocene (Walker et al, 2009). In calendar years, the base of the Holocene would be close to 11,700 yrs BP (Before Present refers to before 1950 CE). Table 1 gives the current status of Holocene with respect to the Quaternary.

Other Terms Used for Holocene

Flandrian and Versilian

The geological time scale has always been flooded with introduction of terminologies based on local or regional events without much acknowledgement to the standard stratigraphic code. The Flandrian interglacial Stage has been frequently used by the scientists in northern Europe as a synonym for Holocene. The term Flandrian has been derived from transgressive sediments of the Flanders Coast of Belgium (Hein Zelin and Tavernier, 1957). This was considered as the only Stage of Holocene and thus was often used as a substitute for the Holocene in British Isles (Nilsson, 1983). It was adopted by authors who consider that the last 10,000 years should have the same stage status as previous interglacial stages (Holstenian, Emian) (Geological Time Scale, 2004). In that sense, the Flandrian becomes a designated stage within the Pleistocene, in which case the later extends to the present day (Walker et al., 2009; West, 1968, 1977, 1979; Woldstedt, 1969). In the Mediterranean, the same transgressive event has been named as Versilian derived from Versilia, the alluvial coastal plane of Eastern Liguria, near La Spezia (Blanc, 1966; Pamerol, 1980). Now both the terms "Flandrian" and "Versilian" have local significance and conceptually encompass the entire Holocene Epoch. There are some advantages of using the local stage names as this gives a physical picture to the conceptual frame work. Simultaneously, applying these local stage names in other areas creates confusion and thus use of the standard terms practiced by international geological time scale should be encouraged.

Status of Quaternary

In the last few years, there has been much heated debate on the status of Quaternary (Gibbard et al., 2010; Van couvering et al., 2009). The International Commission on

Stratigraphy, a body of the International Union of Geological Sciences (IUGS) proposed earlier to eliminate the Quaternary as a formal chronostratigraphic unit and extend the Neogene System to the present (see GTS, 2004 in Gradstein et al., 2004). A large number of Quaternary scientists represented to ICS that it was done without proper consultation and following International Geological Congress in 2004 in Florence, INQUA (International Union of Quaternary Research) and ICS made a task force to look into this mater. The task force made certain recommendations after meeting in Leuven, Belgium in September, 2005 which are in brief as (1) Status of Quaternary should be formal chronostratigraphic Unit, (2) Base of Quaternary should be stretched up to Base of Gelasian Stage (2.6 Ma), (3) That the Quaternary will have the rank of either (a) System/Period at the top of Neogene System/ Period with its lower boundary marking top of a shortened Neogene or (b) Sub-Erathem/Sub Era correlative with the upper part of the Neogene period.

This was followed by ICS decision that the Quaternary should be recognized as a formal chronostratigraphic/geochronologic unit with a lower boundary coinciding with the base of Gelasian Stage and defined by the Gelasian GSSP. John Clague, then President of the INQUA through the INQUA executive committee sought the opinion of the scientists about the status of the Quaternary. Several opinions were expressed and are available on the internet. Following several recommendations for the base of the Quaternary has been redefined at the base of Gelasian Stage (Table 1) with an absolute age of 2.6 Ma BP and the Quaternary has been given the status of a formal Period/System.

Subdivisions of the Holocene

Till date, no serious attempt has been made to formally subdivide the Holocene and no proposal has yet been made by the International Stratigraphic Commission. As a result, much subjectivity has been involved in loosely using the terms and several authors have expressed as 'Early Holocene', 'Mid Holocene' or 'Late Holocene' when referring to climate changes during various time intervals for the last 11,700 years (Ramesh et al., 2010). One of the earliest subdivisions of the Holocene was proposed by Lowe and Walker (1997) based on the pollen stratigraphy of Swedish peat bogs. The Holocene climatic changes can thus be considered as perturbations in the overall warmer period which began~ 11,700 yr BP after the short spell of the cold Younger Dryas interval (Ramesh et al., 2010). The scarce record of Younger Dryas in many parts of the world itself puts constraints on recognizing the Pleistocene–Holocene boundary. Since the chronostratigraphic units are Time–Rock units, they must be defined by certain facies. Whenever a standard Global Boundary Stratotype Section and Point (GSSP) is proposed, it is based on some record of characteristic facies, which often differ from the one that is being studied and correlated with the type locality (Ramesh et al., 2010). As the GSSP for the Holocene is located in the Ice, marking the base of the Holocene, we must find the evidence of Younger Dryas episode (event-wise) or simply get ages close to 10,000 yrs BP (chronometrically). The demarcation of the Holocene for any marine sedimentary or limnic sequence in tropical countries would be based on chronometrically determined level rather than a strong signal of termination of the Younger Dryas (Ramesh et al., 2010).

Proposed Subdivisions of the Holocene

The earliest subdivision of the Holocene, popularly known as the Blytt-Sernander Sequence, also owes its origin to a climatostratigraphic concept. This classification based on the Sweedish peat layers, first noticed by Dau (1829), was given by Blytt (1876) who interpreted that the darker peat layers were deposited in drier times while the lighter layers were deposited in moist times. Blytt (1876) used the term "Atlantic" for warm and moist and "Boreal" for cool and dry. Sernander (1908) added late-glacial chronology and included temperature in the classification of the peat layers. Though the subdivisions did not get any universal acceptance, yet the attempt to subdivide the Holocene Series based on climate change was applauded.

At present, we regard the Holocene as a warm interglacial stage, marked by abrupt phases of cooling as recorded in the Greenland Ice core and ice rafted debris in the marine sediments of the North Atlantic (Bond et al., 1997). These authors observed that the ice-rafted debris (IRD) events exhibit a distinct pacing on millennial scales, with peaks at about 1400, 2800, 4200, 5900, 8100, 9400, 10300, and 11100 yr BP (Bond et al., 1997). The cooling events were supported by influx of *Globigerina quinqueloba* and *Neogloboquadrina pachyderma* (sinistral), typical polar water planktonic foraminiferal species. The Bond events have been found to be in good correlation with weakening of summer monsoon (Gupta et al., 2003). Thus, there is a scope of classifying the Holocene either based on the major changes in North Atlantic climate (Bond events) or on the major changes in the strength of the Indian monsoon which also has a correlation with the North Atlantic climate change and can be located in the marine as well as well dated records of the Himalayan fluvial sequences.

Conclusion

It is concluded that (1) Holocene, the youngest Epoch/Series in the formal Geological Time Scale (2009) has status of Series/Epoch and continues till today and it lies above the Pleistocene Series/Epoch and the two form the Quaternary System/Period. The Quaternary has now been formally recognized as a Period/System in the Geological Time Scale (ratified by ICS) whose base is marked at the base of Gelasian Stage with an absolute age of 2.6 Ma BP (close to Matuyama/ Gauss Boundary). Further, the base of the Holocene is fixed in the North Greenland Ice Core Project (NGRIP) ice-core record at the horizon showing the clearest signal of climatic warming, an event that marks the end of the last cold episode (Younger Dryas Stadial/Greenland Stadial 1) of the Pleistocene. In calendar years, the base of the Holocene would be close to 11,700 yr BP (Before Present refers to before 1950 CE). We also conclude that the Holocene is only interval in the Geological Time Scale whose base (Pleistocene/Holocene boundary) is defined on the climatostratigraphic consideration in contrast to other boundaries of the Phanerozoic. Keeping this in view, further formal subdivisions of the Holocene should be made on the basis of the major climatic changes, i.e., the cooling events within this warm interglacial interval. There is scope of finding the signals of Bond Events and related monsoon strength variations in marine as well as fluvial records beyond the North Atlantic. A universal usage of the terms "Early", Middle and Late Holocene should be encouraged based on its subdivisions.

ACKNOWLEDGMENTS

DKS thanks Prof. B. S. Kotlia for inviting him to contribute this chapter. DKS and AKS thank the ISRO-GBP Project and DST Fast Track Project respectively for support. KRM thanks the Department of Geology, Utkal University. VPS is grateful to the CSIR for support in the form of Junior Research Fellowship. All the authors thank the Department of Geology, Delhi University for technical facilities.

REFERENCES

Alley, R. B., Mayewski, P. A. Sowers, T., Stuiver, M., Taylor, K. C., Clark, P. U., 1997. Holocene climatic instability: A prominent, widespread event 8200 yr ago. *Geology* 25, 483-486.

Blytt, A., 1876. Essay on the Immigration of the Norwegian Flora during Alternating Rainy and Dry Periods. *Alb. Cammermeyer*, Christiana, Oslo, Norway, 89 p.

Bond, G., Showers, W., Cheseby, M., Lotti, R., Amasi, P., deMenocal, P., Priore, P., Cullen, H., Hajdas, I., Bonani, G., 1997. A pervasive millennial-scale cycle in North Atlantic Holocene and glacial climates: *Science* 278, 1257–1266.

Curry, W.B., Ostermann, D.R., Gupta, M.V.S., Ittekkot, V., 1992. Foraminiferal production and monsoonal upwelling in the Arabian Sea: evidence from sediment traps. In: Summerhays, C.P., Prell, W.L., Emeis, K.C., (Eds.), Upwelling Systems: Evolution since Early Miocene. *Geological Society Special Publication*, 93–106.

Dau, Allerunterthänigster Bericht an die Königliche Dänische Rentekammer über die Torfmoore Seelands nach einer im Herbste 1828 deshalb unternommenen Reise. (usually simply Über die Torfmoore Seelands) Copenhagen and Leipzig, 1829.

Gary, N., 1999. Sedimentology and Stratigraphy, second edition,Wiley-Blackwell, ISBN 978-1-4051-3592-4.

Geological Time Scale, 2004. Gradstein, F. Ogg, J., Smith, A. (Eds.), *A Geologic Time Scale.* Cambridge University Press. pp. 589.

Gervais, P., 1867-9. *Zoologie et paleontologie générales. Nouvelles recherches sur les animaux vertétebrés et fossiles* . Paris, pp. 263.

Gibbard, P. L., Head, M. J., Walker, M. J. C. and the Subcommission on Quaternary Stratigraphy. 2010. Formal ratification of the Quaternary System/Period and the Pleistocene Series/Epoch with a base at 2.58 Ma. *Journal of Quaternary Science* 25, 96–102.

Gupta, A.K., Anderson, D.M., Overpeck, J. T., 2003. Abrupt changes in the Asian southwest monsoon during the Holocene and their links to the North Atlantic Ocean. *Nature* 421, 354-357.

Heinzelin J.de., Tavernier, R., 1957. Flandrian. In: Pruvost, P. (Eds.), Lexique stratigraphique international Vol.1 Europe. Paris, Centre National de la Recherche Scintifique, p.32 *International Commission on Stratigraphy* ICS.2009.

Lowe J. J., Walker M. J. C., 1997. Reconstructing Quaternary Environments, (2nd edn). Addison-Wesley-Longman: London.

Mayewski, P. A., Meeker, L. D., Whitlow, S., Twickler, M. S., Morrison, M. C., Alley, R. B., Bloomfield, P., Taylor, K., 1994, The atmosphere during the Younger Dryas. *Science* 263, 1747–1751.

Pamerol, 1982. The Cenozoic Era, Tertiary and Quaternary, Ellis Horwood Limited. pp. 272.

Ramesh, R., Tiwari, M., Chakraborty, S., Managave, S. R., Yadava M. G., Sinha, D. K., 2010. Retrieval of south Asian monsoon variation during the Holocene from natural climate archives. Current Science 99, 12.

Reineck H.E., Singh I. B., 1973. Depositional Sedimentary Environments. With Reference to Terrigenous Clastics. Corrected Reprint of the First Edition. Berlin, Heidelberg, New York: Springer Verlag, pp. 439, ISBN 3-540-07377-9.

Rio, D., Sprovieri, R., Castradori, D., Distefano, E. , 1998. The Gelasian Stage (Upper Pliocene): A new unit of the global standard chronostratigraphic scale. *Episodes* 21, 82-87.

Rosenzweig, C., G. Casassa, D.J. Karoly, A. Imeson, C. Liu, A. Menzel, S. Rawlins, T. L. Root, B. Seguin, P. Tryjanowski, 2007: Assessment of observed changes and responses in natural and managed systems. *Climate Change 2007:* Impacts, Adaptation and Vulnerability.Contribution of Working Group II to the Fourth Assessment Report of the Intergovernmental Panel on Climate Change, M.L. Parry,n O.F. Canziani, J.P. Palutikof, P.J. van der Linden and C.E. Hanson, Eds., Cambridge University Press, Cambridge, UK, 79-131.

Sernander, R., 1908. On the evidence of postglacial changes of climate furnished by the peat-mosses of northern Europe. *Geolgie Foren Stockholm Forh* 30, 465-473.

Sinha, D.K., Singh, A.K., 2008. Late Neogene planktic foraminiferal biochronology of the ODP site 763A, Exmouth Plateau, southeast Indian Ocean. *Journal of Foraminiferal Research* 38(3), 251–270.

Van Couvering, J., Aubry, Marie-Pierre., Berggren, W. A., Gradstein, F. M., Hilgen, F. J., Kent, D. V., Lourens, L. J., McGowran, B. 2009. What, if anything, is Quaternary? *Episodes* 32 (2), 1-2.

Walker, M., Johnsen, S., Rasmussen, S. O., Popp, T., Steffensen, J.-P., Gibbard, P., Hoek, W., Lowe, J., Andrews, J., Bjorck, S., Cwynar, L. C., Hughen, K., Kershaw, P.,Kromer, B., Litt, T., Lowe, D. J., Nakagawa, T., Newnham, R., Schwander, J., 2009. Formal definition and dating of the GSSP (Global Stratotype Section and Point) for the base of the Holocene using the Greenland NGRIP ice core, and selected auxiliary records. *Journal of Quaternary Science* 24, 3–17.

Walker, J. D., Geissman, J.W., compilers, 2009. Geologic Time Scale: *Geological Society of America*, doi: 10.1130/2009.CTS004R2C. © 2009.

West, R. G., 1968. *Pleistocene Geology and Biology*. Longman, London. 379pp.

West, R. G., 1977. *Pleistocene Geology and Biology*, (2nd Edn.). Longman: London.

West, R. G., 1979. Further on the Flandrian. *Boreas* 8, 126p.

Woldstedt, P. 1969. Handbuch der Stratigrdphischen Ceologie. Ed. Fr. Lotzc, Bd. II. viii + 263 p., 77 figs. Ferdinand Enke Vcrlag, Stuttgart.

In: Holocene
Editor: Bahadur Singh Kotlia

ISBN: 978-1-62257-722-4

Chapter 2

HYDROGRAPHIC CHANGES IN THE BAY OF BENGAL DURING THE HOLOCENE AS INFERRED FROM CARBON AND OXYGEN ISOTOPES OF FOSSIL FORAMINIFERA

S. M. Ahmad, T. Raza, N. Sagar and G. Suseela
National Geophysical Research Institute (CSIR), Uppal Road, Hyderabad, India

ABSTRACT

Stable carbon and oxygen isotopic records from the Bay of Bengal sediment cores show significant hydrographic changes during the Holocene. The current chapter deals with the $\delta^{13}C$ and $\delta^{18}O$ records from the Holocene foraminifera to infer surface and deep water changes at different locations in the Bay of Bengal. The oxygen isotopes in surface dwelling foraminifera (*Globigerinoides ruber)* in the western Bay of Bengal and the Andaman Sea cores are significantly more depleted relative to the cores in the central and southern Bay of Bengal. This depletion in planktonic $\delta^{18}O$ of the western Bay of Bengal and the Andaman Sea suggest low salinity surface water compared to the central and southern Bay of Bengal. Planktonic $\delta^{18}O$ values in the central and southern Bay of Bengal cores are relatively more enriched in $\delta^{18}O$ because of the diminished effect of riverine fresh water fluxes. However; a marked depletion in $\delta^{18}O$ values is seen at 8-6 ka BP in all cores from the Bay of Bengal, indicating inflow of low salinity water. It appears that the Indian monsoon precipitation at 8-6 ka BP may have been very strong to influence the salinity structure in the entire Bay of Bengal. Also a marked enrichment in $\delta^{18}O$ at ~ 13-11 ka BP indicates a 'Younger Dryas' cooling event in the Bay of Bengal and the Andaman Sea.

Oxygen and carbon isotopic records of benthic foraminifera (*Cibicidoides* spp.) show little variations during the Holocene. Benthic $\delta^{13}C$ values in *Cibicidoides* species of the Holocene sections are similar to the modern deep water $\delta^{13}C$ values at the studied locations. Holocene benthic $\delta^{13}C$ values suggest greater influx of North Atlantic Deep Water (NADW) into the deep Bay of Bengal.

Keywords: Indian summer monsoon, $\delta^{18}O$, $\delta^{13}C$, Bay of Bengal

INTRODUCTION

The Indian summer (southwest) monsoon results from the differential heating and pressure gradients between the tropics and Inter Tropical Convergence Zone (ITCZ). The low pressure over Asia and the high pressure over the ocean gives rise to the strong low-level atmospheric pressure gradient that, in turn, generates the Indian summer monsoon (Overpeck et al., 1996). The sea surface circulation in the Bay of Bengal changes its direction and intensity from season to season in response to the seasonal rhythm of Indian summer and winter monsoons (Clemens et al., 1991; Prell et al., 1992; Colin et al., 1999; Gupta et al., 2003). The heavy precipitation during the summer monsoon over the land is collected by major rivers such as the Ganga, Brahmaputra, Salween, Irrawaddy etc. These rivers drain their freshwater and sediment fluxes into the northern Bay of Bengal and the Andaman Sea. Consequently, north-south salinity gradient forms in the Bay of Bengal, which is reflected in the distribution pattern and oxygen isotopes of the planktonic foraminifera (Cullen, 1981; Duplessy, 1982; Sarkar et al., 1990; Kudrass et al., 2001). It has been demonstrated that the sea surface salinity in the glacial northern Bay of Bengal was significantly higher relative to the modern conditions (Cullen, 1981; Duplessy, 1982; Rashid et al., 2011). On the other hand, the salinity in the southern Bay of Bengal during the Last Glacial Maximum (LGM) was slightly higher or similar to today. This glacial increase in salinity in the northern Bay of Bengal and the Andaman Sea has been explained as due to the decreased Indian monsoonal precipitation over the Indian subcontinent and North Indian Ocean (Cullen, 1981; Duplessy, 1982; Fontugne and Duplessy, 1986; Sarkar et al., 1990; Kudrass et al., 2001; Rashid et al., 2007). Several paleoceanographic investigations from the Arabian Sea and the Bay of Bengal have shown large variations in the Indian summer monsoon precipitation since the last ice age (Clemens et al., 1991; Prell et al., 1992; Sirocko et al., 1993; Colin et al., 1998; Schulz et al., 1998; Ahmad et al., 2008; Anand et al., 2008, 2000; Rashid et al., 2011). However, the maximum monsoon intensity has reported a time lag of several thousand years from the peak insolation forcing (Kutzbach, 1987). A number of mechanisms have been proposed to explain this time lag in maximum monsoon intensity and solar forcing (Kutzbach, 1987; Prell and Kutzbach, 1987; Clemens et al., 1991). It has been suggested that the Indian monsoon responded more linearly to insolation forcing after the disappearance of glacial boundary conditions (Overpeck et al., 1996). Prell and Kutzbach (1987) have argued that both orbital forcing and glacial boundary conditions are necessary to explain the major regional features of monsoonal activity. In a recent study, Zhisheng et al. (2011) have suggested a strong interhemispheric connection in controlling the Indian monsoon activity. The sediments from the northern Bay of Bengal and the Andaman Sea have recorded these changes in the Indian monsoon activity (Duplessy, 1982; Colin et al., 1998; Ahmad et al., 2000, 2008; Rashid et al., 2007, 2011). It has been suggested that the monsoon-induced salinity changes in the glacial northern Bay of Bengal were more pronounced compared to the central and southern Bay of Bengal (Cullen, 1981; Duplessy, 1982). Moreover, the relative strength of the Indian southwest and the northeast (winter) monsoons have changed even on short time scales and such monsoonal variability have a relationship with cold (stadial) and warm (inter-stadial) events of the North Atlantic (Sarkar et al., 1990; Prell et al., 1992; Sirocko et al., 1993; Colin et al., 1998; Schulz et al., 1998; Kudrass et al., 2001; Gupta et al., 2003; Anand et al., 2008). The distribution of $\delta^{13}C$ in deep waters is related to the oxygen and nutrient contents and the circulation pattern of the water masses (Curry et al., 1988). Benthic foraminiferal $\delta^{13}C$ values are widely used in tracing past distribution of carbon in the deep and mid-depth seawater and provide clues to the variations in atmospheric CO_2 concentrations on glacial-interglacial time scale (Boyle, 1988; Duplessy et al., 1988; Charles and Fairbanks, 1992; Yu et al., 2010). Under modern conditions, deep water in the northeast Indian Ocean is filled with water originating in the circumpolar region and transported northward by two western boundary currents (You, 2000). Previous isotopic investigations on benthic foraminifera have also shown significant changes in chemical characteristics of the deep Bay of Bengal since the last glaciation (Kallel et al., 1988; Naqvi et al., 1994; Ahmad et al., 2008). However, the available isotopic records from the Bay of Bengal, particularly on benthic foraminifera, have a very low temporal resolution of the investigated sample sets. In this study we present both surface and deep water records pertaining to the Holocene from the well-dated sediment core of the Bay of Bengal.

METHODOLOGY

We have generated carbon and oxygen isotopic records from two deep sea sediment cores from the central and the southern Bay of Bengal cores (SK157-18 and SK157-14 respectively) (Figure 1). These records are compared with previously published isotopic records from the western Bay of Bengal (VM29-19) and the Andaman Sea (RC12-344) in Figure 2 (Rashid et al., 2007, 2011). The isotopic measurements on planktonic foraminifera *(Globigerinoides ruber)* and benthic foraminifera (mostly *Cibicidoides wuellerstorfi*) from cores SK157-14 and SK157-18 were carried out at the National Geophysical Research Institute, Hyderabad, India. The microscopic examination showed that the foraminiferal tests were well preserved with no apparent dissolution. All the tests were cleaned ultrasonically in methanol to remove adhering fine particles. The oxygen ($\delta^{18}O$) and carbon ($\delta^{13}C$) isotopic measurements were performed on a Delta^{Plus} Advantage Isotope Ratio Mass Spectrometer (IRMS) in conjunction with Kiel-IV automatic carbonate device. The foraminifera tests were reacted with saturated H_3PO_4 at ~ 70°C in a vacuum system and evolved carbon-di-oxide was then analyzed by mass spectrometer. Isotopic compositions are reported in δ notation as per mil deviation from VPDB standard. The analytical precision was better than 0.10‰ for $\delta^{18}O$ and 0.05‰ for $\delta^{13}C$. The calibration to the VPDB standard was achieved by repeated measurements of international reference standards NBS-19 and NBS-18. *Cibicidoide*s oxygen isotope data was converted into the *Uvigerina* scale by adding scale by adding 0.64‰ and carbon isotopic values of two levels of *Uvigerina*were brought to *Cibicidoides* scale by adding 0.90‰ (Mix et al., 1991). The age models of all sediment cores presented here are mainly based on AMS radiocarbon ages. A reservoir age correction of 400 years is applied to all the radiocarbon ages (Stuiver and Reimer, 1993). This correction is a reasonable estimate of the reservoir age for the studied location in the Bay of Bengal (Dutta et al., 2001). The calibration to calendar year ages for cores SK157-14 and SK157-18 was performed using 'Fairbanks0107' programme (Fairbanks et al., 2005).

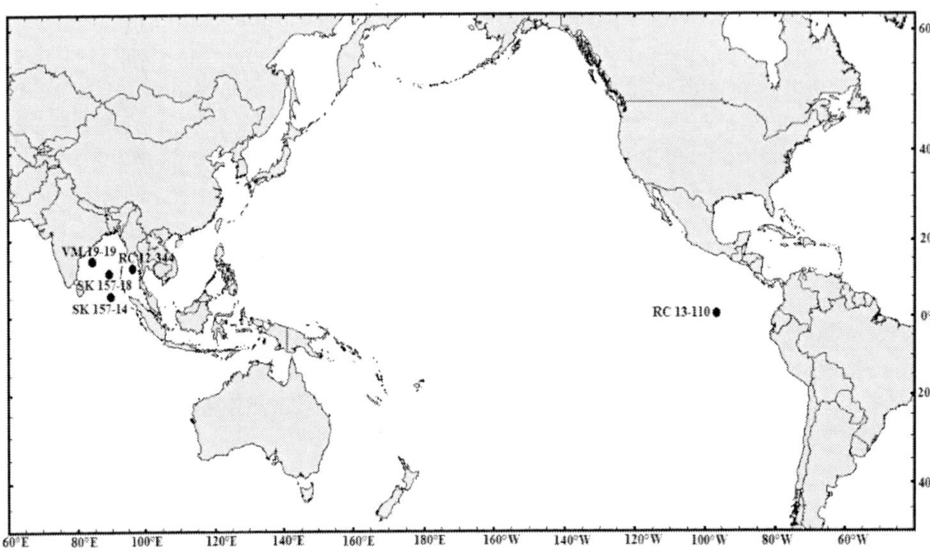

Figure 1. Location map of deep sea sediment cores discussed in text.

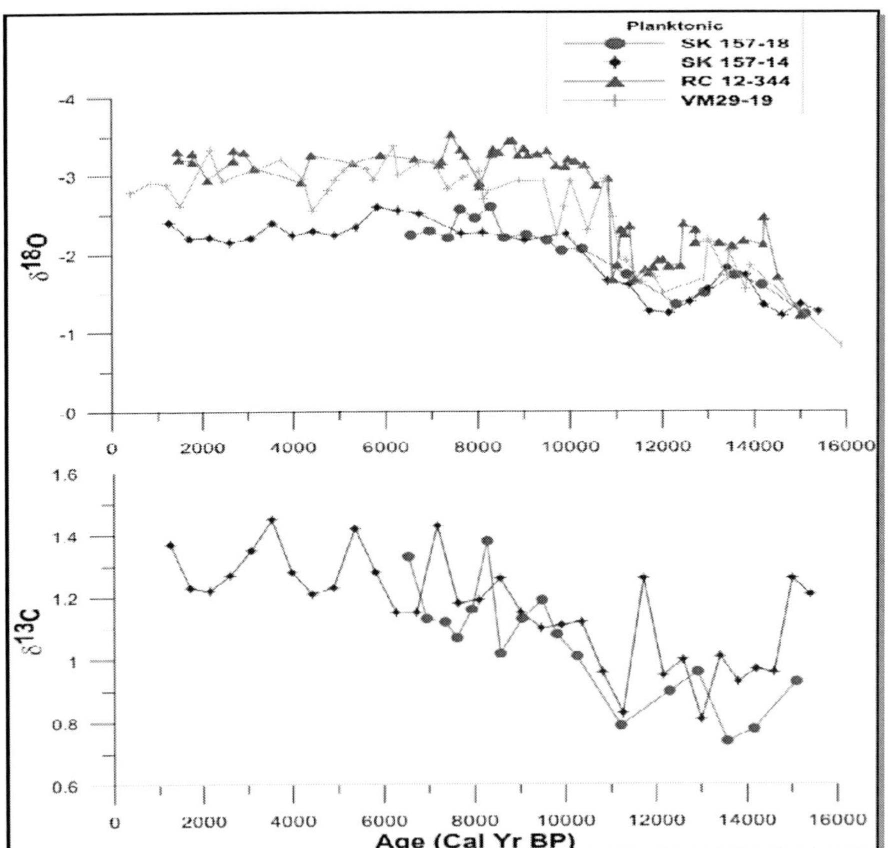

Figure 2. Planktonic $\delta^{18}O$ and $\delta^{13}C$ records of the Bay of Bengal cores SK157-14 and SK157-18 plotted with cores from the western Bay of Bengal (VM29-19) and the Andaman Sea (RC12-344) (Rashid et al., 2007, 2011).

Results and Discussion

Holocene Evolution of Surface Water in the Northeast Indian Ocean

Previous studies from the Bay of Bengal cores have clearly demonstrated a distinct north-south $\delta^{18}O$ gradient in core-top and Holocene planktonic foraminifera (Cullen, 1981; Duplessy, 1982; Fontugne and Duplessy, 1986). These workers have suggested an intensification of Indian southwest monsoon (ISM) and discharge of fresh water fluxes into the northern Bay of Bengal during the Holocene. Consequently a north-south $\delta^{18}O$ gradient forms in the Bay of Bengal, which is reflected in oxygen isotopes of planktonic foraminifera. Therefore, our planktonic $\delta^{18}O$ records from the central (SK157-18) and southern Bay of Bengal (SK157-14) cores show higher $\delta^{18}O$ values compared to the cores in the north or western Bay of Bengal (Kudrass et al., 2001; Rashid et al., 2011) and the Andaman Sea (Rashid et al., 2007) (Figure 2). Both these cores show considerable changes in $\delta^{13}C$ and $\delta^{18}O$ values during the Holocene. Despite the fact that the late Holocene section is missing from the core SK157-18, some fluctuations in both $\delta^{13}C$ and $\delta^{18}O$ values are recorded in the early Holocene section of this core. The maximum $\delta^{18}O$ depletion in core SK157-14 occurs at 7-6 ka BP, whereas maximum $\delta^{18}O$ depletion in core SK157-18 occurs at 8-7 ka BP. This discrepancy in the timing of the negative $\delta^{18}O$ pulse may be due to age constraints of the two cores. The observed negative $\delta^{18}O$ excursion at ~ 8-6 ka BP in these cores could arise either due to a decrease in surface salinity as a result of intensification of Indian monsoonal precipitation or increase in sea surface temperature (SST) or a combination of both. However previously published other proxy records from the northeast Indian Ocean, such as the planktonic Mg/Ca ratios and alkenone records, have demonstrated a negligible SST change during the Holocene (Sonzogni et al., 1998; Saraswat et al., 2005; Rashid et al., 2007). Therefore, the observed negative $\delta^{18}O$ excursion in the central and southern Bay of Bengal at 8-6 ka BP is probably caused by the intensified Indian monsoon precipitation.

The oxygen isotopic record of planktonic foraminifera (*G. ruber*) from the western Bay of Bengal core (VM29-19) shows most depleted $\delta^{18}O$ values during the early Holocene suggesting strengthening of the ISM (Rashid et al., 2011). Another core, SK117-GC8 from the eastern Arabian Sea has also shown maximum ISM at ~ 8 ka BP (Banakar et al., 2010).

These results are consistent with several other terrestrial and marine records of the enhanced Indian monsoon activity at ~ 8 ka BP (Fleitmann et al., 2003; Gupta et al., 2003; Rashid et al., 2007). However, this Holocene maximum in the Indian summer monsoon activity lags behind solar (orbital) forcing by a few thousand years, reflecting either a large response time or additional forcing mechanisms in determining the time of maximum response (Clemens et al., 2010), In addition, the core VM29-19 also exhibits enriched $\delta^{18}O$ values between 4.8 and 3.1 ka BP indicating a dryer climate. A similar enrichment in $\delta^{18}O$ values at 4.8-1.4 ka BP interval has been reported from another core in the Andaman Sea (Rashid et al., 2007). Our planktonic oxygen isotopic record of core SK157-14 also shows enriched $\delta^{18}O$ values during the 5-2 ka BP, indicating a decrease in the Indian monsoon precipitation (Figure 2). Moreover, these records are consistent with the planktonic $\delta^{18}O$ records off Indus delta (Staubwasser et al., 2003) and northern Bay of Bengal (Chauhan,

2000) supporting a drastic reduction in Indian monsoon activity at ~ 5 ka BP interval, coinciding with the disappearance of the Indus valley civilization.

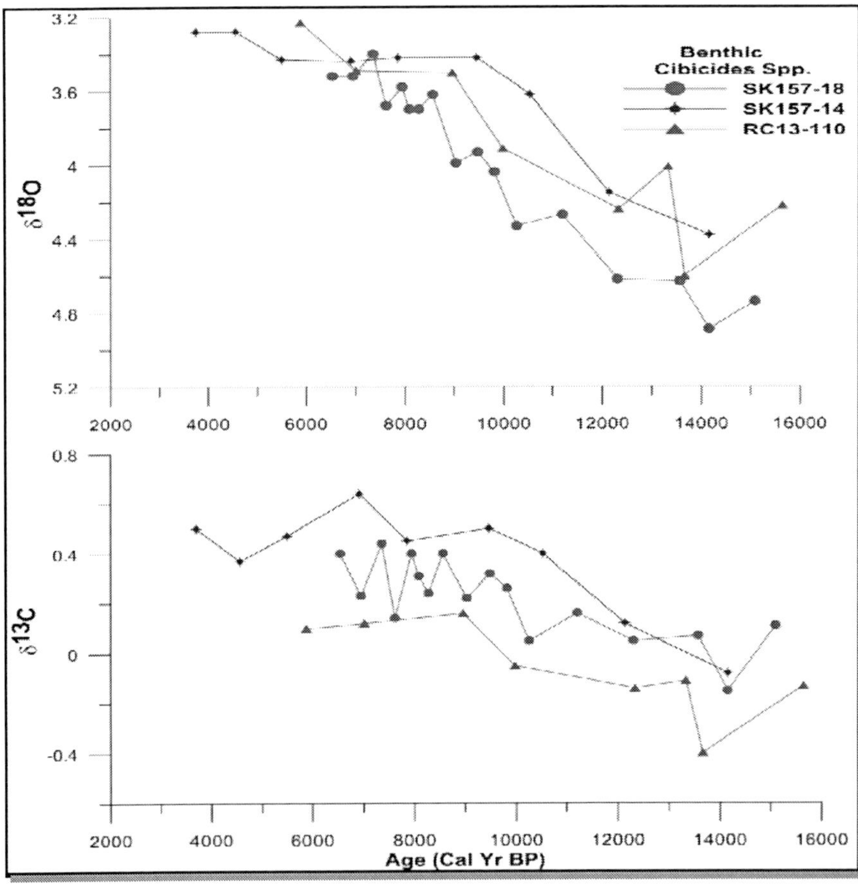

Figure 3. Benthic $\delta^{18}O$ and $\delta^{13}C$ records from cores SK157-18, SK157-14 and RC13-110 (Mix et al., 1991), plotted against calibrated ages.

Interestingly, a systematic pattern is observed in the planktonic $\delta^{13}C$ records of SK157-14 and SK157-18 cores (Figure 3). An increasing $\delta^{13}C$ trend is seen between 12 and 8 ka BP. We attribute this $\delta^{13}C$ shift to the variations in carbon isotopic budget of surface waters due to wind-induced variability. During the times of enhanced wind, sub-surface water with low $\delta^{13}C$ may have been brought to surface from sub-surface, causing depletion in planktonic $\delta^{13}C$ and vice-versa. It may be noted that water mass from a depth of ~100 m is generally depleted in $\delta^{13}C$ by ~ 0.5‰ (Kroopnick, 1985).

Deep Water Evolution During the Holocene

Under modern conditions, large scale variations in deep water $\delta^{13}C$ have been reported from different ocean basins with values ranging from ~ 1‰ in the North Atlantic (Raymo et al., 2004), 0.2-0.4‰ in the Weddell Sea (Mackensen et al., 2001), ~ 0.4‰ for Circumpolar

Deep Water (CDW) (Charles and Fairbanks, 1992) and - 0.5 to 0‰ in mid-depth water masses of the North Pacific (Charles and Fairbanks, 1992). The most depleted $\delta^{13}C$ values, as low as - 0.9‰, are reported from the Southern Ocean deep water (Curry and Oppo, 2005). The deep waters in the Bay of Bengal are of circumpolar origin and are transported northward by two western boundary currents (Warren, 1981). The northward movement of this circumpolar deep water (CDW) is mainly through two paths, one in the western Indian Ocean and another in the central Indian Ocean and beyond into the Arabian Basin through passage in the Chagos-Laccadive Ridge, the other branch of the later propagates northward at the southern tip of the Chagos-Laccadive Ridge, feeding the deep Bay of Bengal. The CDW has a $\delta^{13}C$ value of ~ 0.4‰, reflecting a balance North Atlantic Deep water (NADW) and deep water formed in Southern Ocean (Charles and Fairbanks, 1992; You, 2000). The CDW is characterized by its relatively high salinity and oxygen and low nutrients, whereas North Indian Deep Water (NIDW) has high nutrients and low oxygen (Wyrtki, 1971; You, 2000; Schmiedl and Leuschner, 2005). The low oxygen and high nutrients in NIDW could also be due to the introduction of water from the marginal Seas (Mantyla and Reid, 1995).

To compare deep water evolution in the Bay of Bengal and the Pacific Ocean, we have plotted $\delta^{18}O$ and $\delta^{13}C$ data sets from three different cores; two from the Bay of Bengal, SK157-18 and SK157-14 and one from the Pacific Ocean (RC13-110) in Figure 3 (Mix et al., 1991). All these cores were retrieved from the similar water-depths (3000-3300 m) and have been analyzed for same benthic foraminifera (*Cibicidoides* spp.). A close examination of $\delta^{18}O$ and $\delta^{13}C$ values in Figure 3 reveals the subtle variations in both $\delta^{18}O$ and $\delta^{13}C$ values since the last deglaciation. Interestingly, the Holocene $\delta^{18}O$ values in *Cibicidoides* spp. in core RC-110 from the deep Pacific are quite similar to that of deep Bay of Bengal cores. The deep water production in the North Atlantic has increased drastically since the last deglaciation. The Maximum NADW fluxes have been reported during interglacials and minimum during glacial times (Duplessy et al., 1988; Oppo and Fairbanks, 1990; Bickert and Mackensen, 2003; Curry and Oppo, 2005). The increase in the $\delta^{13}C$ values during the Holocene in all three cores from the deep Bay of Bengal and deep Pacific suggests greater contribution from the NADW (Figure 3). In contrast, the glacial deep water in the Bay of Bengal was depleted by ~ 0.7‰ (Ahmad et al., 2008). The mean $\delta^{13}C$ values for the Holocene section of cores SK157-14 and SK157-18 (0.3‰) are close to the modern deep water $\delta^{13}C$ (0.2‰) at these core locations (Kroopnick, 1985), indicating a similar deep water $\delta^{13}C$ during the Holocene and today. Closer inspection of our data suggest that the benthic $\delta^{13}C$ values respond to the source water variation in the deep Bay of Bengal. The minor changes in $\delta^{13}C$ values of the Holocene benthic foraminifera in the Bay of Bengal cores reflect local variability in deep water chemistry.

ACKNOWLEDGMENTS

We are thankful to Waseem Raza, K. Venkatesham, Mahjoor Lone, Sadia Farnaaz and P.S. Srikanth for their help during the course of this study. The financial support provided by the Indian Space Research Organization (ISRO-GBP), Department of Science and Technology (DST) and Ministry of Earth Sciences (MoES-INCOIS), Govt. of India is thankfully acknowledged. We are grateful to the Director, N.G.R.I. for his permission to submit this chapter.

REFERENCES

Ahmad, S.M., Patil, D. J., Rao, P.S., Nath, B. N. , Rao, B. H., Rajagopalan, R., 2000. Glacial-interglacial changes in the surface water characteristics of the Andaman Sea: Evidence from the stable isotopic ratios of planktonic foraminifera. *Proceedings Indian Academy of Sciences* (Earth Planetary Science) 109 (1), 153-156.

Ahmad, S.M., Anil Babu, G., Padmakumari, V.M., Waseem Raza, 2008. Surface and deep water changes in the northeast Indian Ocean during the last 60 ka inferred from carbon and oxygen isotopes of planktonic and benthic foraminifera. Palaeogeography Palaeoclimatology Palaeoecology 262, 182-188, doi: 10.1016/j.palaeo.2008.03.007.

Allison, M.A., Khan, S.R., Goodbred, S.L., Kuehl, S.A., 2003. Stratigraphic evolution of the late Holocene Ganges-Brahmaputra lower delta plain. *Sedimentary Geology* 155, 317-342.

Banakar, V. K., Mahesh, B. S., Burr, G., Chondakar, A. R., 2010. Climatology of the eastern Arabian Sea during the last glacial cycle reconstructed from paired measurement of foraminiferal $\delta^{18}O$ Mg/Ca. *Quaternary Research*, 73, 535-540, doi:10.1016/j.yqres.2010.02.002.

Bickert, T., Mackensen, A., 2003. Last Glacial to Holocene changes in South Atlantic deep water circulation. In: Wefer, G., Mulitza, S., Rathmeyer, V. (Eds.), The South Atlantic in the Late Quaternary – Reconstruction of Material budget and Current Systems , Springer, New York, pp. 671-693.

Boyle, E.A., 1988. Vertical oceanic nutrient fractionation and glacial-interglacial CO_2 cycles. Nature 331, 55-56.

Charles,C.D., Fairbanks, R.G., 1992. Evidence from Southern Ocean sediments for the effect of North Atlantic Deep water flux on climate. *Nature* 355, 416-419.

Chauhan, O.S., 2000. Past 2,000 years history of Himalayan aridity: Evidence from oxygen isotope records in the Bay of Bengal. *Current Science* 84 (1), 90-93.

Clemens, S., Prell, W.L., Howard, W.R., 1991. Forcing mechanisms of the Indian Ocean monsoon. *Nature* 353, 720-725.

Clemens, S., Prell, W.L., Sun, Y., 2010. Orbital scale timing and mechanisms driving Late Pleistocene Indo-Asian summer monsoons: Reinterpreting cave speleothem $\delta^{18}O$. *Paleoceanography* 25, PA4207, doi:10.1029/ 2010 PA001926.

Colin, C., Kissel, C., Blamart, D., Turpin, L., 1998. Magnetic properties of sediments in the Bay of Bengal and the Andaman Sea: Impact of rapid North Atlantic Ocean climatic events on the strength of Indian monsoon. *Earth and Planetary Science Letters* 160, 623-635.

Colin, C., Turpin, L., Bertaux, J., Desprairies, A., Kissel, C., 1999. Erosional history of the Himalayan and Burman ranges during the last two glacial-interglacial cycle. *Earth and Planetary Science Letters* 171, 647-660.

Cullen, J., 1981. Microfossil evidence for changing salinity patterns in the Bay of Bengal over the last 20,000 years. *Palaeogeography Palaeoclimatology Palaeoecology* 85, 315-356.

Curry, W.B, Duplessy, J.C., Labeyrie, L.D., Shackleton, N.J., 1988. Changes in the distribution of $\delta^{13}C$ of deep water $\sum CO_2$ between the last glaciation and the Holocene. *Paleoceanography* 3, 317-341.

Curry W. B., Oppo, D.W., 2005. Glacial water mass geometry and the distribution of $\delta^{13}C$ of ΣCO_2 in the western Atlantic Ocean, *Paleoceanography* 20, PA1017, doi:10.1029/2004PA001021.

Duplessy, J.C., 1982. Glacial to interglacial contrast in the northern Indian Ocean. *Nature* 295, 494-498.

Duplessy, J.C., Shackleton, N.J., Fairbanks, R.G. Labeyrie, L.D., Oppo, D., Kallel, N., 1988. Deepwater source variations during the last climatic cycle and their impact on global deepwater circulation. *Paleoceanography* 3, 343-360.

Dutta, K, Bhushan, R., Somayajulu, B.L.K., 2001. R correction values for the northern Indian Ocean. *Radiocarbon* 43, 483-488.

Fairbanks, R.G., Mortlock, R.A., Chiu, T.C., Li, Kaplan, A., Guilderson, T.P., Fairbanks, T.W., Bloom, A.L., Grootes, P.M., Nadeau, M.J., 2005. Radiocarbon calibration curve spanning 0 to 50,000 years BP based on paired $^{230}Th/^{234}U/^{238}U$ and ^{14}C dates on pristine corals. *Quaternary Science Reviews* 24, 1781-1786.

Fleitmann, D., Burns, S. J., Mudelsee, M., Neff, U., Kramers, J., Mangini, A., Matter, A., 2003. Holocene forcing of the Indian monsoon from southern Oman. *Science* 300, 1737-1739.

Fontugne, M.R., Duplessy, J.C., 1986. Variations in the monsoon regime during the upper Quaternary: Evidence from carbon isotopic record of organic matter in North Indian Ocean sedimentary cores. *Palaeogeography Palaeoclimatology Palaeoecology* 56, 69-88.

Gupta, A.K., Anderson, D.M., Overpeck, J.T., 2003. Abrupt changes in Asian southwest monsoon during the Holocene and their links to the North Atlantic Ocean. *Nature* 421, 354-356.

Kallel, N., Labeyrie, L.D., Juillet-Leclerc, A., Duplessy, J.C., 1988. A deep hydrological front between intermediate and deep water masses in the glacial Indian Ocean. *Nature* 333, 651-655.

Kroopnick, P., 1985. The distribution of carbon-13 in the world oceans. *Deep Sea Research* 32, 57-84.

Kudrass, H. R., Hofmann, A., Doose, H., Emeis, K., Erlenkeuser, H., 2001. Modulation and amplification of climatic changes in the Northern Hemisphere by the Indian summer monsoon during the past 80 k.y.. *Geology* 29, 63-66.

Mackensen, A., Rudolph, M., Kuhn, G., 2001. Late Pleistocene deep-water circulation in the subantarctic eastern Atlantic. *Global and Planetary Change,* 30, 197–229, doi:10.1016/S0921-8181(01)00102-3.

Mantyla, A.W., Reid, J.L., 1995. On the origins of deep and bottom waters of the Indian Ocean. *Journal of Geophysical Research* 100, 2417-2439.

Mix, A.C., Pisias, N.G., Zahn, R., Rugh, W., Lopez, C., Nelson, K., 1991. Carbon 13 in Pacific deep and intermediate waters, 0–370 ka: implications for ocean circulation and Pleistocene CO_2, *Paleoceanography* 6 (2), 205–226, doi:10.1029/90PA02303.

Naidu, P.D., Malmgren, B.A., 1995. A 2,200 years periodicity in the Asian monsoon system. *Geophysical Research Letters* 22 (17), 2361-2364.

Naqvi, W.A., Charles, C.D., Fairbanks, R.G., 1994. Carbon and oxygen isotopic records of benthic foraminifera from the Northeast Indian Ocean: Implications on glacial-interglacial CO_2 changes. *Earth and Planetary Science Letters* 121, 99-110.

Oppo, D.W., Fairbanks, R.G., 1990. Atlantic Ocean thermohaline circulation of the last 150,000 years: Relationship to climate and atmospheric CO_2. *Paleoceanography* 5, 277-288.

Overpeck, J., Anderson, D., Trumbore, S., Prell, W., 1996. The southwest Indian Monsoon over the last 18000 years. *Climate Dynamics* 12, 213-225, doi:10.1007/BF00211619.

Prell, W.L., Kutzbach, J.E., 1987. Monsoon variability over the past 150000 years. *Journal of Geophysical Research D-7 92*, 8411-8425.

Prell, W.L., Murray, W.W., Clemens, S.C., Anderson, D.M., 1992. Evolution and variability of the Indian Ocean Summer Monsoon: Evidence from western Arabian Sea Drilling Program. *Geophysical Monograph* 70, 447-469.

Rashid, H., Flower, B.P., Poore, R.Z., Quinn, T.M., 2007. A ~ 25 ka Indian Ocean monsoon variability record from the Andaman Sea. *Quaternary Science Reviews* 26, 2586-2597, doi:10.1016/j.quascirev.2007.07.002.

Rashid, H., England, E., Thomson, L., Polvak, L., 2011. Late glacial to Holocene Indian summer monsoon variability based upon sediment records taken from the Bay of Bengal. *Terra Atmosphere Ocean Science* 22, doi:10.3319/TAQ2010.09.17.02(TibXS).

Raymo, M. E., Oppo, D.W., Flower, B.P., Hodell, D.A., McManus, J.F., Venz, K.A., Kleiven, K.F., McIntyre, K., 2004. Stability of North Atlantic Water masses in face of pronounced climate variability during the Pleistocene. *Paleoceanography* 19, PA 2008, doi:10.1029/2003PA000921.

Saraswat, R., Nigam, R., Weldeab, S., Mackensen, A., Naidu, P.D., 2005. A first look at past sea surface temperatures in the equatorial Indian Ocean from the Mg/Ca in the foraminifera. *Geophysical Research Letters* 32, doi:10.1029/2005GL024093, 2005.

Sarkar, A., Ramesh, R., Bhattacharya, S.K., Rajagopalan, G., 1990. Oxygen isotope evidence for a stronger winter monsoon current during the last glaciations. *Nature* 343, 549-551.

Schmiedl, G., Leuschner, D.C., 2005. Oxygenation changes in the deep western Arabian Sea during the last 190,000 years: Productivity versus deep water circulation. *Paleoceanography* 20, PA2008, doi:10.1029/ 2004PA001044.

Schulz, H., Von Rad, U., Erlenkeuser, H., 1998. Correlation between Arabian Sea and Greenland climate oscillations of the past 110,000 years. *Nature* 393, 54-57.

Sirocko, F., Sarnthein, M., Erlenkeuser, H., Lange, H., Arnorld, M., Duplessy, J.C., 1993. Century scale events in monsoon climate over the past 24,000 years. *Nature* 364, 322-32.

Sonzogni, C., Bard, E., Rostek, F., 1998. Tropical sea surface temperatures during the last glacial period: A view based on alkenones in Indian Ocean sediments. *Quaternary Science Reviews* 17, 1185-1201.

Staubwasser, M., Sirocko, F., Grootes, P.M., Segl, M., 2003. Climate change at 4.2 ka BP termination of the Indus valley civilization and Holocene south Asian climate variability. *Geophysical Research Letters* 30, 1425, doi:10.1029/2002GLO16822.

Stuiver, M., Reimer, P.J., 1993. Extended ^{14}C data base and revised CALIB 3.0 ^{14}C age calibration program. *Radiocarbon* 35, 215-230.

Warren, B.A., 1981. Transindian hydrogrphic section at Lat. 18°S. Properly distribution and circulation in the south Indian Ocean. *Deep-sea Research* 28, 759-788.

Wyrtki, K., 1971. Oceanographic Atlas of the International Indian Ocean Expedition. National Science Foundation, Washington, D.C., pp. 531.

You, Y., 2000. Implications of the deep circulation and ventilation of the Indian Ocean on the renewal mechanism of North Atlantic Deep Water. *Journal of Geophysical Research* 105, 23,895-23,926.

Yu, J., Broecker, W.S., Elderfield, H., Jin, Z., McManus, J., Zhang, F., 2010. Loss of carbon from the deep sea since the last glacial maximum. *Science* 330, 1084-1087.

Zhisheng, A., Clemens, S.C., Shen, J., Qiang, X., Jin, Z., Sun, Y., Prell, W.L., Juo, J., Wang, S., Xu, H., Cai, Y., Zou, W., Liu, X., Liu, W., Shi, Z., Yan, L., Xiao, X., Chang, H., Wu, F., Ai, L., Lu, F., 2011. Glacial-Interglacial Indian summer monsoon dynamics. *Science* 333, 719-723, doi:10.1126/science.1203752.

In: Holocene
Editor: Bahadur Singh Kotlia

ISBN: 978-1-62257-722-4

Chapter 3

SPATIAL AND TEMPORAL VARIABILITY IN HOLOCENE RIVER FLOW CONDITIONS DEDUCED FROM FLUVIAL TERRACE SYSTEMS IN THE MAHI RIVER BASIN, GUJARAT, INDIA

Alpa Sridhar and L. S. Chamyal*

Department of Geology, Faculty of Science,
M. S. University of Baroda, Vadodara, India

ABSTRACT

Riverbank stratigraphy and palaeofluvial style of the Mahi River provides a framework to examine the variations in the flow conditions through time. The exposed sediment records from the upland to the lower reaches have been studied in order to understand the hydrologic regime during which these were deposited. The regime based palaeohydrological and palaeocompetence analyses have been employed to estimate the flow energy and discharge conditions. The nature and dimensions of the various sediment units have altered in response to the quantity of sediment supply, hydrological conditions and tectonic disturbances. The exposed sequences exhibit fining of sediment facies from the upland to the alluvial-estuarine zones. However, the gravel terrace from the upland reach provides evidence of braided gravel-bed river that switched to a gravel bed wandering fluvial style during early Holocene; the mid–late Holocene channel form changed from wide, large-amplitude meanders to smaller meanders with decreased width/depth ratio, unit stream power and bed shear stresses. Based on the available chronology on these sediments it can be inferred that an abrupt change in the hydrological conditions in the Mahi River Basin occurred post 9 ka which led to the incision of gravel deposits in the uplands and movement of the gravels further downstream following which, the extensive sedimentation occurred in the lower reaches till about 2 ka BP. The results suggest that discharge varied spatially and temporally during Holocene, the incision and aggradation of the sediments was controlled by fluctuations in sediment influx and the river responded to these changes by switching stream patterns.

* E mail: lschamyal@yahoo.com.

Keywords: River flow, Fluvial terrace, Mahi river, Gujarat

Introduction

Fluvial valleys in western India have well-developed terrace systems - bedrock surfaces overlain by a fluvial sediment cover of various thicknesses. The role of Quaternary climate oscillations and tectonic uplift in the terrace formation in the alluvial rivers of Indian fluvial systems and alluvial rivers in particular have been briefly discussed by many authors (Chamyal et al., 2003; Chamyal and Juyal, 2008; Sridhar and Chamyal, 2010 and references therein). However, the process-response relationship of sediment erosion and accumulation to the climate and tectonic pulses, presented in the current chapter is not well understood. In the evolution of a valley, well-marked phases of erosion and accumulation, separated by periods of lateral erosion occur in a cycle (Hancock and Anderson, 2002). Erosion and aggradation depend directly on the discharge into the river channels and on the supply of slope material to the channel (Bogaart and van Balen, 2000). When the two processes are so balanced that discharge is high enough to remove all slope material supplied from the valley bottom, the river does not erode or aggrade (Blum and Törnqvist, 2000). An increase in the supply of slope material, a change in river regime and pattern (Hancock and Anderson, 2002; Starkel, 2003) or a decrease in river gradient leads to the widening and steepening of a valley. Therefore, analyses of riverbank stratigraphy and palaeofluvial style will reveal the variations in the flow conditions through time. Here, an attempt has been made to decipher the relation between the flow conditions and aggradation-incision phases in a river basin during the Holocene.

Mahi River Basin is one of the major alluvial river basins in western India and the valley has preserved distinct terrace systems. The exposed terrace sediments from the upland to the lower reaches have been studied in order to understand the hydrologic regime during which these were deposited based on palaeohydrological and palaeocompetence analyses. This paper discusses the aggradation and incision of the terrace system in response to change in the river regime and pattern during the Holocene.

Study Area and Regional Setting

The Mahi River is the third largest river in Gujarat having a drainage basin area of ~ 32 000 km^2; originates in Malwa region of Madhya Pradesh and flows 180 km along the eastern flank of the intracratonic Cambay graben (Biswas, 1987) before meeting the Gulf of Cambay.

The East Cambay Basin Margin Fault (ECBMF) cuts across the Mahi River, which is manifest in the change in orientation of the river. Several N-S and NNW-SSE trending fractures that developed parallel to the ECBMF have strongly influenced the drainage pattern of the river (Figure 1). Subsurface data indicate the presence of pre-existing step faults parallel to the ECBMF that acted as the depocenter of late Quaternary sedimentation. The Mahi basin lies in a horst segment and has over 300 m of Quaternary sediments overlying the Tertiary rocks (Maurya et al., 1995). Quaternary alluvium in the basin has source in the Precambrian and Cretaceous highlands located in the north and northeast.

Varying thickness of Quaternary alluvium (10–40 m) exposed along the river course indicate that sedimentation in the Mahi basin was influenced by tectonics, climate and the sea level (Pant and Chamyal, 1990; Raj et al., 1998; Juyal et al., 2000; Chamyal et al., 2003). The Mahi River flows through the Aravalli orogenic mountains and the Cambay graben, both of which have been tectonically active domains. Based on morphological and morphometric characters of the river basin, the role of tectonics in the geomorphic evolution of the Mahi basin has been emphasized (Maurya et al., 1998; 2000; Raj et al., 1999). The degree of tectonic activity is indicated to increase from the alluvial tract towards the uplands. Tectonic activity is considered to be at its peak between 10-6 ka in the study area (Maurya et al., 2000).

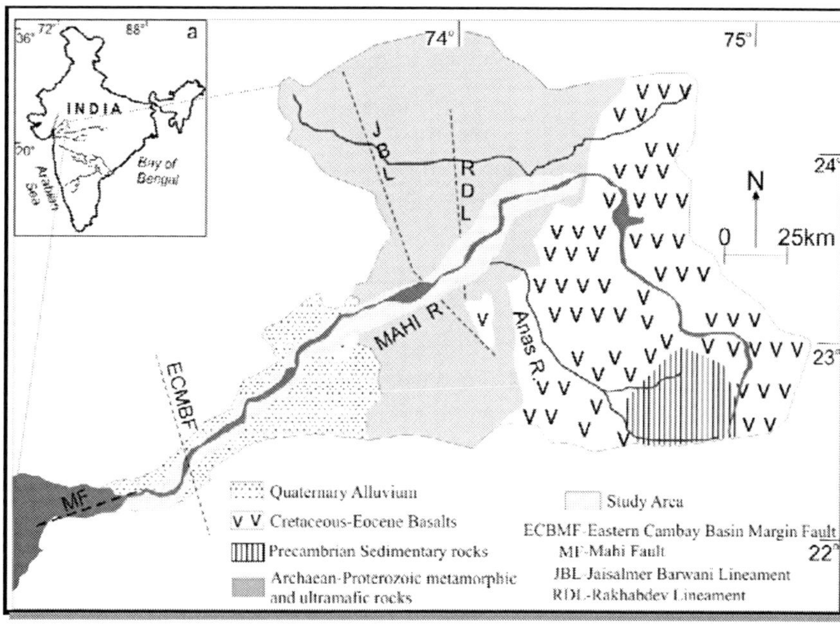

Figure 1. Location of the study area, the generalized geology and major lineaments in the Mahi River Basin (after Raj et al., 1999).

The extensive incision in the river valleys of the arid western India, and the formation of intruded valleys such as that along the Mahi River, is supposed to be related to this phase of tectonic uplift. Two phases of uplift, the early Holocene and post 2 ka have been proposed for the study area, and Maurya et al., (2000) suggest that development of the present landscape is a result of both. Maurya et al. (1998) reported soft sediment deformation structures from the lower Mahi valley as being attributable to a seismic event bracketed between 3,300 and 2,800 yr BP. Agarwal et al. (1996) identified a palaeo-delta of ancestral Mahi River towards the west of present day river and suggested that the palaeo Mahi had 5–10 km wide flood plain.

METHODOLOGY

The fluvial terraces were studied by detailed geomorphologic field mapping of the Mahi River valley. The mapping consisted of the delineation of the preserved terraces and their characteristics: elevation above the present-day river channel, extent and elevation of bedrock surfaces, thickness of fluvial deposits, and state of preservation. The terraces were marked in the area between the river mouth and the piedmont zone beyond which well developed terraces do not exist.

Table1. Equations used for palaeohydrological estimations

Equations used	Developed by
$\omega = 0.079d^{1.27}$ ($10 \leq d \leq 1500$mm)	Williams (1984)
$W_b = 3/2$ pointbar width	Williams (1988)
$D_b = 0.027\ L_m^{0.66}$	Williams (1988)
$Q=0.027\ W_b^{1.69}$	Osterkamp and Hedman (1982)
$Q_b=4.0A_b^{1.21}S^{0.28}$	Williams (1978)

The sediments were collected for comparison of the size fractions for various terraces. Channel cross sections as well as the terrace dimensions were measured wherever necessary for palaeohydrological estimations. The channel dimensions, palaeodischarge and palaeocompetence were estimated using the empirical relations (Table 1). Based on the empirical relations given by Williams (1988), estimates of channel width and depth were made from parameters such as present and palaeo meander wavelengths and pointbar dimensions. The present and palaeo-planform properties such as the meander wavelength (Lm) and the sinuosity index were determined using topographic maps.

When we employ the relation between the width of the pointbar at its apex and the width of the present channel, the measured width values are similar to those estimated. The estimation of bed shear stress and unit stream power for the channel was made using the formulae by Baker and Costa (1987). The chronology is based on the already existing ages of the sediments through IRSL, OSL and radiocarbon dating for some of the terraces.

GEOMORPHIC, LITHOLOGIC AND PALAEOHYDROLOGICAL DATA OF THE TERRACES

The semi-arid Mahi River channel is bounded by indurated Pleistocene terrace sediments forming 30–40m high cliffs that provide high resistant boundaries and restrict floodplain widths, channel migration and splay formation. If we assume that the extent of the river basin has remained essentially constant since the end of the Pleistocene, changes in channel morphology will be largely the product of climate changes and associated variations in flow regime. The terraces in the Mahi River basin occur along the river banks in the piedmont, alluvial and estuarine zones (Figure 2). Typically, these terraces are observed exposed along the river at elevations between 5-12m above the river level. In the piedmont zone, the terrace comprises mainly gravels that overlie eroded bedrock surfaces, with terrace remnants and exposures present on both river banks, whereas the alluvial zone sections through the discontinuous terrace remnants exhibit typical point bar facies near the base, which in turn are overlain by floodplain facies (Sridhar, 2007).

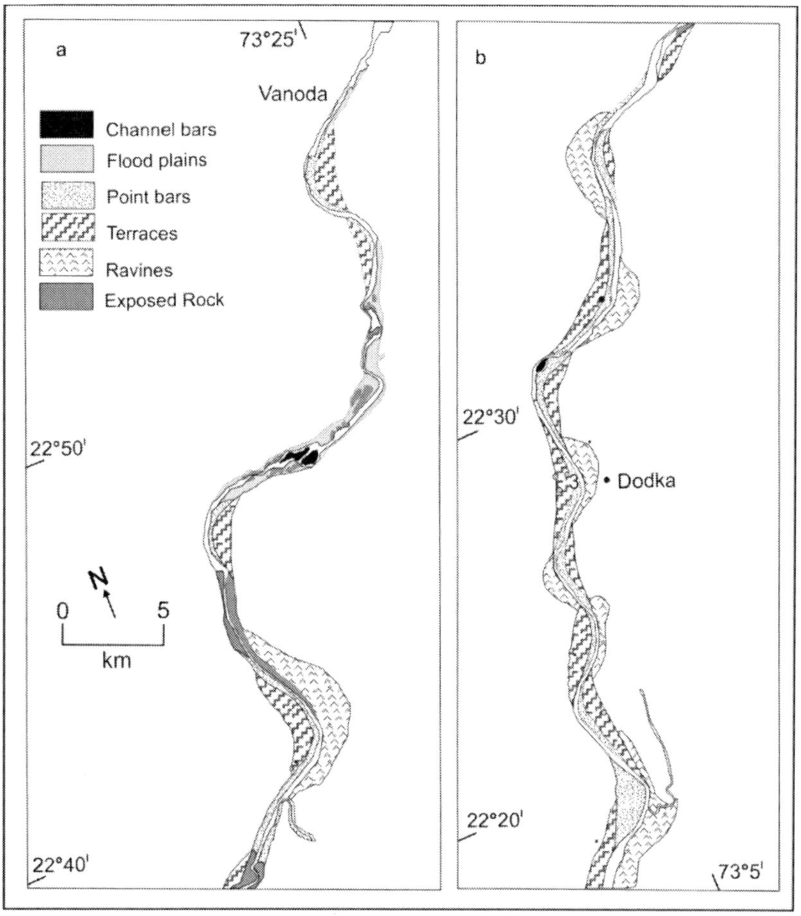

Figure 2. Geomorphic map of the study area. (a) Landforms exposed in the piedmont zone and (b) Landforms exposed in alluvial-estuarine zones (Sridhar, 2007a).

The terraces in the estuarine zone comprise of fine silt and clay intercalations. The gravel has been dated to 9.2 ka in the upper reach (Sridhar et al., comm.) whereas further downstream, the base gravel has been dated to 6,400± 120 yr BP (Maurya et al., 2000).

Terraces recorded in the estuarine zone are interpreted as valley fill features (Maurya et al., 1997) and their basal sediments have been dated to 3660 ± 90 ^{14}C yr BP (Kusumgar et al., 1998). The sediment sequences of the terraces exhibit gravel bar, pointbar and floodplain facies and their elevation above the contemporary river level indicates a higher bankfull stage and enhanced flow regime at the time of deposition.

Palaeohydrological parameters determined for the early Holocene gravel terrace in the piedmont zone are seen to vary with changing facies assemblages. The palaeocompetence of the flow has been estimated using the maximum clast sizes and the unit stream power estimates range between 57 Wm^{-2} - 17.41 Wm^{-2}. The flow velocity estimates range from 2.94 to 1.22 m/s^{-1} based on the clast size. The thickness of the cross bed sets varies between 0.2-0.5 m and the corresponding depth of the channel is between 2-4m. The average and peak flow discharges quantified based on the equation by Williams (1984) are ~17 m^3s^{-1} and ~ 760 m^3s^{-1} respectively. Based on relationships by Osterkamp and Hedman (1982) for mean discharge and that by Williams (1988) for bankfull discharge the estimated mean annual and bankfull discharges for the mid–late Holocene channel is ~3400 and ~55000 m3 s−1 and at present it is ~175 and ~2000 m^3s^{-1} respectively. The shear stresses (τ) at bankfull flows for the mid–late Holocene and present channels are 6.8 and 2.2 N m^{-2} and the unit stream powers (ω) are 15.0 and 2.7W m^{-2}, respectively (Sridhar, 2007).

Discussion

The exposed terrace sequences in the Mahi river basin exhibit fining of sediment facies from the upland to the alluvial-estuarine zones. These terraces provide evidence of braided gravel-bed river to a gravel bed wandering fluvial style during early Holocene and a wide, large-amplitude meandering channel form during the mid–late Holocene. The gravel was deposited in longitudinal bars, the progressive abandonment of which marks the transition between the low sinuosity braided channel and the present day meandering river. The maximum unit stream power for the early Holocene gravel terrace in the piedmont zone is an order of magnitude higher than that for the mid Holocene terrace in the alluvial reach. The early Holocene discharge was almost the same as the present day. However, the average discharge of the mid–late Holocene channel has been estimated to be much greater than that of the present river.

The available chronology of the exposed sediment sections suggests that whereas early Holocene sediments have been preserved in form of terraces in the piedmont zone, the terrace sediments in the alluvial and estuarine zone are mid-late Holocene in age. The river dynamics is a function of climate-related boundary conditions like channel morphology, water flow and sediment transport (Bogaart, 2000).

The reduction in stream power with time is indicative of aggradation of the river from the early Holocene to mid–late Holocene and the present. Phases of incision and aggradation can be explained by changes in sediment supply or discharge and may not be necessarily coupled. The incision/aggradation and channel pattern are a function of the ratio between sediment supply and bankfull discharge. The time lag between changes in river discharge and sediment supply determines the river response to changes. The multi cyclic stacking of the gravels in the early Holocene terrace sediments is an indication of the vertical adjustment of the river suggesting that the discharge and sediment supply were not in phase. The discharge for the early Holocene channel was similar to the present day channel (Sridhar et al., comm.), the competence was much higher and the sediment load was coarser than the present. It can be inferred that the river flow pre 9.2 ka BP had low transport capacities and high sediment supplies resulting into aggradation. However, post 9.2 ka probably the transport capacity of the flow increased related to the enhanced precipitation resulting into the deposition of gravels further downstream. This implies that the increased sediment supply was compensated by an increase in transport capacity and sediment moved faster through the channels. However, as the precipitation further increased, the sediment supply and the discharge went out of phase once again but this time the transport capacity exceeding the sediment supply resulting into the incision of the gravel deposits forming terraces. This implies that either the discharge increased or there was a significant reduction in the sediment supply. An intensified summer monsoon with increased fluvial activity during the early Holocene following the reduced but erratic fluvial activity during LGM has been envisaged

(Sirocko, 1993). Periods of alluviation are marked at 9.8-9.0 cal ka BP and 10.8-8.9 cal ka BP respectively in the Central Ganga Basin and Deccan Peninsular rivers of India (Kale, 2007). Short term abrupt climatic events of reduced monsoon strengths have also been identified at 8.2, 4.7 and 4.2 cal ka BP (Staubwasser et al., 2003). Based on the numerical modeling Bogaart (2000) demonstrated that high transport capacity and low sediment supply lead to maximal incision. It can be therefore inferred that the incision of the gravel deposits was more in response to the increase in the discharge and competence of the flow when the precipitation enhanced. The warming of climate, high precipitation and low sediment supply post ~9 ka resulted into the down cutting. River down cutting has been associated to warming transitions (Bridgland and Westaway, 2008) and also cooling transitions (Lewin and Gibbard, 2010) in glacial areas. The down cutting related to climate warming and reduced sediment supply has also been reported (Olszak, 2011). Prevalence of erosion in the fluvial systems has been recorded mainly in the transitional phases (Vandenberghe, 2002; Starkel, 2003).

The transition in the Mahi valley from a phase of low precipitation and high sediment supply to the phase of high precipitation and low sediment discharge was marked by erosion resulting into the widening and the deepening of the valley. This actually created accommodation space in the lower alluvial reaches where extensive aggradation occurred at a later date. The braiding pattern of the channel can also cause lateral erosion of the valley slopes. The widening of the valley has been attributed to change in river regime and pattern (Hancock and Anderson, 2002) and a reduced river gradient (Starkel, 2003). In this case, the early Holocene channel has been inferred as braided with low discharge and hence the lateral erosion because of the braiding would be minimal. As spatially the discharge increases downstream as compared to the upstream, lateral erosion may have occurred in these reaches. According to Bogaart (2002), given equal changes in the forcing factors, changes in discharge have the most impact on incision/aggradation, and changes in sediment supply have the highest impact on channel pattern. The river channels are constantly meandering in the downstream reach and are always braided type in the upper reaches whereas in the middle reaches the river behavior is more dynamic and the river switches from meandering to braided and back (Vandenberghe, 1995) as in the present case.

The wide terraces in the alluvial reach are a testimony to large sedimentation that occurred post ~6 ka as the gravel at the bottom of the terrace has been dated to 6,400 yr BP (Maurya et al., 2000). The river was meandering but less sinuous than the present, had high discharge and low competence (Sridhar, 2007a). The high discharge led to the erosion of the late Pleistocene sediment terraces exposed along the river banks. The mid Holocene terraces therefore comprise finer sediments as compared to the early Holocene terrace. The fine sediments were deposited in the widened valleys; periodic deepening and frequent shifting of channels occurred related to the high magnitude flood events between ~5 ka and 2 ka (Sridhar, 2007b). As inferred from the age of the top surface sediments of the terraces (Kusumgar et al., 1998) and the palaeoflood deposits (Sridhar, 2007b), the deposition ceased around 2 ka. However, the palaeohydrological estimates indicate a further reduction in the discharge and the channel width also decreased around this time (Sridhar, 2007a). Therefore the incision of these sediments may be considered to be independent to the discharge conditions. The possible reason for the incision may be then related to a much reduced sediment supply, the river response to the change in sea level or tectonic activity.

The inferences drawn here have implications for the reconstruction of climate events based on documented river response in the sediment record. The phases of incision and aggradation cannot be possibly related to changes in sediment supply without considering discharges as well. The link between reconstructed incision/aggradation events and palaeohydrology based on channel width and related bankfull discharge can be used as a proxy. Aggradation and incision occur only when the discharge and the sediment supply are not in phase and there is a time lag.

The results suggests that discharge varied spatially and temporally during Holocene, the incision and aggradation of the sediments was controlled by fluctuations in sediment influx and the river responded to these changes by switching of stream patterns.

CONCLUSION

There have been phases of aggradation followed by incision in the Mahi River Basin during the Holocene and palaeohydrological estimations have helped in inferring the relation between the flow conditions and the aggradation -incision phases. Whereas the aggradation of the gravel terrace in the peidmont zone during the early Holocene was controlled by the large sediment influx, the incision that followed was in response to the increase in the discharge and competence of the river flow. On the other hand, aggradation in the alluvial zone during the mid-Holocene occurred in response to the fluctuating discharge conditions related to the fluctuating monsoon. The phase of incision that followed however was possibly independent of the discharge conditions.

ACKNOWLEDGMENTS

The financial support from the Department of Science and Technology under WOS (A) scheme (DST/WOS-A/ ES-35/2003) to Alpa Sridhar is thankfully acknowledged.

REFERENCES

Agarwal, R. P., Dotiwala, S., Mitra, D. S., Bhoj, R., 1996. The palaeodelta of the "Proto" Vatrak and "Proto" Mahi rivers of northeastern Gujarat, India: A remote sensing interpretation. *Geomorphology* 15, 67–78.

Baker, V.R., Costa, J.E., 1987. Flood power. In: Mayer, L., Nash, D. (Eds.), *Catastrophic Flooding*. Allen and Unwin, London, pp. 1–21.

Biswas, S. K. 1987. Regional tectonic framework and structural evolution of western marginal basins of India. *Tectonophysics* 135, 307–327.

Blum, D.M., Törnqvist, T.E., 2000. Fluvial responses to climate and sea-level change: a review and look forward. *Sedimentology* 47 (1), 2–48.

Bogaart, P.W., van Balen, R.T., 2000. Numerical modelling of the response of alluvial rivers to Quaternary climate change. *Global and Planetary Change* 27, 147–163.

Bridgland, D. R.,Westaway, R., 2008. Climatically controlled river terrace staircases: a worldwide Quaternary phenomenon. *Geomorphology* 98, 288-315.

Chamyal, L. S., Maurya, D. M., Rachna Raj., 2003. Fluvial systems of the drylands of western India: a synthesis of Late Quaternary environmental and tectonic changes. *Quaternary International* 104, 69–86.

Chamyal L.S, Juyal, N. 2008. Late Quaternary continental studies in parts of India: implications for monsoon variability. *Journal of Geological Society of India* 71, 611-629.

Hancock, G.S., Anderson, R.S., 2002. Numerical modelling of the fluvial strath-terrace formation in response to oscillating climate. *Geological Society of America Bulletin* 114, 1131–1142.

Juyal, N., Rachna, R., Maurya, D. M., Chamyal, L. S., Singhvi, A. K., 2000. Chronology of Late Pleistocene environmental changes in the lower Mahi basin, western India. *Journal of Quaternary Science* 15, 501–508.

Kale, V S., 2007. Fluvio-sedimentary response of the monsoon-fed Indian rivers to Late Pleistocene-Holocene changes in Monsoon strength: reconstruction based on existing ^{14}C dates. *Quaternary Science Reviews* 26, 1610-1620.

Kusumgar S., Rachna R., Chamyal L.S., Yadav M.G., 1998. Holocene palaeoenvironmental changes in the lower Mahi basin, western India. *Radiocarbon* 40, 819-823.

Lewin, J., Gibbard, P.L., 2010. Quaternary river terraces in England: forms, sediments and processes. *Geomorphology* 120, 293–311.

Maurya, D. M, Malik, J. N, Rachna Raj, Chamyal., L. S., 1997. Soft sediment deformation in the Quaternary sediments of the lower Mahi river basin, western India. *Current Science* 72, 519–522.

Maurya, D. M., Chamyal, L. S., Merh, S. S., 1995. Tectonic evolution of the Central Gujarat plain, western India. *Current Science* 69, 610–613.

Maurya, D. M., Malik, J. N., Rachna R., Chamyal, L.S., 1998. Seismically induced deformational structures from Mid-Late Holocene terraces in lower Mahi valley, Gujarat. *Journal Geological Society of India* 51, 755-758.

Maurya, D., Rachna, M., Chamyal, R., 2000. History of tectonic evolution of Gujarat Alluvial Plains, Western India during Quaternary: a review. *Journal of Geological Society of India* 55,343–366.

Osterkamp, W.R., Hedman, E.R., 1982. Perennial-streamflow characteristics related to channel geometry and sediment in Missouri River basin. *USGS Professional Paper* 1242, 847–851.

Olszak J 2011. Evolution of fluvial terraces in response to climate change and tectonic uplift during the Pleistocene: Evidence from Kamienica and Ochotnica River valleys (Polish Outer Carpathians) *Geomorphology* 129, 71–78.

Pant, R. K., Chamyal, L. S., 1990. Quaternary sedimentation pattern and terrain evolution in the Mahi river, Gujarat, India. *Proceedings of Indian National Science Academy* 56, 501–511.

Raj R, Maurya, D. M., Chamyal, L. S., 1998. Late Quaternary sea level changes in western India: evidence from Lower Mahi valley. *Current Science* 79, 910– 914.

Raj R., Maurya, D. M., Chamyal, L.S., 1999. Tectonic geomorphology of the Mahi river basin, Western India. *Journal of Geological Society of India* 54, 387-398.

Sirocko, F., Sarnthein, M., Eriekeuser, E., Larnge, H., Arnold, M. and Duplessy, J. C. 1993. Century scale events in monsoonal climate over the past 24,000 years. *Nature* 364, 322–324.

Sridhar, A., 2007a. Mid-late Holocene hydrological changes in the Mahi River, arid western India. *Geomorphology* 88, 285-297.

Sridhar, A., 2007b. A mid-late Holocene flood record from the alluvial reach of the Mahi River, western India, *Catena* 70, 330–339.

Sridhar, A., Chamyal, L.S., 2010. Sediment records as archives of the Late Pleistocene–Holocene hydrological change in the alluvial Narmada River basin, western India. *Proceedings of Geologists' Association* 121, 195-202.

Sridhar, A., Chamyal, L.S., Bhattacharya, F., Singhvi, A.K., (comm.) Early Holocene fluvial activity from palaeodischarge and palaeocompetence estimations of a gravel terrace in the Mahi River Basin, India.

Starkel L., 2003. Climatically controlled terraces in uplifting mountain areas. *Quaternary Science Reviews* 22, 2189–2198.

Staubwasser, M. F., Sirocko, F., Grootes, P.M., Segl, M., 2003. Climate change at the 4.2 ka BP termination of the Indus valley civilization and Holocene south Asian monsoon variability. *Geophysical Research Letters* 30, 1425.

Vandenberghe, J., 1995. Timescales, climate and river developments. *Quaternary Science Reviews* 14, 631–638.

Vandenberghe, J., 2002. The relation between climate and river processes, landforms and deposits during the Quaternary. *Quaternary International* 91, 17–23.

Williams, G.P., 1988. Palaeofluvial estimates from dimensions of former channels and meanders. In: Baker, V.R., Kochel, R.C., Patton, P.C., (Eds.), *Flood Geomorphology*. Wiley, New York, 321-334.

In: Holocene
Editor: Bahadur Singh Kotlia
ISBN: 978-1-62257-722-4

Chapter 4

THE HOLOCENE EVENTS: DATABASE ON THE KONKAN-KERALA COAST

C. P. Rajendran[*]
Centre for Earth Sciences, Indian Institute of Science,
Bangalore, India

ABSTRACT

In this chapter, we evaluate the regional database on the Holocene studies from the west coast of India that stretches from Kanyakumari at the southern extremity to the vicinity of Mahim, located to the north of Mumbai. The database provides insights into the morphological changes and also brings out the timings of the regional Holocene events including the sea level still stands and climatic events. During the last glacial maximum (LGM; 18,000 yr BP) much of the western continental shelf of India was aerially exposed and the coast extended up to the present shelf edge. The changes in the existing LGM coastal morphology were set in motion by the beginning of the Holocene when the sea started transgressing the shelf region, punctuated with still stands, until it reached the present stage, profoundly transforming the landscape. The radiocarbon dates from the region indicate a major transgression during 8,000-9,000 yr BP, which impacted the then existing luxuriant coastal mangrove vegetation. The transgression event coincided with a strong monsoonal regime, helping a deciduous evergreen vegetation to flourish in the mountainous region of the Western Ghats. The brief spell of strong monsoon lasted until 6,000 yr BP., ending in a cycle of a weak summer monsoon. The consequent aridity that ensued around 3,500 yr BP resulted in the destruction of the remaining coastal mangrove vegetation. The arid climatic conditions were succeeded by a stronger spell of monsoon that also coincided with a phase of marine transgression (4,000-2,500 yr BP). The present architecture of the Konkan-Kerala Coast characterized by barrier-island system was stabilized during this interval.

Keywords: Holocene, Database, Kerala Coast

---

[*] E mail: cprajendran@ceas.iisc.ernet.in.

INTRODUCTION

The Holocene Epoch is the last and latest major subdivision of the geological record, extending to the present day. Identified as a warm period, it is considered to be an interglacial that began around 10,000 years ago. The most characteristic feature of this period is the frequency of climatic changes and oscillations in an extremely short time span. The effects of these climatic changes are drastic, with phases of aridity alternating with periods of increased precipitation, having a consequent impact on weathering rates and pedogenic processes.

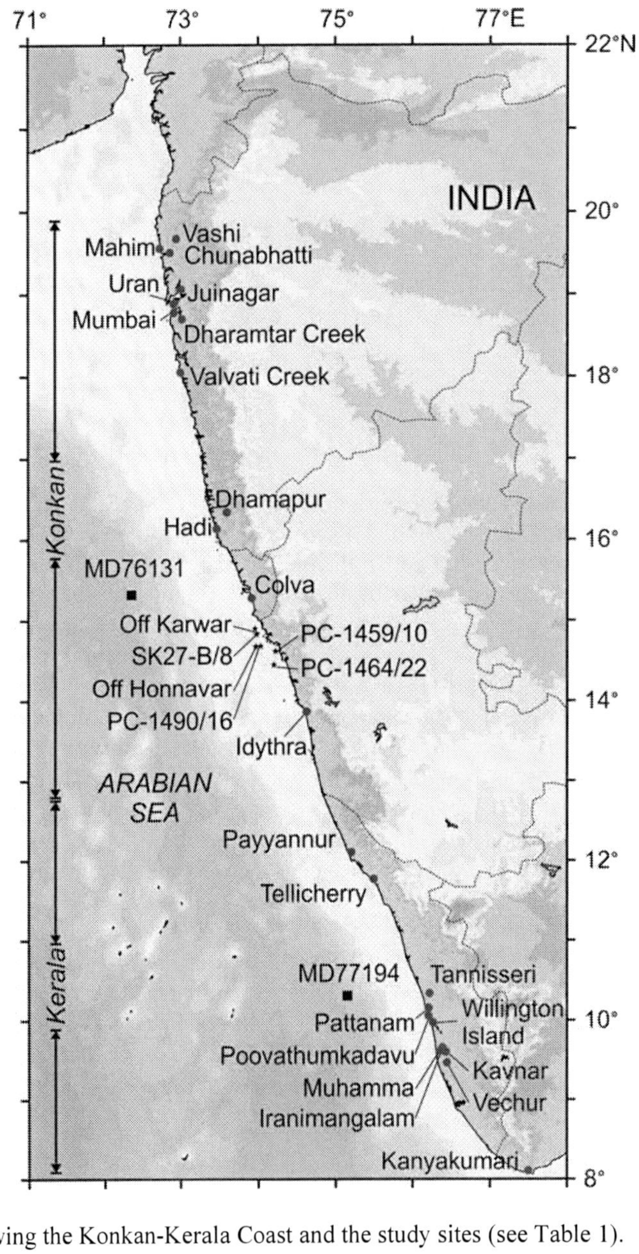

Figure 1. Map showing the Konkan-Kerala Coast and the study sites (see Table 1).

The river regimes and sea levels fluctuated markedly while plant and animal populations, including the human species, were forced to migrate or perish as the rapidly changing environment was thrust upon them. The dynamic nature of the Holocene environment is reflected in the complex mosaic of landforms, sedimentary sequences, faunal and floral elements. For the Konkan-Kerala region, the Holocene (and the Pleistocene) record is highly fragmented and absent in many areas due to high precipitation and the erosional activity of streams, resulting in detailed sequences being preserved only locally. However, local depositional centers, like the lakes, the backwaters and the offshore regions, can profitably be explored for sedimentary sequences that archive the major natural events. In the following sections, we present the existing database to discuss the Holocene events like the sea level variations and the changing climatic conditions that prevailed along the Konkan-Kerala coast, an area extending from Mumbai in the north to Cape Comorin (Kanyakumari) in the south (Figure 1).

Sea Level Changes

The Holocene evolution of the continental shelf off Konkan-Kerala is closely associated with the intermittent eustatic sea level variations. For example, the submerged terraces found at -92, -85, -75 and -55 m below the present sea level have been attributed to four still stands of the sea during the Holocene (Nair, 1974; 1975). Warmer climate and low terrestrial run-off were considered to be other attributes for the Holocene Epoch, evidenced by the formation of oolitic limestone on continental shelf off Kerala (Nair and Hashimi, 1980). Recent studies have provided further clarity to the complex spatio-temporal variations of sea level, which also lead to emergence of a regional picture extending toward the Konkan part of the continental shelf. In a more recent study, Wagle and Veerayya (1996) identified parallel rows of morphologically conspicuous submerged sand ridges on the mid and outer continental shelf, 100 to 200 km off Mumbai. They also identified inter-ridge depressions that may have been paleo-lagoons. They suggest that these are signatures of transgression in an episodic fashion (still stands) occurring in the late Pleistocene to the early Holocene. The dates obtained from the carbonate sand and limestone around the ridges range from 8,340±185 to 11,330±350 yr BP (Wagle and Veerayya, 1996). The offshore data also revealed the presence of a buried coral reef at the edge of the western continental shelf (-136 to -85 m) extending for a distance 1300 km from Saurashtra to Kanyakumari (Vora et al., 1996). This feature essentially suggests an exposed shelf during the Pleistocene, which got submerged by a rise in the sea level at the beginning of the Holocene. The coral reef samples provided ages ranging from 11,330±350 to 9,200±135 yr BP (Vora et al., 1996). The sea level curve for the west coast of India by Wagle et al. (1994) suggests a still stand at around 12 ka, followed by a rapid rise in the sea level at 10-11 ka. The sand ridges and shelf-edge reefs validate that assumption. Remains of shallow water carbonate deposits have also been identified along the outer part of the western continental shelf from the Gulf of Kachchh to Kanyakumari, which provide further evidence for the Pleistocene-Holocene sea level variations (Rao et al., 2003). All the aforementioned studies suggest a subsidence of the continental shelf region during the beginning of the Holocene. Although subsidence can be attributed to tectonism, the question remains as to how this can be explained for a region that forms a part of a stable continental craton. In an earliest attempt to study the sea level variations within the Holocene on the western Indian shelf, Nair (1974) provided a curve based on the available ages of samples from the continental shelf. Later, Kale and Rajaguru (1983, 1985) have inferred a lowering of the sea level to the tune of 1.5 to 2.5 m at about 6.5 ka BP, and have also obtained a sea level curve at the closing phase of the Pleistocene to the late Holocene. In their work on the Neogene and Quaternary marine transgressional and regressional history along the west coast of India, they attributed the sea level variations through time to global eustatic changes as well as to tectonic movements. Agrawal (1990) considered the west coast of India to be of the emergent type and identified a sea level rise to the order of 0.3 m. Merh (1992) observed that the Last Glacial Stage was a period of regression and the sea surface during that period was about -150 m, below the present level. The sea level had risen since then and reached its maximum during 4,000 – 6,000 yr BP, and the rise was gradual with several fluctuations and still stands. According to Rao and Ramanujam (1976) sea level changes in the region are purely the product of climatic, eustatic or isostatic phenomena, while Ahmed (1972) supports the theory of regional uplift. Thrivikramji and Ramasarma (1983) inferred a 4-5 m fall in sea level during the late Holocene. The peat beds, that are prominent units within the shallow coastal sections in Kerala, are considered to represent the submergence of coastal forests (Power et al., 1993; Rajendran et al., 1989), although it was not clear what processes led to submergence.

Lack of reliable geological proxies is a major hindrance in developing well-calibrated sea level curves for the region. For example, the Paleolithic sites in Kerala provide no convincing stratigraphic control and are devoid of any organic material to provide any meaningful dates. The Mesolithic sites occur on the lateritic surfaces, more than 65 m above MSL, are no help either. Some age controls, however, are derived from the stratified microliths discovered within the 'Teri' dune sands, which were dated at 6 ka BP, and therefore were assigned to the late Atlantic phase of post-glacial transgression.

This lends support to the theory of rise in sea level (2 to 3 m) during the mid-Holocene (Gardner and Martingell, 1990). The Teri dunes were generally considered to have formed during the Last Glacial Maxima (LGM) when the sea level was the lowest and the shelf sands were exposed and reworked by the wind (Rajendran et al., 1986; Thrivikramji et al., 2008).

However, Thrivikramji et al. (2008) obtained a date of 3,680±110 yr BP from rhizolithic calcrete from the Teri sands, which suggests arid condition in southern Tamil Nadu during the late Holocene. These authors suggest an initial semi- arid phase in late Pleistocene (18 ka BP), an intervening humid phase and late Holocene hot arid climate. The luminescence dating methods helped Singhvi et al. (1986) to confirm that the red sand beds of Sri Lanka to be of last glacial period.

As compared to Kerala, the Konkan Coast fares slightly better as the onshore information regarding eustatic changes can be derived from various preserved geomorphic features that are suggestive of emergence and submergence of the coast due to sea level fluctuations (Prudhi Raju and Vaidyanathan, 1977).

The radiocarbon dates obtained from the carbonized wood and continental sediments from Mangalore indicate sea level changes around 7,000 yr BP (Agrawal et al., 1970). Nigam and Nair (1989) attempted to study the monsoonal variations during late Quaternary based on foraminiferal analyses from core samples off Karwar. Agrawal and Guzder (1972) and Agrawal et al. (1970) identified different stages of transgression and regression along the Konkan Coast. Sukhantar and Pandian (1990) identified various geomorphic features related to marine regression along the Maharashtra Coast during the Holocene, and highlighted the role of neotectonism in moulding the coast line of this region. Age data of the shell deposits of the Konkan Coast suggest that the Holocene transgression was maximum around 2,000-2,500 yr BP (Brückner, 1989). An independent shell date of 3940 ± 100 yr BP of a beach rock 5 m above the MSL from Mahim (Kelva beach), north of Mumbai obtained from our studies suggests that the transgression must have started much earlier (Figure 1; Table 1).

In Kerala the Holocene geology is mostly remians under soil cover and there are few surface exposures. The occurrences of peat sequences in the low-lying areas around the Vembanad Lake in the Kottayam district of Kerala suggest a submerged forest in this area (Pawar et al., 1983). Later studies and the available age data also indicate the development of a dense coastal forest not only along the Kerala coast, but extended all along the western coast of India, including the Konkan part. The age data of the peat beds suggest an episode of marine transgression around 8,000 yr BP.

The thick development of peat sequence in some parts of the coastal areas of Kerala is a conspicuous feature of the subsurface Holocene geology of the region (Rajendran et al., 1989; Kumaran et al., 2005; Table 1). These peat and carbonized wood deposits are dated at 9-6 ka BP. Evidence of thick mangrove vegetation along the west coast at the beginning of the Holocene is validated also by a recent set of dating results.

We report here an independent radiocarbon date of a carbonized wood, collected at the bottom-most part of a 5-m-deep well section at Pattanam located on a branch of Periyar River on the coastal part of the Ernakulam district, Kerala. The site is known for archeological artifacts that suggest this to be an early inland port town (Muziris?) dating to BC 2nd to AD 2nd centuries (Cherian et al., 2009).

The open dug well was located near the excavation sites in Pattanam. The well-section consisting of thick layers of fine grained sand is suggestive of a paleobeach environment, and the carbonized wood occurs as a distinct layer under this sand. The age of the sample obtained in our study (10,490-11,200 or 9,240-8,470 yr BP; Table 1) is comparable to ages of peat /carbonized wood from onshore and offshore parts of the Kerala Coast. In a seminal work, Van Campo (1986) suggested a very humid period around 11 ka BP with heaviest rainfall on the basis of pollen content derived from an ocean core on the southwest coast of India at 10°N latitude (Figure 1 for location of the offshore coring sites- MD 76131 and 77194). The pollen in the core essentially represented the coastal mangrove vegetation.

Another interesting aspect of the Holocene stratigraphy of the region is the widespread occurrence of shell beds. In many localities in Kerala, the shell deposits occur above the peat beds or carbonaceous clay, as reported from the Konkan Coast. Rajendran et al. (1989) reported shell dates from various parts of the Kerala Coast that range from 4,500 to 2,500 yr BP (Table 1). At Payyanur in the north Kerala, the shell deposit occurs about 3 meters above MSL whereas the shell deposit at the Vembanad Lake in the central Kerala occurs two meters below the lake level (Rajendran et al., 1989).

These dates match with that of the shell deposits reported by Brückner (1989) and other workers from the Konkan Coast that implied a more recent marine transgression. Thus the results suggest that this transgression event must have been regional and affected the entire western coast.

Table 1. Radiocarbon ages of peat and shell deposits from the Konkan-Kerala Coast

Conventional Radiocarbon Age (BP)	Reference	Material dated, sample / lab ID Sample Location	Calibrated Age Range (cal. yr. BP)
10760 ± 130	Pandarinath et al., 2001	Organic matter in sediment; 14°40'N 74°E, off Honnavar, Karnataka	12400 - 12970 12685 ± 285
9790 ± 120	Pandarinath et al., 2001	Carbonized wood peat; 14°40'N 74°E, off Honnavar, Karnataka	10760 - 11610 11185 ± 425
9630 ± 120	Nambiar et al., 1991	Carbonized wood, RVSS-PC-1464/2; 14°25.62'N 74°12.59'E, Arabian Sea	10660 - 11240 10950 ± 290
9510 ± 130	This study	Carbonized wood; 10°09.43'N 76°12.56'E, Pattanam, near Kochi, Kerala	10490 - 11200 10845 ± 355
9280 ± 150	Pandarinath et al., 2001	Peat and organic matter in sediment; 14°50.4'N 74°E, off Karwar, Karnataka	10180 - 10880 10530 ± 350
8620 ± 300	Nambiar et al., 1991	Carbonized wood, (RVSS)PC-1490/16; 14°40.14'N 73°59.16'E, Arabian Sea	8980 - 10440 9710 ± 730
8315 ± 125	Agrawal et al., 1970	Wood; Willington Island, Kochi, Kerala	9020 - 9520 9270 ± 250
7820 ± 120	Kumaran et al. 2004	Carbonaceous clay; Juinagar, Mumbai, Maharashtra	8420 - 8990 8705 ± 285
7820 ± 120	Pawar et al., 1983	Peat; Iranimangalam, 12 km ESE of Alleppey, Kerala	8420 - 8990 8705 ± 285
7450 ± 110	Shajan, 1998	Peat; Poovathumkadavu, 15 km NNW of Kochi, Kerala	8020 - 8430 8225 ± 205
7230 ± 120	Rajendran et al., 1989	Peat, BS711; Tellicherry / Thalassery, SSE of Kannur, Kerala	7830 - 8330 8080 ± 250
7230 ± 120	Caratini et al., 1990	Peat, Tellichery / Thalassery, SSE of Kannur, Kerala	7830 - 8330 8080 ± 250
7220 ± 110	Kumaran et al., 2004	Carbonaceous clay; 16°08'N 73°28'E, Hadi, Malvan, Maharashtra	7840 - 8220 8030 ± 190
7210 ± 110	Kumaran et al., 2004	Black clay; 16°20'N 73°36'E, Dhamapur, Malvan, Maharashtra	7820 - 8220 8020 ± 200
7090 ± 100	Nair et al., 2006	Black clay; Kavnar, 18km WNW of Kottayam, Kerala	7690 - 8060 7875 ± 185
6980 ± 170	Limaye, 2004	Carbonaceous clay; 19°40'N 72°57'E, Vashi, Thane, Maharashtra	7560 - 8160 7860 ± 300
6760 ± 200	Limaye, 2004	Carbonaceous clay; 18°69'N 72°55'E, Uran, Thane, Maharashtra	7270 - 7970 7620 ± 350
6720 ± 70	Shajan, 1998	Peat; Poovathumkadavu, 15 km NNW of Kochi, Kerala	7460 - 7680 7570 ± 110
6430 ± 110	Kale and Rajaguru, 1983	Peat; 15°N 74°E, Colva, Goa	7160 - 7570 7365 ± 205
6430 ± 120	Caratini et al., 1990	Peat; Idythra, Karnataka	7160 - 7570 7365 ± 205
6420 ± 120	Rajendran et al., 1989	Peat, BS689; Tannisseri, Irinjalakuda, Kerala	7160 - 7570 7365 ± 205
5230 ± 100	Limaye, 2004	Carbonaceous clay; 18°69'N 72°55'E, Uran, Thane, Maharashtra	5840 - 6220 6030 ± 190

Conventional Radiocarbon Age (BP)	Reference	Material dated, sample / lab ID Sample Location	Calibrated Age Range (cal. yr. BP)
4610 ± 110	Limaye, 2004	Carbonaceous clay; 19°30'N 72°52'E, Chunabhatti, Mumbai, Maharashtra	5030 - 5590 5310 ± 280
4490 ± 90	Rajendran et al.,1989	Limeshell, BS673; Payyannur, Kerala	4210 - 4810 4510 ± 300
4370 ± 100	Rajendran et al., 1989	Limeshell, *BS674;* Payyannur, Kerala	3980 - 4660 4320 ± 340
3940 ± 100	This study	Shell-rich rock; Mahim (Kelva beach), Mumbai, Maharashtra	3470 - 4040 3755 ± 285
3710 ± 90	Rajendran et al., 1989	Limeshell, BS701; Vechur, Vembanad Lake, Kerala	3200 - 3760 3480 ± 280
3460 ± 100	Kumaran et al.,2004a	Carbonaceous clay; 16°08'N 73°28'E, Hadi, Malvan, Maharashtra	3480 - 3980 3730 ± 250
3410 ± 90	Nambiar et al., 1991	Shells, (RVSS)PC-1459/10; 14°35.99'N 74°13.83'E, Arabian Sea	2730 – 3230 2980 ± 250
3130 ± 100	Rajendran et al., 1989	Limeshell, BS717; Muhamma, Vembanad Lake, Kerala	2420 - 3080 2750 ± 330
2800 ± 110	Agrawal and Guzdar, 1972	Peat; Valvati Creek, north of Srivardhan, Maharashtra	2750 - 3170 2960 ± 210
2600 ± 60	Brückner, 1989	Shells; 18°41.82'N 73°01.83'E, Dharamtar Creek, near Mumbai, Maharashtra	1940 - 2300 2120 ± 180
2305 ± 95	Agrawal and Guzdar, 1972	Peat; Valvati Creek, north of Srivardhan, Maharashtra	2110 - 2550 2330 ± 220

The CALIB Radiocarbon Calibration program (version 6.1.1) that uses the 2009 international calibration datasets (INTCAL09 and MARINE09 ^{14}C curves) was used here for calibration (Stuiver and Reimer, 1993). ΔR values for marine carbonate samples from near Karwar and off the coast of Kerala are taken as 252 ± 51 and 138 ± 64 respectively. For the marine shell samples from in and around (Bombay) Mumbai, Maharashtra , the ΔR value is taken as 136 ± 36, averaged from 113 ± 50 and 165 ± 57 (Southon et al., 2002; http://radiocarbon.pa.qub.ac.uk/marine). The calibrated age ranges are the highest probability 2 sigma confidence intervals rounded off to the nearest decade. *BS*: Samples dated at the Birbal Sahani Institute of Paleobotany, Lucknow, India; RVSS- PC*:* Piston Core from Research Vessel Samudra Shaudhikama.

Paleoenvironmental and Climatic Signatures

The cores obtained about 100 km off Cochin and Karwar are probably the most representative archives of onshore environmental changes as they are closest to the Konkan-Kerala Coast (Van Campo, 1986). This study, based on the oxygen isotope records and pollen analyses, indicates three paleoclimatic events: the arid interval around the Last Glacial Maximum (LGM; 18,000 yr BP); the humid and wet climate during 11-10 ka BP (beginning of the Holocene); and the drier period from 4,000-2,000 yr BP. A later paper on cores from off Karwar (Nambiar et al., 1991; Figure 1) reporting the age of carbonized wood from the 9000-8000 year interval in fact provides further validation to the isotopic event intervals identified by Van Campo.

The paleoclimatic studies on the peat bogs of the Nilgiri Hills are equally revealing about the fluctuating environmental conditions during the last 20,000 years. The stable carbon isotope records from peats in the Nilgiri Hills (Figure 1) support arid conditions during the LGM (Sukumar et al., 1993). The stratigraphic interval representing the LGM showed the dominance of C4 plants indicating aridity. The relative proportions of C3 and C4 plants from the Nilgiri peats further suggest that from 16 ka BP, the C3 plants started dominating as more negative $\delta^{13}C$ values were detected, indicating the onset of stronger summer monsoon (Sukumar et al., 1993; Geetha et al., 1997). The humid interval peaked at the beginning of the Holocene. These findings agree with the paleoclimate interpretations derived from distant ocean cores from the Arabian Sea (Gupta et al., 2003; Gupta et al., 2006; Thamban et al., 2007).

Further, Sukumar et al. (1993) were able to capture the possible signatures of weaker monsoon starting from 6,000 yr BP, including a more recent Medieval Warm Period (AD 1200-1400). Cores collected from near-shore waters off Karwar provided evidence of drastic vegetation changes at around 3,500 yr BP as indicated by the increased proportion of pollen from savanna-type C4 plants (Caratini et al., 1991). This drastic vegetation change is also supplemented by the reduction in the sedimentation rate (18 cm/100 years during 4,500-3,500 years BP to 7 cm/100 years during 3,500 yr BP to Recent) implying the reduced erosional rates from the continent owing to a weaker monsoon.

Conclusion

The coast of the western India extended all the way to the current shelf edge during LGM (18 ka BP). The arid climatic regime possibly resulted in desert conditions with intermittent saline lagoons. The sea level had a gradual rise since then, with various still stands. A major transgression may have occurred around 9 ka BP, which probably flooded the mangrove vegetation. A strong monsoonal regime set in during the same period and mangrove vegetation thrived all along the Konkan-Kerala coast and deciduous evergreen vegetation flourished in the Western Ghats. The wet period probably lasted until 6 ka BP. Another cycle of arid conditions and a weak summer monsoon may have peaked around 3,500 yr BP. By this time most of the mangrove vegetation had been destroyed. Maximum landward marine transgression along the Konkan-Kerala Coast had occurred between 4,000- 2,500 yr BP. The sea probably regressed a little after 2,000 yr BP leading to emergence of the coast. The present barrier-island system of the region was developed during this interval. Typical broad strand plains and cliffed shorelines with or without a beach characterize the coastal tract of Kerala. The strand plains consist of series of sub parallel ridges formed by successive accretion of low narrow beach ridges parallel to the shorelines.

Acknowledgments

I thank Prof. B. S. Kotlia for inviting me to contribute to this volume, and Ms. Vanessa Andrade for helping with the figure and the text.

REFERENCES

Agrawal, D.P., Gupta, S.K., Kusumgar, S., 1970. Radiocarbon dates of Quaternary Samples. *Current Science* 30, 219-222.

Agrawal, J.M., 1990. Sea level variation-through bathymetric data example-Azhikkal on west coast of India. In: Sea level variation and its impact on coastal environment. Tamil University Press, Thanjavur, 1-6.

Agrawal, D.P., Guzder, S., 1972. Quaternary studies on the western coast of India: Preliminary observations. *Paleobotanist* 21, 216-222.

Ahmed, E., 1972. Coastal Geomorphology of India. Orient Longman, New Delhi, 222p.

Banerjee, M., Sen, P.K., 1987. Paleobiology in understanding the change of sea- level and coastal line in Bengal basin during Holocene period. Indian Journal of Earth Science 14 (3-4), 307-320.

Borole, D.V. Kameshwara Rao, K. Krishnamoorthy, R.V., Somayajulu, B.L.K., 1982. Late Quaternary faunal change in coastal Arabian Sea sediments. *Quaternary Research* 18, 236-239.

Brückner, H., 1989. New data on the evolution of Konkan (western India). In Datye, V.S., Diddee, J., Jong, S. R., Path, C. (Eds.), Exploration in Tropics. Professor K. R. Dikshit Felicitation Volume, Pune, 173-184. In: Gunnell Y., Radhakrishna, B. P. (Eds.) (2001), *Memoirs of the Geological Society of India* 47, 845-854.

Brückner, H., 1987. Late Quaternary shorelines in India. In: Scott D.B., et al. (Eds.), *Late Quaternary Sea-level Correlation and Applications*. Kluwer, London, pp. 160-194.

Caratini, C., Delibrias, G., Rajagopalan, G., 1990. Palaeomangroves of Kanra coast, Karnataka, India and their implications on Late Pleistocene sea-level changes. *The Palaeobotanist* 38, 370–378.

Caratini, C., Fontugne, M., Pascal, J.P., Tissot, C., Bentaleb, I., 1991. A major change at ca. 3500 years BP in the vegetation of the Western Ghats in north Kanara, Karnataka. *Current Science* 61 (9-10), 669-672

Chamyal, L.S., Merh, S.S., 1995. The Quaternary formations of Gujarat. *Memoirs of the Geological Society of India* 32, 246-257.

Cherian, P.J. Prasad, G.V. Ravi, Dutta, K., Ray, D. Kr., Selvakumar, V., Shajan, K.P., 2009. Chronology of Pattanam: a multi-cultural port site on the Malabar Coast, *Current Science* 97(2), 236-240.

Gardner, R., Martingell, H., 1990. Microlithic sites and paleoenvironmental setting, southeast India; A reevaluation. *Geoarchaeology* 5(1), 1-13.

Geetha R., Sukumar, R., Ramesh, R., Pant, R.K., Rajagopalan, G. 1997. Late Quaternary vegetational and climatic changes from tropical peats in southern India- an extended record up to 40,000 years BP. *Current Science* 73, 60-63.

Gupta, A.K., Anderson, D.M., Overpeck, J.T. 2003. Abrupt changes in the Asian southwest monsoon during the Holocene and their links to the North Atlantic Ocean. *Nature* 423, 354-357.

Gupta, A.K., Anderson, D.M., Pandey, D.N., Singhvi, A.K. 2006. Adaptation and human migration, and evidence of agriculture coincident with changes in the India summer monsoon during the Holocene. *Current Science* 90, 1082-1090.

Gupta, A.K., Singh, R.K., Thomas, E., Joseph, S. 2004. Indian Ocean high productivity event (10-8 Ma): Linked to Global Cooling or to the Initiation of the Indian monsoons? *Geology* 32, 753-756.

Hashimi, N.H., Nigam, R., Nair, R.R., Rajagopalan, G. 1993. Holocene sea level fluctuations on western Indian continental margin: An update. *Journal of the Geological Society of India* 46, 157-162.

Kale, V.S., Rajaguru, S.N., 1983. Mid-Holocene fossil woods from Colva, Goa. *Current Science* 52, 778-779.

Kale, V.S., Rajaguru, S.N., 1985, Neogene and Quaternary transgressive and regressional history of the west coast of India- an overview. *Bulletin of the Deccan College Research Institute* 44, 154-163.

Kumaran, K.P.N., Nair, K.M., Shindikar, M., Limaye, R. B., Padmalal, D., 2005. Stratigraphical and palynological appraisal of the Late Quaternary mangrove deposits of the west coast of India. *Quaternary Research* 64, 418-431.

Kumaran, K.P.N., Shindikar, M.R., Mudgal, T.R., 2004. Floristic composition, palynology and sedimentary facies of Hadi mangrove swamp (Maharashtra). *Journal of the Indian Geophysical Union* 8, 55–63.

Kunte, P.D., 1995. On some aspects of barrier islands of the west coast, India. *Journal of Coastal Research* 11(2), 508-515.

Manjunatha, B.R., Shankar, R., 1992. A note on the factors controlling the sedimentation rate along the western continental shelf of India. *Marine Geology* 104(1-4), 219-224.

Merh, S.S. 1992. Quaternary sea level changes along Indian coasts. *Proceedings of the Indian National Science Academy* 58, 461-472.

Merh, S.S., 1987. Quaternary sea level changes: The present status vis-a-vis records along coasts of India. *Indian Journal of Earth Sciences* 14, 235-251.

Nair, R.R., 1974. Holocene sea levels of the western continental shelf of India. *Proceedings of the Indian Academy of Science* 79(B), 197-203.

Nair, K.M., Padmalal, D., Kumaran, K.P.N., 2006. Quaternary geology of south Kerala sedimentary basin – an outline. *Journal of the Geological Society of India* 67, 165-179.

Nair, R.R. 1975. Nature and origin of small scale topographic prominence on the western continental shelf of India. *Indian Journal of Marine Sciences* 4, 25-29.

Nair, R.R., Hashimi, N.H. 1980. Holocene climatic inferences from sediments of the western Indian continental shelf. *Proceedings of the Indian Academy of Science* (Earth Planetary Science) 89(1), 299-315.

Nambiar, A.R., Rajagopalan, G., Rao, B.R.J., 1991. Radiocarbon dates of sediment cores from inner continental shelf off Karwar, west coast of India. *Current Science* 61, 353-354.

Nigam, R., Nair, R.R., 1989. Cyclicity in monsoons. *Bulletin of Sciences* 5, 42-45.

Pandarinath, K., Shankar, R., Yadava, M.G., 2001. Late Quaternary changes in sea level and sedimentation rate along the seacoast of India: evidence from radiocarbon dates. *Current Science* 81, 594–600.

Pawar, S.D., Venkataramana., Thomas Mathai., Mallikarjuna, C., 1983. Systematic geological mapping around Shertallai, Vikkom, Alleppey, Kottayam and Pandalam areas in parts of Alleppy and Kottayam Districts. Unpublished Report of the Geological Survey of India.

Rajendran, C.P., Narayanaswamy, Chitraprabha, J.P., 1986. Grain size analysis and surface textures of coastal red sands around Muttom, Kanyakumari District. and their implications on environment of deposition. *Current Science* 55(4), 1241-1244.

Rajendran, C.P., Rajagopalan, C., Narayanaswamy, 1989. Quaternary geology of Kerala: Evidence from radiocarbon Radiocarbon dates. *Journal of the Geological Society of India* 33, 218-222.

Rao, V. P., Rajagopalan, G. Vora, K.H., Almeida, F. 2003. Late Quaternary sea level and environmental changes from relic carbonate deposits of the western margin of India. *Proceedings of the Indian Academy of Science* (Earth Planetary Science) 112(1), 1-25.

Rao, V.P., Veerayya, M., 1996. Submarine terrace limestone from the continental slope off Saurashtra-Bombay: Evidence of Late Quaternary neotectonic activity. *Current Science* 71, 36-41.

Shajan, K.P., 1998. Studies on the Late Quaternary sediments and sea level changes of central Kerala coast, India. PhD thesis, *Cochin University of Science and Technology*, Kochi. 230p.

Singhvi, A.K., Deraniyagala, S.U., Sengupta, D., 1986. Thermoluminicence dating of Quaternary red-sands beds: a case study of coastal dunes in Sri Lanka. *Earth and Planetary Science Letters* 80(1-2), 139-144.

Southon, J.R., Kashgarian, M., Fontugne, M., Metivier, B., Yim, W.W-S., 2002. Marine reservoir corrections for the Indian Ocean and Southeast Asia. *Radiocarbon* 44(1), 167-180.

Stoddart, D.R., 1971. Environment and history in Indian Ocean reef morphology. In: Stoddart, D.R., Yonge, M., (Eds.), Regional Variations in Indian Coral Reefs. *Symposia of the Zoological Society of London* 28, 3-38.

Stoddart, D.R., 1973. Coral reefs of the Indian Ocean. In: Jones, O.A., Endean, R. (Eds.), *Biology and Geology of Coral Reefs*. Academic Press, New York, 51-92.

Stuiver, M., Reimer, P.J., 1993. Extended ^{14}C Data Base and Revised CALIB 3.0 ^{14}C Age Calibration Program. *Radiocarbon* 35(1), 215-230.

Subbaraju, L.V., Wagle, B.G., 1996. Shallow seismic and geomorphic expression of buried wave-cut terrace and erosional valleys off Redi, west coast of India. *Journal of Coastal Research* 12, 205-210.

Sukhtandar, R.K., Pandian, R.S. 1990. Evolution of sea level rise on the shore zone areas of the Maharashtra Coast. In: Sea level variation and its impact on coastal environment, Tamil Nadu. University Press, Thanjavur, pp. 329-338.

Sukumar, R., Ramesh, R., Pant, R.K., Rajagopalan, G. 1993. A δ^{13}C record of late Quaternary climate change from tropical peats in southern India. *Nature* 364, 703-705.

Thamban, M., Kawahata, H., Rao, V.P. 2007. Indian summer monsoon variability during the Holocene as recorded in sediments of the Arabian Sea, timing and implications. *Journal of Oceanography* 63, 1009-1020.

Thrivikramji, K.P., Ramasarma, M. 1983. Implications of sedimentary structural facies of a calcareous sandstone body at Kovalam, Kanyakumari District, Tamil Nadu. *Journal of the Geological Society of India* 24, 203-207.

Thrivikramji, K.P., Sabu Joseph, Anirudhan, S., 2008. Teris of southern Tamil Nadu: Saga of Holocene climate change. *Memoirs of the Geological Society of India* 74, 351-359.

Vaidyanandhan, R., 1981. Sea-levels. Information Bulletin, *IGCP Project* 61 (5), 1-5.

Van Campo, E., 1986. Monsoon fluctuations in two 20,000-Yr B.P. oxygen-isotope/pollen records off southwest India. *Quaternary Research* 26, 376-388.

Vora, K.H., Wagle, B.G., Veerayya, M., Almeida, F., Karisiddaiah, S.M., 1996. 1300 km long late Pleistocene-Holocene shelf edge barrier reef system along the western continental shelf of India: occurrence and significance. *Marine Geology* 134, 145-162.

Wagle, B.G., Vora, K.H., Karisiddhaiah, S.M., Veerayya, M., Almeida, F., 1994. Holocene submarine terraces on the western continental slope of India: Implications for sea-level changes. *Marine Geology* 117, 207-225.

Wagle, B.G., Veerayya, M., 1996. Submerged sand ridges on the western continental shelf off Bombay, India: Evidence for Late Pleistocene Holocene sea-level changes. *Marine Geology* 136, 79-95.

In: Holocene
Editor: Bahadur Singh Kotlia

ISBN: 978-1-62257-722-4

Chapter 5

CATASTROPHIC VOLCANIC EVENTS IN THE EARLY HOLOCENE: SUCCESSIVE LARGE-SCALE CALDERA-FORMING ERUPTIONS IN JAPAN

Takeshi Hasegawa*
Department of Earth Sciences, College of Science,
Ibaraki University, Japan

ABSTRACT

One of the most catastrophic geological events on the Earth was the caldera-forming explosive eruption. This chapter deals with the review and reconstruction of the impact of two caldera-forming eruptions in Japan in the early Holocene. These eruptions occurred in the Kikai and Mashu calderas, which are located in southwestern and northeastern Japan respectively. The sequence of both of the large-scale eruptions involved initial plinian fallout, climactic pyroclastic flow (ignimbrite) related to caldera collapse, and a widespread covering of co-ignimbrite ash. The catastrophic eruption from Kikai caldera at 7.3 ka BP caused severe damage to the southern Kyushu mainland, resulted in a change in vegetation and pre-historic culture. The hazardous lahar occurred during the 7.5 ka BP Mashu caldera-forming eruption just after the deposition of voluminous ignimbrite.

Keywords: Volcanic events, Caldera, Pyroclastic flow, Holocene, Japan, lahar

INTRODUCTION

One of the most catastrophic geological events on Earth is caldera-forming explosive volcanism. The historic examples of this phenomenon include the eruption of Mt. Pinatubo in the Philippines in 1991, the eruption of Novarupta crater in Alaska in 1912, and the eruption of Krakatau in Indonesia in 1883 (e.g., Simkin and Fiske, 1983; Lipman, 2000; Hildreth and

* E mail: hasegawt@mx.ibaraki.ac.jp

Fierstein, 2000). These explosions ejected about 5-10 km^3 DRE (dense rock equivalent) of pyroclastic material into the atmosphere and surface of the ground, and formed subsided calderas, which typically have diameters of more than 2 km. These events caused substantial numbers of deaths and/or changes to the global climate. During the Holocene epoch, there were larger calderas (>8 km in diameter) and greater explosive eruptions (>10 km^3 DRE) compared to any that were witnessed by human beings, such as Taupo in New Zealand (1.8 ka BP) and Crater Lake in Oregon (7.7 ka BP) (e.g., Bacon, 1983; Wilson, 1993). Additionally, during the Miocene to Pleistocene epochs, even larger calderas (sometimes referred to as super-volcanoes) existed, which produced explosive eruptions that ejected more than 100 km^3 of pyroclastic material (e.g., 73 ka BP Toba super-eruption, Indonesia (Rampino and Self, 1992). The geological features of the large-scale pyroclastic flow deposits (ignimbrite), which are composed of silicic pumice and ash, suggest that a large-scale silicic molten magma chamber was situated in the upper crust just before the caldera-forming eruption. Although, there are no historical records of large-scale (>10 km^3 DRE) caldera-forming eruptions, we need to study these eruptions in order to evaluate the current situation where many large calderas are increasingly recognized as potentially active on the basis of geophysical data for subvolcanic magma systems [e.g. Yellowstone (Lowenstern et al., 2006); Campi Flegrei (Yokoyama and Nazzaro, 2002); Santorini (Stiros et al., 2005); Aira (Kriswati and Iguchi, 2003)].

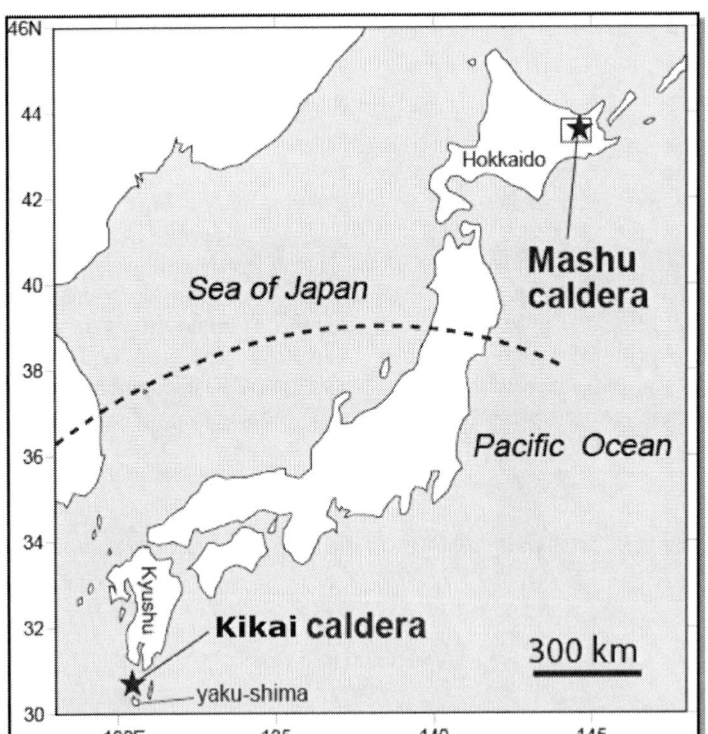

Figure 1. Index map showing the locations of the Holocene Kikai and Mashu calderas. The dashed line encloses the dispersal areas of the Akahoya Ash.

Approximately fifty Quaternary calderas are located in Japan, which is a part of the Ring of Fire in the Northern Hemisphere. At least two large-scale caldera-forming eruptions have occurred during the Holocene in Japan. These are temporarily closed, occurring at about 7.4 ka BP. One is located in the Kikai volcano in southwestern Japan, and another is located in the Mashu volcano in northeastern Japan (Figure 1). This paper will review and reconstruct the successive events that had a catastrophic impact on prehistoric human activities in the early Holocene in Japan, with a particular focus on Mashu caldera, and discuss the magma plumbing systems of large-scale eruptions.

KIKAI CALDERA, SOUTH OF KYUSHU, JAPAN

General Geology

The Kikai caldera (17 x 20 km in diameter) is a Holocene submerged caldera that is located in the Ohsumi Islands of the Kagoshima prefecture in southern Kyushu, Japan (Figure 1). The caldera forms a part of the volcanic-tectonic depression (Kagoshima graben) that extends to the southern part of the Kyushu mainland, where the Philippine Sea Plate subducts under the Eurasian Plate. The caldera-forming eruption occurred in shallow seawater at approximately 7.3 ka BP. The total eruptive volume is estimated to be approximately 170 km^3, which is the largest eruption in Japan during the Holocene epoch (Machida and Arai, 2003). Maeno and Taniguchi (2007) studied the eruption and conducted a systematic facies analysis on the eruption's deposits. The products of the eruption can be roughly categorized into three units; the lowermost plinian pumice fallout (Koya pumice), a climactic pyroclastic flow (Koya ignimbrite), and the uppermost co-ignimbrite ash-fall (Akahoya Ash). There is no significant time interval (often marked by paleosol layers) between these eruption units. Although the dominant component of the juvenile materials from the eruption deposits is silicic white pumice, the Koya ignimbrite characteristically includes banded pumice with streaks of andesitic dark and silicic white pumice. All of the units are traceable from their source to the southern Kyushu mainland. The Koya ignimbrite has been interpreted as a low-aspect ratio ignimbrite (Walker et al., 1980) as a result of its wide distribution. The ignimbrite flowed more than 80 km over the sea and was deposited on southern Kyushu and the neighboring islands, including Yaku-shima, which is a World Heritage Site. The Akahoya (which means reddish in color) Ash was dispersed more than 1,000 km away from its source. The ash is characterized by orange-colored, glassy, cohesive facies. This ash covers most of southwest and central Japan and is a Holocene key-marker tephra in Japan (Figure 1). The evidence of at least two intense earthquakes related to the caldera-forming eruptions, such as clastic dykes and sand boiling, were also found with the eruption deposits at archaeological sites in southern Kyushu (Naruo and Kobayashi, 2002). The timing of the ejection of the two clastic dykes associated with the two large earthquakes has been determined on the basis of the precise correlation between the tephra and liquefied material. The first earthquake occurred just before the eruption and was much more severe, affecting a wide area of the southern Kyushu mainland and some of the neighboring islands. The second earthquake is thought to have occurred during the co-ignimbrite ash (Akahoya ash) deposition and affected only the southern mainland of Kyushu.

The Sequence of the 7.3 ka BP Caldera-Forming Eruption and Its Impact

The sequence and impact of the 7.3 ka BP Kikai caldera-forming eruption can be reconstructed on the basis of not only volcanological studies (e.g., Ui, 1973; 1982; Walker et al., 1984; Maeno and Taniguchi, 2007, 2009) but also chronological, archaeological and paleobotanical studies of the deposits (e.g., Takahashi, 2008). The Kikai eruption began with a large earthquake and was followed by the plinian eruption, which provided 40 km^3 of Koya pumice fallout from a submarine vent system. The column height and duration of the plinian eruption is estimated to be a maximum of approximately 40 km and over a minimum of 28 hours, respectively, according to clast dispersal data. The pumice fallout reached the southern Kyushu mainland by a southwesterly wind at a velocity of 40 m/s and was deposited as a layer that was tens of centimeters thick. Soon after the deposition of the Koya pumice fallout, large-scale pyroclastic flow known as Koya ignimbrite (30 km^3) was generated. The pyroclastic flow is characterized by a low-aspect ratio and was probably rich in vapors because of the heating of seawater. This flow traveled across the sea and caused devastating damage to the prehistoric human settlements of the southern Kyushu mainland. The collapse of the caldera (which is the collapse of the magma chamber roof) may have occurred before and during the deposition of Koya ignimbrite. The collapse may have caused intense earthquakes and Tsunamis. The maximum tsunami wave height was more than 20 m according to a numerical simulation (Maeno et al., 2006). In addition, the co-ignimbrite ash cloud (Akahoya Ash) resulting from the Koya ignimbrite was widely spread and was deposited in western Japan as a layer that was more than 50 cm thick. The total volume of the Akahoya Ash has been estimated to be about 100 km^3.

The devastating Kikai caldera-forming eruption event changed natural circumstances. The types of vegetation on the southern Kyushu mainland shifted from laurel forests, such as Castanopsis, to a barren grassland after the catastrophic event. The forests needed 600~900 years to recover after the accumulation of the 7.3 ka BP eruption deposits, according to studies of Phytolith in the soil layers (Sugiyama, 2002). Pre-historical Jomon (which means cord patterned pottery) peoples could not gather food and live in the area, which resulted in the loss of their pottery culture. After the fertile vegetation recovered, another pre-historic group of people with a culture of using different archaeological material (e.g., Sobata type pottery) seems to have migrated from eastern Japan and the Korean Peninsula and settled in this area (Kuwahara, 2002).

The Magma System of the Kikai Caldera-Forming Eruption

The 7.3 ka BP caldera-forming eruption represents the evacuation of the silicic magma system, which has an estimated magma volume of 70–80 km^3 (DRE). The magma chamber depth of the eruption is estimated to be 3–7 km from the gas-saturation pressure of the melt inclusions (Saito et al., 2001). The presence of banded pumice showing heterogeneous textures of both silicic and andesitic types of magma in the Koya ignimbrite indicates that magma mixing (e.g., Sakuyama, 1979; Bacon, 1986) had occurred. Andesitic magma could have co-existed or intruded into a main, voluminous silicic magma chamber deeper than approximately 7 km during the 7.3 ka BP eruption (Maeno and Taniguchi, 2007).

MASHU CALDERA, EASTERN HOKKAIDO, JAPAN

General Geology

The Mashu caldera is located in eastern Hokkaido, Japan, where westward oblique subduction of the Pacific plate under the North American plate occurs (Kimura, 1985). The Mashu, Akan and Kutcharo calderas formed a Quaternary caldera cluster in this area (Figure 2).

The Mashu caldera (7.5 x 5.5 km in diameter) was formed on the edifice of a pre-caldera stratovolcano, like the Somma-Vesuvius in Italy. The caldera is famous for its beautiful Mashu Lake, which has a maximum depth of approximately 210 m. The Lake is known as one of the most transparent lakes in the world. On the southeastern rim of the Mashu caldera, a post-caldera cone of the Kamuinupuri (which means a mountain of god, in Ainu) exists. Voluminous pyroclastic deposits from the caldera have been widely distributed to the southeast of the volcano.

The Mashu volcano became active at 35 ka BP. The activity has been divided into three stages: the stratovolcano building stage, the Caldera forming stage, and the central cone building stage (Katsui et al., 1975).

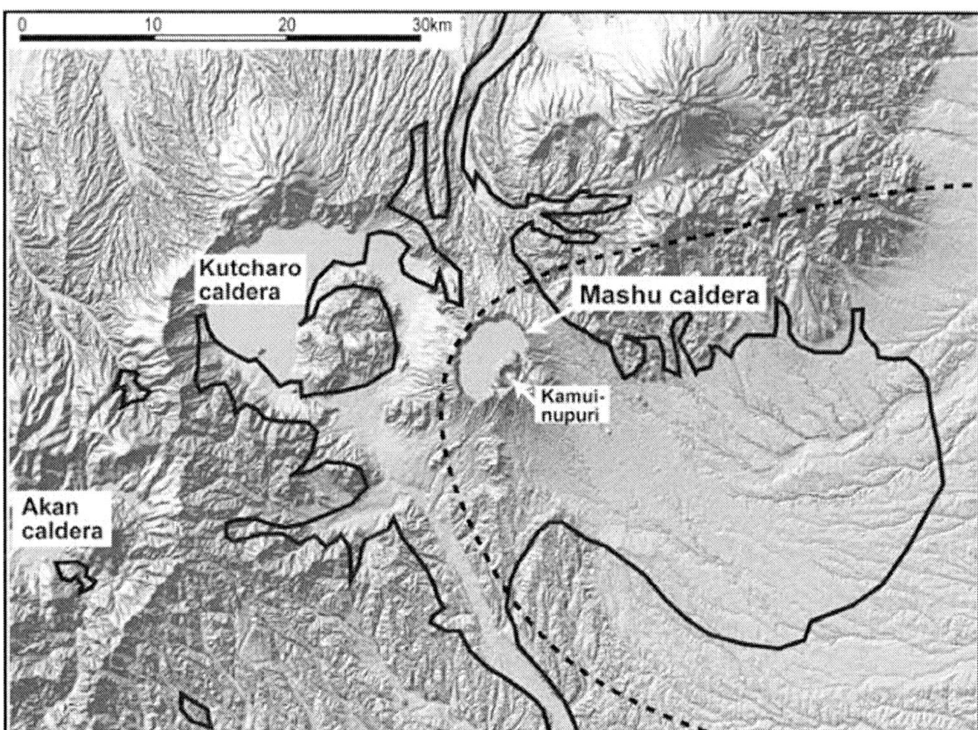

Figure 2. Shaded relief map demonstrating the digital topography of the eastern Hokkaido caldera cluster, which is illuminated by light from the northwest. This area is shown in Figure 1. Solid lines and a dashed line enclose the distributions of Unit 2 and Unit 1-b of the Mashu caldera-forming eruption, respectively.

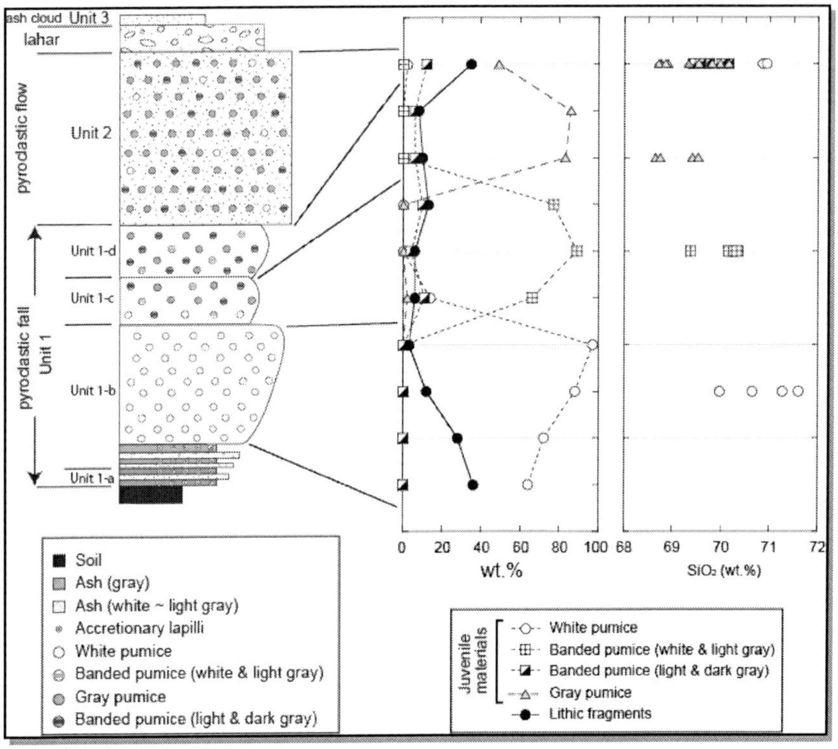

Figure 3. Stratigraphic variations of the rock-type proportions (left), and the whole-rock SiO_2 content for the juvenile materials (right) during the Mashu caldera-forming eruption.

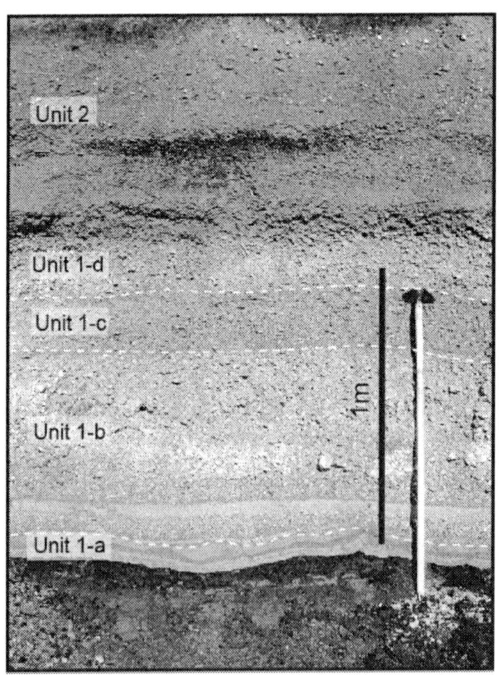

Figure 4. A field occurrence of the caldera-forming eruption deposits from the Mashu caldera. Unit 3 is not presented in this photograph.

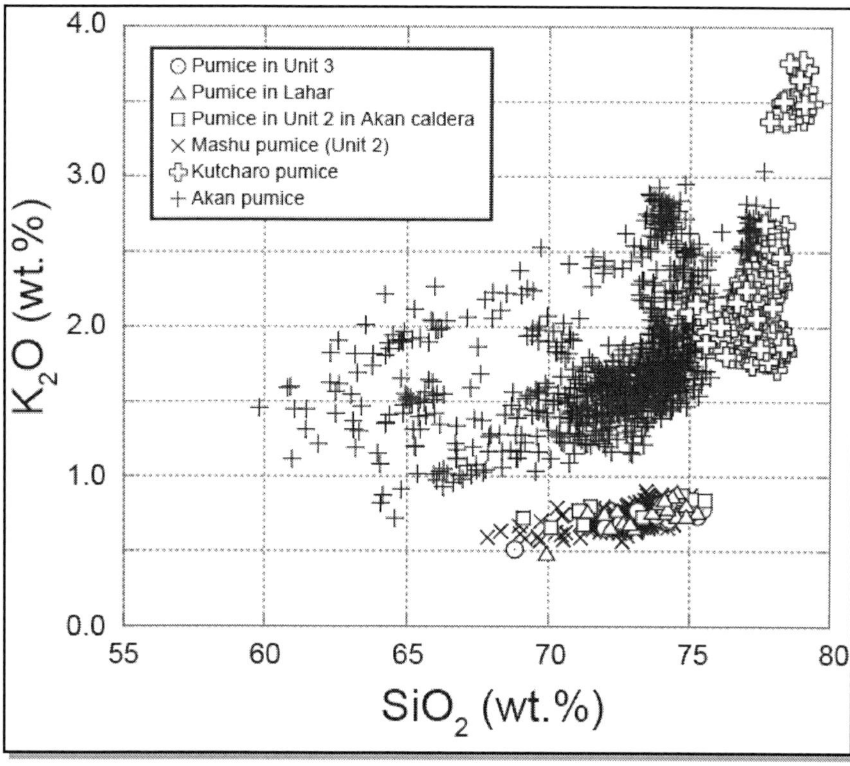

Figure 5. Glass compositions (SiO_2-K_2O) of pumices from the Mashu, Kutcharo and Akan calderas. Mashu pumice is characterized by a low-potassium content for a given SiO_2. Compositional fields of pumices in Unit 3 and Lahar from the Mashu eruption overlap with those of Unit 2.

The Kamuinupuri crater (1.2 x 1.5 km) was formed in the latest eruption, which occurred approximately one thousand years ago during the central cone building stage. The most catastrophic eruption event of the volcano occurred at 7.5 ka BP, during the Caldera-forming stage. The eruption deposits are also divided into three units: the lowermost plinian pumice fallout (Unit 1), the large-scale pyroclastic flow deposit (Unit 2) and the uppermost co-ignimbrite ash (Unit 3, see Figure 3). The total volume of the eruptive products is approximately 20 km^3. Within the eruption deposits, there are no soil layers that indicate significant dormancy before the eruption. Additionally, no archaeological materials were found in these deposits.

Unit 1 can be further subdivided into four subunits (Unit 1-a through 1-d, in ascending order) by their grain size and by the ratio of types of pyroclastic materials (Figure 4). Unit 2 is the thick and voluminous ignimbrite (the eruption volume is approximately 10 km^3), which traveled over 30 km and was deposited in the adjacent Akan caldera. Unit 3 is a widely distributed fine ash layer in eastern Hokkaido, which has an average thickness of about 10 cm.

In addition, lahar deposits were recognized between Unit 2 and Unit 3. This deposit is poorly sorted and contains a considerable amount of round and angular grayish pumice clasts and lithic fragments. Juvenile materials of the 7.5 ka BP Mashu eruption deposits are composed of white pumice, gray pumice and banded pumice that show streaks of white and light gray or dark and light gray pumice. These Juvenile materials are commonly aphyric

(1~6 %) two-pyroxenes dacites (SiO_2=68-72%, K_2O=0.5-0.8% in whole-rock composition). Eruptive products of the Mashu volcano characteristically have an extremely low K_2O content and can be clearly distinguished from the products of the adjacent Akan and Kutcharo calderas (Figure 5).

The Sequence of the Mashu Caldera-Forming Eruption

The sequence of the 7.5 ka BP caldera-forming eruption will be reconstructed here on the basis of a facies analysis and the temporal changes in the lithic/juvenile ratio of pyroclastic deposits during the eruption (Figure 3). The hazardous caldera-forming event of the Mashu volcano was initiated by a relatively small explosive eruption from Unit 1-a.

This unit is dominant in lithic fragments, which is indicative of a vent-opening event. Small amounts of the juvenile materials consisted of fine sand-sized, poorly vesiculated blocky ash with accretionary lapilli, suggesting that the phreatomagmatic eruption style was caused by the interaction of water and magma (e.g., Wohletz, 1983).

The next layer, Unit 1-b, is relatively thick and has fewer lithic fragments. The ratio of juvenile/lithic fragments increases with time and reaches nearly 100% at the top of this unit, which suggests that the vent system was becoming more stable and that conditions were moving from wet to dry. The eruption style of this Unit 1-b is typically described as plinian. The eruption column height and wind speed at that time are estimated to be more than 30 km and 70 m/s, respectively, according to the methods of Carey and Sparks (1986) in analyzing clast dispersal data. In Units 1-c and 1-d, the plinian eruption style demonstrated that a lower lithic/juvenile ratio continued and that the catastrophic pyroclastic flow (Unit 2) had finally occurred. The pyroclastic flow discharged voluminous magmas and a considerable amount of lithic fragments because of the caldera collapse.

The Unit 2 ignimbrite traveled over the 570 m high wall of the Akan caldera and was deposited in the caldera basin (Hasegawa et al., 2009). The maximum velocity of the ignimbrite was calculated to be more than 100 m/s, as determined by the law of conservation of energy ($v = \sqrt{2gh}$; $g=9.8\ m/s^2$, $h=570\ m$). This value is the higher than any other directly observed pyroclastic flow in the world. The Unit 2 ignimbrite was accompanied with a hazardous lahar event. The composition of the gray pumice clasts in the lahar deposits are the same as those from the juvenile pumice in the Unit 2 ignimbrite (Figure 5), which suggests that the lahar eroded and the pyroclastic debris of Unit 2 were introduced. Either during or after the deposition of the lahar, the co-ignimbrite ash cloud resulting from the Unit 2 ignimbrite began to settle across the broad area of eastern Hokkaido.

The dispersal of fine-grained ash (Unit 3) and sulfur aerosols into the upper atmosphere could have modified the climate, possibly resulting in anomalously cold weather. The successive threats of the large-scale eruption including the lahar event might have endangered the pre-historic Jomon people, who most likely lived in eastern Hokkaido during this period.

Magma System

The existence of banded pumice and a single linear chemical trend of juvenile materials in the Harker diagram (Figure 5) indicate that a magma integration of the two end-members was the main magmatic process during the caldera-forming eruption. The SiO_2 content is the highest (70~72%) in the white pumice and the lowest (68-70%) in the gray pumice (Figure 3). On the other hand, banded pumice shows intermediate compositions of SiO_2 (69-71%). During the 7.5 ka BP eruption, white pumice is rich in the early units (such as Unit 1-a and 1-b) and gray pumice increases in the later units (Unit 1-d and Units 2, 3). Banded pumice is easily recognized in middle unit (Unit 1-c). These temporal changes in types of juvenile materials suggest that there was a zoned magma chamber composed of more silicic magma (providing white pumice) and less silicic magma (providing gray pumice) just before the caldera-forming eruption. The steep decline in the ratio of white pumice at Unit 1-c indicates that most of the upper silicic magma was ejected during the early units (Unit 1-a and 1-b). In the middle (Unit 1-c), the two end member magmas (white and gray) mingled and were ejected as banded pumice. Finally, climactic pyroclastic flow primarily discharged gray pumice from the remaining less silicic magma system.

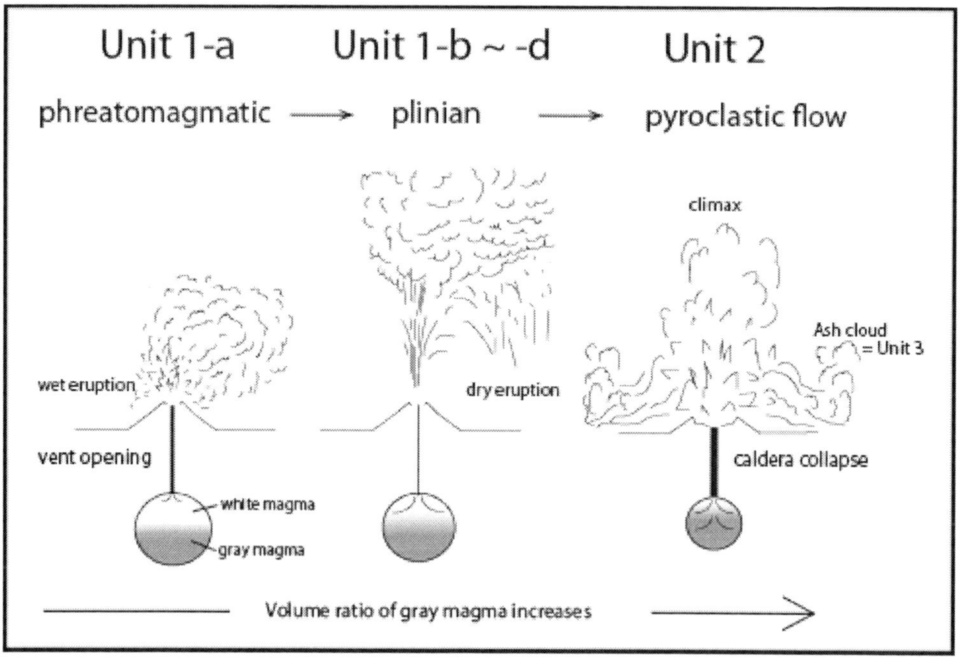

Figure 6. Schematic illustrations showing the eruptive sequence of magma systems during the Mashu caldera-forming eruption.

Comprehensive Discussion: The Clustering of Caldera-Forming Eruptions

The most interesting feature of the two largest Holocene caldera-forming eruptions at Kikai and Mashu in Japan is their close approximate time of eruption. The time of eruption at Kikai has been determined to be 7.3 ka BP by ^{14}C dating method and by counting varied sediments with Akahoya Ash (Okuno, 2002). Furthermore, the time of eruption at Mashu has been determined by ^{14}C dating to be between 7.7 ka BP and 7.3 ka BP (Yamamoto et al., 2010).

Thus, the two catastrophic events may have occurred simultaneously, considering the range of error of the carbon dating from the Mashu eruption. The author is aware of another instance of closely paired ignimbrite eruptions between Kyushu and Hokkaido at approximately 90 ka BP. At the some outcrop in eastern Hokkaido, we observed that the distal ash layer from the Aso caldera (Aso-4) in central Kyushu is directly covered by ignimbrite from the Kutcharo caldera (Kp III) with no geological time gap (Hasegawa et al., 2012).

Similarly coupled large-scale eruptions have been reported at the 240 ka BP Mamaku and Ohakuri ignimbrite at the Taupo Volcanic Zone (TVZ) in New Zealand, on the basis of detailed age data and geological field evidence (Gravley et al., 2007). Gravley et al. (2007) emphasized that such close pairing of large-scale caldera-forming eruptions may be more common than we think because the pairs are overlooked and/or treated as complex events as a result of poor geological information. In the case of the simultaneous eruptions of Mamaku and Ohakuri ignimbrites, which are located ~30 km apart, the paired events could have been controlled by the regional tectonics associated with the TVZ rifting processes. However, it is difficult to explain the mechanisms of the paired ignimbrite eruptions between Kyushu and Hokkaido, which are located over 1,500 km apart and are on different tectonic settings. In this case, we need to consider the effects not only of the tectonic triggering but also of the regional events.

CONCLUSION

The sequence of two successive caldera-forming eruptions, the 7.3 ka BP Kikai eruption in Kyushu and the 7.5 ka BP Mashu eruption in eastern Hokkaido, that damaged Japan during the early Holocene was reviewed and reconstructed in this paper. The sequence of both large scale eruptions can be summarized as (1) an initial plinian fallout, (2) climactic pyroclastic flow related to caldera collapse and (3) a covering of widespread co-ignimbrite ash. In addition, a large tsunami may have been generated in the case of the formation of a submerged Kikai caldera. The catastrophic eruption event from the Kikai caldera severely damaged the southern Kyushu mainland, represented by changes in vegetation and pre-historic culture. Additionally, hazardous lahar occurred during the Mashu eruption just after the deposition of voluminous ignimbrite. The silicic magma systems under both of the Kiaki and Mashu calderas are not homogeneous and endured some mixing processes by more mafic magmas, that might be the trigger of the eruptions. The most outstanding feature of the two largest eruptions during the Holocene epoch in Japan is their close approximate time of eruption. However, this synchronism cannot only be explained by the effects of tectonic control because the two calderas have different geological settings.

ACKNOWLEDGMENTS

I thank Professor B. S. Kotlia of the Centre of Advamced Study in Geology, Kumaun University, Nainital (India) for reviewing the manuscript. This study was partly supported by the KAKENHI (Grant-in-Aid for Young Scientists: #23710201).

References

Bacon, C.R., 1983. Eruptive history of Mount Mazama and Crater Lake caldera, Cascade Range, U.S.A. *Journal of Volcanology and Geothermal Research* 18, 57–115.

Bacon, C.R., 1986. Magmatic inclusions in silicic and intermediate volcanic rocks. *Journal of Geophysical Research* 91, 6091-6112.

Carey, S., Sparks, R.S.J., 1986. Quantitative models of the fallout and dispersal of tephra from volcanic eruption columns. *Bulletin of Volcanology* 48, 109–125.

Gravley, D.M., Wilson, C.J.N., Leonard, G.S., Cole, J.W., 2007. Double trouble: Paired ignimbrite eruptions and collateral subsidence in the Taupo Volcanic Zone, New Zealand. *GSA Bulletin* 119, 18-30.

Hasegawa, T., Yamamoto, A., Kamiyama H., Nakagawa M., 2009. Gravity structure of Akan composite caldera, eastern Hokkaido, Japan: Application of lake water corrections. *Earth Planets and Space* 61, 933-938.

Hasegawa, T., Nakagawa, M., Kishimoto, H., 2012. The eruption history and silicic magma systems of caldera-forming eruptions in eastern Hokkaido, Japan. *Journal of Mineralogical and Petrological Sciences* 107, 39-43.

Hildreth, W., Fierstein, J., 2000. Katmai volcanic cluster and the great eruption of 1912. *GSA Bulletin* 112, 1594–1620.

Katsui, Y., Ando, S., Inaba, K., 1975. Formation and Magmatic evolution of Mashu Volcano, East Hokkaido, Japan, *Journal of the Faculty of Science,* Hokkaido University 16, 533-552.

Kimura, G., 1986. Oblique subduction and collision: Forearc tectonics of the Kuril Arc. Geology 14, 404-407.

Kriswati, E., Iguchi, M., 2003. Inflation of the Aira Caldera prior to the 1999 Eruptive Activity at Sakurajima Volcano detected by GPS network in South Kyushu. *Disaster Prevention Research Institute Report* 46, 817–825.

Kuwahata, M., 2002. Age and Cultural Influence of the Kikai-Akahoya Eruption as Seen from Archaeological Material in South Kyushu, Japan. *Quaternary Research* (Daiyonki–Kienkyu) 41, 317-330 (in Japanese with English abstract).

Lipman, P.W., 2000. Calderas. In: Sigurdsson, H. (Eds.), Encyclopedia of Volcanoes, Academic Press, London, pp. 643–662.

Lowenstern, J.B., Smith, R.B., Hill, D.P., 2006. Monitoring Super-Volcanoes: Geophysical and Geochemical signals at Yellowstone and other caldera systems. *Philosophical Transactions of the Royal Society* A364, 2055-2072.

Machida, H., Arai, F., 2003. Atlas of Tephra in and around Japan. Tokyo University Press, Tokyo, pp. 336 (in Japanese).

Maeno, F., Imamura, F., Taniguchi, H., 2006. Numerical simulation of tsunami generated by caldera collapse during the 7.3 ka Kikai eruption, Japan. *Earth Planets Space* 58, 1013–1024.

Maeno, F., Taniguchi, H., 2007. Spatiotemporal evolution of a marine caldera-forming eruption, generating a low-aspect ratio pyroclastic flow, 7.3 ka, Kikai caldera, Japan: implication from near-vent eruptive deposits. *Journal of Volcanology and Geothermal Research* 167, 212-238.

Maeno, F., Taniguchi, H., 2009. Sedimentation and welding processes of dilute pyroclastic density currents and fallout during a large-scale silicic eruption, Kikai caldera, Japan. *Sedimentary Geology* 220, 227-242.

Naruo, H., Kobayashi, T., 2002. Two large-earthquakes triggered by a 6.5 ka BP eruption from Kikai caldera, southern Kyushu, Japan. Quaternary Research (Daiyonki–Kienkyu) 41, 287–299 (in Japanese with English abstract).

Okuno, M., 2002. Chronology of Tephra Layers in Southern Kyushu, SW Japan, for the Last 30,000 Years Many tephra. Quaternary Research (Daiyonki–Kienkyu) 41, 225-236 (in Japanese with English abstract).

Rampino, M.R., Self, S., 1992. Volcanic winter and accelerated glaciation following the Toba super-eruption. *Nature* 359, 50-53.

Saito, G., Kazahaya, K., Shinohara, H., Stimac, J., Kawanabe Y., 2001. Variation of volatile concentration in a magma system of Satsuma–Iwojima volcano deduced from melt inclusion analyses. *Journal of Volcanology and Geothermal Research* 108, 11–31.

Sakuyama, M., 1979. Evidence for magma mixing: Petrological study of Shirouma-Oike calc-alkaline andesite volcano, Japan. *Journal of Volcanology and Geothermal Research* 5, 179-208.

Simkin, T., Fiske, R.S., 1983. Krakatau 1883: The Volcanic Eruption and Its Effects. Smithsonian Institution Press, Washington, DC., pp.464.

Stiros, S., Chasapis, A., Kontogianni, V., 2005. Geodetic evidence for slow inflation of the Santorini caldera. *Developments in Volcanology* 7, 205-210.

Sugiyama, S., 2002. The Impact of the Kikai-Akahoya Explosive Eruption on Vegetation in Southern Kyushu, Japan, Clarified by Phytolith Studies. Quaternary Research (Daiyonki–Kienkyu) 41, 311-316 (in Japanese with English abstract).

Takahashi, M., 2008. Catastrophic Eruption (Hakyoku-funka). Shodensha Shinsho, Tokyo, pp. 244 (in Japanese).

Ui, T., 1973. Exceptionally far-reaching, thin pyroclastic flow in Southern Kyushu, Japan, *Bulletin of Volcanological Society of Japan* 18, 153–168.

Walker, G.P.L., McBroome, L.A., Caress M.E. (Eds.), 1984. Products of the Koya Eruption from the Kikai Caldera, Japan. A progress report of the U.S.–*Japan Cooperative Science Program*, 4–8.

Yamamoto, T., Itoh, J., Nakagawa, M., Hasegawa, T., Kishimoto, H., 2010. ^{14}C ages for the ejecta from Kutcharo and Mashu calderas, eastern Hokkaido, Japan. *Bulletin of Geological Survey of Japan* 61, 161-170 (in Japanese with English Abstract).

Yokoyama, I., Nazzaro, A., 2002. Anomalous crustal movements with low seismic efficiency– Campi Flegrei, Italy and some examples in Japan. *Annales Geophysicae* 45, 709–722.

Walker, G.P.L., Heming, R.F., Wilson C.J.N., 1980. Low-aspect ratio ignimbrite. *Nature* 283, 286–287.

Wilson, C.J.N., 1993. Stratigraphy, chronology, styles and dynamics of late Quaternary eruptions from Taupo volcano, New Zealand. *Philosophical Transactions of the Royal Society of London A* 343, 205-306.

Wohletz, K.H., 1983. Mechanisms of hydrovolcanic pyroclastic formation: grain-size, scanning electron microscopy, and experimental studies. *Journal of Volcanology and Geothermal Research* 17, 31-63.

In: Holocene
Editor: Bahadur Singh Kotlia

ISBN: 978-1-62257-722-4

Chapter 6

RECONSTRUCTION OF MIDDLE TO LATE HOLOCENE PALAEOENVIRONMENTS OF POOKODE LAKE, SOUTH INDIA: SEDIMENT TEXTURE AND GEOCHEMISTRY

***M. P. Veena*[1], *Hema Achyuthan**[1], *Christopher Eastoe*[2] *and Navin Shankar*[1]**

[1]Department of Geology, Anna University, Chennai, India
[2]Department of Geosciences, University of Arizona, Tucson, AZ, US

ABSTRACT

Lake sediment layers are often studied for understanding the hydrodynamic characteristics of the catchment area around and reconstruction of palaeoenvironments. Pookode Lake in Kerala is a fresh water lake situated at an altitude of ~770 masl and receives rain predominantly from the south-west (SW) monsoon. One sediment core of 125 cm was raised from this lake, sub-sampled and analysed for sediment texture, radiocarbon dating and geochemistry. Radiocarbon dates obtained range in age from 6240 to 565 yr BP. Major oxides, trace elements and chemical weathering index (CWI) values indicate warm and dry conditions during the intervals ~6200 to ~3900; ~1900 to ~1400 and ~ 760 to ~420 yr BP while the period ~3900 to ~1900, ~1400 to ~760, and ~420 to ~140 yr BP, strengthening of the SW monsoon occurred. These short wetter interludes resulted in rising of the lake water level and expansion of the lake margin for brief periods.

Keywords: Mid-Holocene, South-west monsoon, Lake sediment core, late Holocene dry conditions, Chemical weathering intensity (CWI)

* E mail: hachyuthan@yahoo.com.

INTRODUCTION

The lakes preserve sediment layers that can be analyzed to reconstruct palaeoenvironments of the recent past. Lake histories provide information about precipitation and temperature trends as sediments are deposited at different rates in different aquatic environments (Roberts, 2004). In South Asia, palaeoclimate studies indicate that both south-east and south-west (SW) summer monsoons were stronger in the early Holocene (10,000–7,000 yr BP) than today, resulting in a wetter climate over the Asian continent (Faure et al., 1963; Roberts, 2004). These reconstructions were based on a wide variety of proxies such as pollen and diatoms (Prell and Kutzbach, 1987; Van Campo and Gasse, 1993; Jarvis, 1993; Van Campo et al., 1996), oxygen isotopes of marine sediments (Sirocko et al., 1993), mineral magnetics and magnetic susceptibility of loessic soils (Maher et al., 1994), alluvial and fluctuating terrace levels in river basin (Porter et al., 1992), lake levels (Pachur and Wünnemann, 1995), vegetation dynamics (Miehe, 1996), and foraminifera (Prell and van Campo, 1986; Overpeck et al., 1996). The observed monsoonal variations in the Holocene have been attributed to greater insolation during the early Holocene, which caused greater land–sea temperature and pressure gradients (Overpeck et al., 1996).

A number of studies on palaeoenvironmental changes in India are based on lacustrine deposits in the Himalaya, western Rajasthan and Gujarat (Bhattacharyya, 1989; Kusumgar et al., 1992; Prasad et al., 1997; Chauhan et al., 2000; Enzel et al., 1999; Sinha et al., 2006; Achyuthan et al., 2007), tree rings and speleothems (Tiwari et al., 2005, 2006, 2009; Fleitmann et al., 2003; Yadava et al., 2004; Yadava and Ramesh, 2005, 2006; Managave et al., 2010), marine sediment cores (Rangarajan and Sant, 2000), and mangrove and peat (Kumaran et al., 2005; Farooqui et al., 2010). These studies indicate three distinct climatic events since the last glacial maximum: 19,000-13,500 yr BP, 13,500-10,000 yr BP and 3,400 yr BP to present. In north-west India, studies carried out on the lake sediments (Enzel et al., 1999; Achyuthan et al., 2007) reveal that a more intense monsoon lead to the filling of the fresh water bodies during the early Holocene, followed by desiccation of lakes about 5,300 yr BP. The late Holocene was characterized by multiple wet events of shorter duration (Prasad and Enzel, 2006). The palaeoclimatic reconstructions are also available from the Himalaya (Sharma and Owen, 1996; Phadtare et al., 2000), Ganga Plain (Srivastava et al., 2010) and the Nilgiris (Sukumar et al., 1993).

In Kerala, continuous palaeoclimatic records are available from sediment cores collected from mangrove swamps (Narayana et al., 2002; Nair et al., 2010). However, the data for the last 6,000 years are sparse. The lakes of Kerala offer better sites to reconstruct Holocene climatic changes as variations in the intensity of the SW monsoon should be reflected in the depositional events.

The changes in the lake sediment characteristics, e.g., particle size and geochemical parameters may help in understanding the fluctuations in the intensity of SW monsoon. In this chapter, the geochemical and textural data of sediments from Pookode Lake, Kerala, are used to reconstruct the palaeoenvironmental records since the Middle Holocene. The main objective is to provide the information from the tropical southern end of the Indian subcontinent, where rainfall is predominantly due to the SW monsoon.

STUDY AREA

The Pookode Lake in Kerala (11°32'35"N, 76°01'40"E) is a fresh water body covering an area of 0.085 km² and situated at an altitude of ~770 masl (Figure 1).

The water column is about 6.5 m deep. The lake experiences a salubrious climate with a mean rainfall of ~4,000 mm^{-yr} (Sandeep et al., 2012). The area receives rainfall from both the SW and NE monsoons, the former is dominating.

The summer temperature rises to a maximum of 35°C (95°F), while winters are cool with minimum temperature around 07°C (45°F). The bed rock in the lake catchment includes highly foliated charnockite and hornblende-biotite gneiss of the Wayanad Group (Archean age), overlain by ferruginous loamy soil of the Quaternary age (Sandeep et al., 2012). The magnetic studies of sediments from the lake reveal that the bedrocks are deeply weathered as a result of the high rainfall (Sandeep et al., 2012).

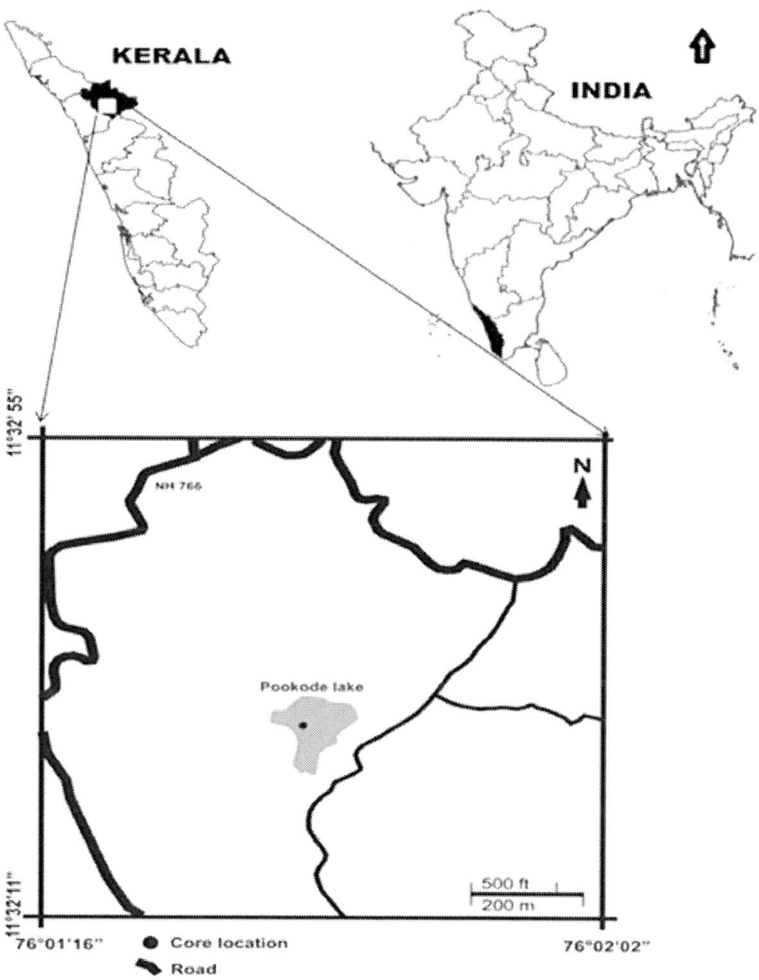

Figure 1. Location map of the study area.

MATERIALS AND METHODS

A sediment core of nearly 125 cm long was raised from the interior of the Pookode Lake (Figure 2). The core lithology was described using the Munsell color chart. The core was sub-sampled at every 2 cm and the samples were sealed, numbered and refrigerated . The textural analyses were carried out by the pipette method of Krumbein and Pettijohn (1938).

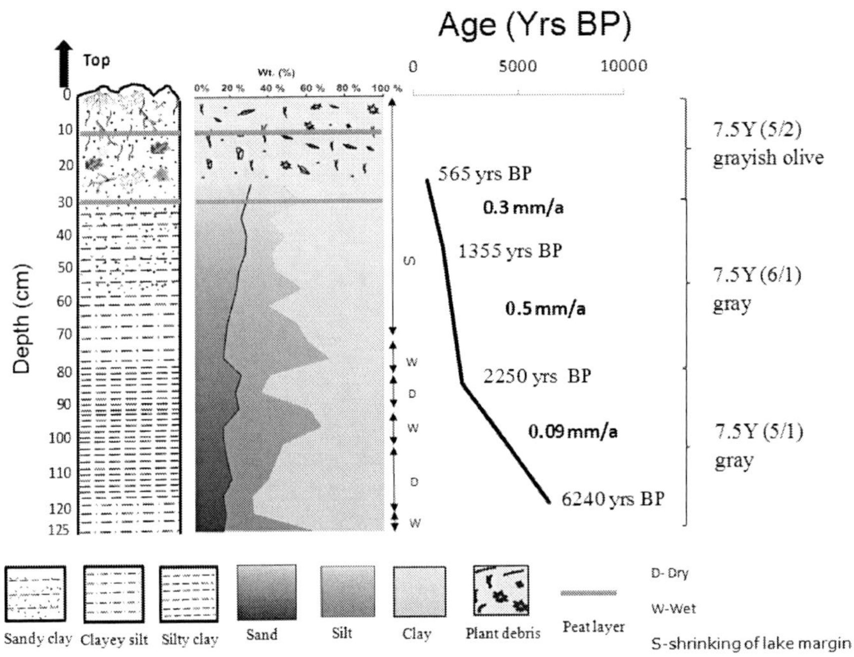

Figure 2. Down sediment core variations in grain texture and radiocarbon ages.

Table 1. Certified PAAS standard values with the present average trace metal data. (n=13 samples)

Elements	Present Study (ppm)	PAAS (ppm)
Co (ppm)	16	23
Ni (ppm)	88	55
Pb (ppm)	30	20
Zn (ppm)	87	85
Cr (ppm)	138	110
Cu (ppm)	41	50

The calcium carbonate ($CaCO_3$%) and total organic carbon content (TOC) were determined using the methods of Loring and Rantala (1992) and Heiri et al. (2001) respectively (Figure 3).

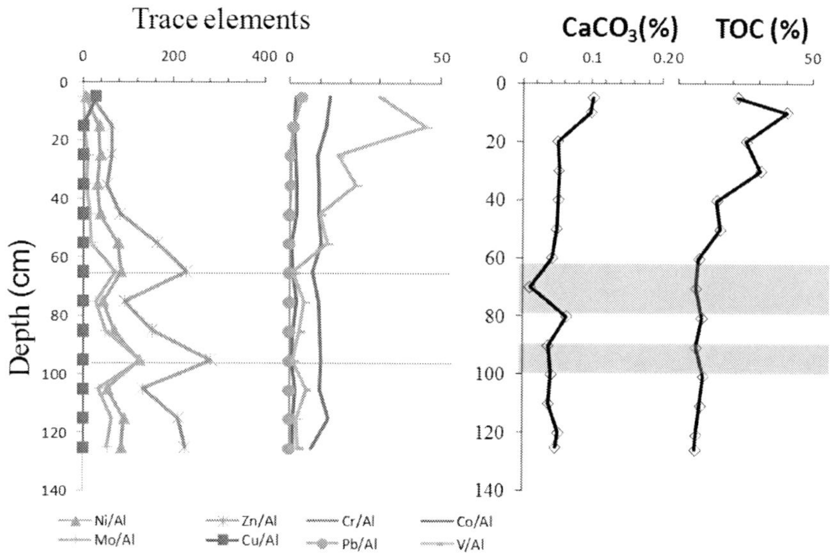

Figure 3. Sediment profile variations in $CaCO_3$, TOC content and trace elements normalized to Al.

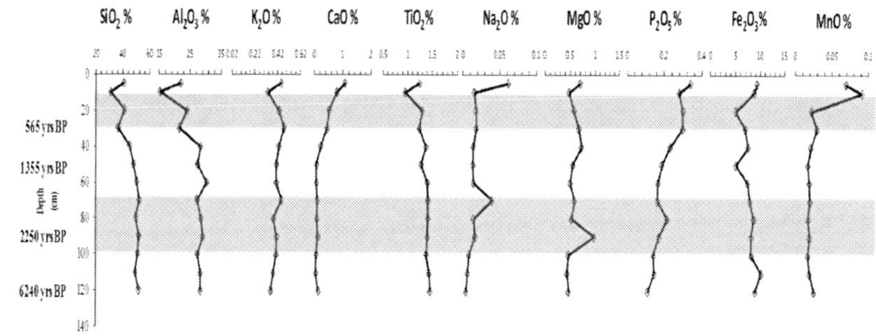

Figure 4. Sediment profile variation in major oxides.

Further, a total of 25 sub-samples covering the entire sediment core were analyzed for major oxides (Figure 4) and trace elements using the XRF Spectro Xepos powder method and Atomic Absorption Spectrometer (AAS GBC 932 PLUS) respectively. Merck standards and certified Post Archaean Australian shale (PAAS) standard were used for each element repeatedly for the accuracy of the analysis. The results obtained were normalized to Al content and are presented in Figures 3 and 5. Four samples were radiocarbon dated by liquid scintillation counting following acid-base-acid pre-treatment at the Department of Geosciences, University of Arizona, Tucson, USA (Table 2).

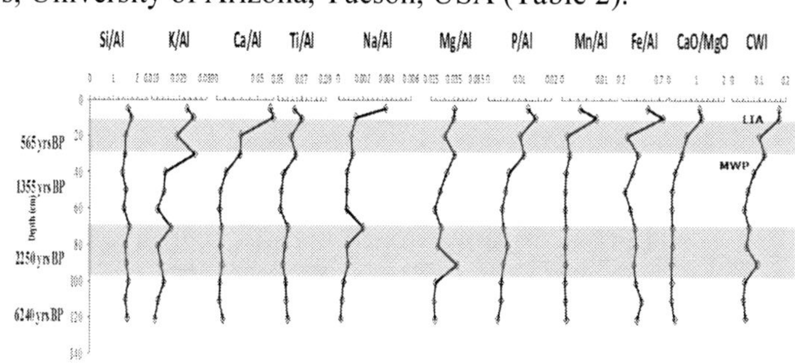

Figure 5. Sediment profile variations in ratios of major elements normalized by Al, CaO/MgO and CWI.

Table 2. Radiocarbon dates of the Pookode Lake core

Number	Depth (cm)	Age (yr BP)	$\delta^{13}C$‰
A 15786	20-30	565 ± 50	-28.6
A 15787	40-50	1,355 ± 75	-27.1
A 15788	80-90	2,250 ± 105	-26.5
A 15790	115-125	6,240± 170	-22.1

RESULTS

Lithology of the Sediment Core

The sediment layers within the core varied both in color and texture (Figure 2). Based on the color, texture and concentration of elements, the core sediments were divided into three units: a). Unit I (125-80 cm, gray in color (7.5Y 5/1) and consists of fine clayey silt from120-125 cm and silty clay from 80-120 cm. This is overlain by Unit II (80-30 cm, gray in color (7.5 Y 6/1), and is composed of thin alternate layers of clayey silt from 60-80 cm and sandy clay from 30-60 cm. Upwards, the Unit III (30-0 cm) is represented by grayish olive colored (7.5Y 5/2) sand which is poorly sorted, mixed with humus and plant debris and consisting of roots and leaves. The layers at depths 30-28 cm and 10-8 cm are rich in organic matter. However, no sedimentary structures were discerned clearly in unit III (Figure 2). In all three units, the heavy minerals were admixed with detritus.

Chronology

Radiocarbon dates range in age from 6,240 to 565 yr BP (Table 2). The average accumulation rate is low; 0.09 mm/a from 120 to 85 cm, 0.55 mm/a from 85 to 45 cm and 0.30 mm/a from 45 to 25 cm (Figure 2). The age/depth plot reveals an increase in the average rate of sedimentation around 6,240 and 2,250 yr BP. The increase in deposition points to changes in one or both of the catchment hydrology and vegetation cover, and reflects probably the initiation of agriculture in the region. The shift to lower values of $\delta^{13}C$ (Table. 2) can be related to such changes. The sedimentation rates may have varied as indicated by the fluctuations in the silt fraction (Figure 2).

Sediment Texture

Of the total clastic fractions, clay ranges from 28.3 % (75 cm) to 68.7 % (115 cm) and is more abundant than silt. The silt content varies from 13.6% (85 cm) to 55.6% (75 cm). The sand content increases steadily from 80 cm (15%) to 25 cm (30%). From 25 to 0 cm, the layers contain sand, mixed with humus of decomposing plant debris with laminae of peat (>40% of TOC) at 10 cm depth (Figure 2).

$CaCO_3$, TOC Content, Major Oxides and Trace Elements

The $CaCO_3$ content is very low while TOC ranges from 6.5 to 40%. The high values of TOC represent occurrence of peaty layers (at depth of 10 cm and 30 cm) (Figure 3). The sediments are rich in SiO_2, varying from 30 to 52 %, Al_2O_3 from 15 to 30 %, Fe_2O_3 from 5 to 10 %, and by TiO_2 from 0.9 to 1.4%. K_2O, CaO, MgO, and MnO are present in minor (<2%) amounts (Figure 4).

Unit I (~6,240 to ~2,100 yr BP) [120-80 cm]

From 125-100 cm (~6,200 to 3,900 yr BP), the major elements, normalized to Al (Si, Ca, Na, Mg, P etc) and $CaCO_3$ percentage are low and the unit is clayey silt and silty clay indicating warm and dry conditions. There is an increase in concentration of Na_2O, CaO, MnO, MgO, K_2O, and ratios of Na/Al, Mg/Al, K/Al, Mn/Al, Ti/Al and Ca/Al from 100-80 cm (~3,900 and ~2,100 yr BP) indicating wetter conditions and also probably bringing in detritus fluxes into the lake (Figure 5). This observation is supported by decrease in the ratios of Si/Al observed between (100-80 cm, ~3,900 and ~2,100 yr BP). Around ~3,000 yr BP (95 cm), the trace elements normalized to Al such as Zn/Al, Ni/Al, Mo/Al and Cr/Al reveal high ratios pointing to a short intense wet phase.

Unit II (~ 2,100 to ~760 yr BP) [80-30 cm]

The clayey silt and sandy clay layers reveal an increase in the concentration of Na_2O, MnO, MgO, K_2O, and ratios of Na/Al, Mg/Al, K/Al and Ca/Al between ~2,100 and ~1,900 yr BP (80-70 cm). Subsequently all soluble elements and oxides show a less concentration from ~1,900 to ~1,400 yr BP (70-50 cm) indicating dry and warm conditions. However, around ~1,800 yr BP (65 cm), the trace element ratios (Zn/Al, Ni/Al, Mo/Al and Cr/Al) show increased values especially V/Al ratio that increases from 60 cm to the surface. From ~ 1,400 to ~760 yr BP (50-30 cm), the major oxides exhibit a mild increase in their trends indicating climatic amelioration.

Unit III (~760 yr BP to Present) [30-0 cm]

From ~760 Yr BP to 420 Yr BP (30-20 cm), the major oxides and the elements normalized to Al show a decreasing trend indicating dry and warm conditions. The increase in TOC content at 30 cm and 10 cm levels and an increasing trend in the major soluble elements from 20-10 cm may point to wetter conditions. This is supported by an increasing trend in ratios of the elements normalized to Al, CaO/MgO ratios and CWI values from 20-10 cm (~420 to 141 yr BP) that point to wet and humid conditions.

DISCUSSION

The lake sediments reflect hydrodynamic environment and hydrology of the lake catchment basin. The textural data exhibit variations in grain size, represented by sand, sandy clay, clayey silt and silty clay layers (Figure 2) that reveal the variations in the hydrological system of the lake basin and the catchment. Occurrence of sand and sandy clay layers indicates wetter conditions, high lake level and expansion of the lake margin. Based on the particle sizes (clay content >50%) and TOC (>20%), some units have been identified which represent alternate dry and wet phases. High clay content (>50%) from 100-120 cm and 70-30 cm indicate a quiescent and dry phase causing shrinking of the lake margin and shallowing of the lake. The TOC content (>20%) from 30 cm to top layers also indicate warm and dry conditions. A major wet phase is observed from 120-125cm and 70-100 cm (clay <50%) with a short dry phase occurring from 80-90 cm (clay >50%). In general, the textural data indicate shallowing of the lake and shrinking of the lake margin from 70 cm to the surface (Figure 2).

Oxides such as CaO, MgO, and Na_2O are considered more soluble and mobile, while Al_2O_3, SiO_2, and TiO_2 are considered more insoluble and resistant oxides (Mackereth, 1966; Engstrom and Wright, 1984). High molar ratios of CaO + MgO + Na_2O)/Al_2O_3 calculated for chemical weathering intensity (CWI) of sediments indicates stronger weathering and presumably wet conditions and a strong monsoon, whereas low CWI reflects a less weathering and drier conditions (Sun et al., 2009). Al is insoluble under both oxic and anoxic conditions and is a common terrestrially derived element (Brown et al., 2000). Thus in this study, the entire major and trace element data were normalized to Al to evaluate chemical weathering intensity in the lake basin (Figure 5). Higher CaO/MgO ratio and high $CaCO_3$ content (Figures 3, 5) generally point towards more carbonate of authigenic origin input under warm and dry climatic conditions (Wang et al., 1990, 1992).

The decrease in CWI values and Na/Al, Mg/Al, Ca/Al and K/Al ratios from ~6,200 to ~ 3,900 yr BP (120-100 cm) indicate a dry period with warmer temperatures which is supported by the overall fluctuating trend of $CaCO_3$ and less concentration of TOC (Hong et al., 2003; Xiao et al., 2006). Distinct fluctuations in TOC observed from ~1,800 yr BP (65 cm) onwards to the present points to an overall dry period with wetter conditions of shorter duration but of greater intensity (Figures 3, 5).

From 100-70 cm (~3,900 to ~ 1,900 Yr BP), an increase in the ratios of Na/Al, Mg/Al, K/Al, Mn/Al and Ca/Al indicate a major wet phase (Figure 5). An intense phase of SW monsoon may have occurred during ~3000 and ~1800 yr BP that is supported by elevated ratios of Ti/Al, Zn/Al, Ni/Al and Mo/Al. Low content of $CaCO_3$ during ~1,900 yr BP (70 cm) also indicates a wet period (Figure 3). V/Al exhibit a fluctuating trend from ~1,800 yr BP (~65 cm to the surface layers) due to the abundance of plant debris, admixed with sand. High organic productivity would have resulted in higher organic matter production, which in turn may have functioned as a substrate for absorbing trace elements which are known to be bound to organics. This scenario of lake expansion implies that the region experienced a humid climate during this period on account of the prevalence of active SW monsoon (Sharma, 2006).

Low ratios of Na/Al, Mg/Al, Ti/Al, K/Al, Mn/Al and Ca/Al from 70-50 cm (~1,900 to 1,400 yr BP) (Figure 5), indicate dry and warm conditions. An increasing trend of CaO/MgO and CWI from 50-30 cm (~1,400–760 yr BP) suggest warm temperature with ameliorated hydrothermal conditions in the lake basin (Sun et al., 2009). This was followed by a dry phase from ~760 to 420 yr BP (30-20 cm), supported by a decreasing trend of CWI and other soluble elements. Further, an increasing trend of all soluble elements, ratios of elements normalized to Al, CaO/MgO and CWI from depth 20–10 cm (~420 to 141 yr BP) indicate a wet and humid period, stronger weathering due to high rainfall combined with warmer conditions. The elevated ratio of CaO/MgO also suggests higher evaporation rate under warmer temperatures from ~420 yr BP to present. High K/Al, Ca/Al, Na/Al and Mg/Al ratios are reflected in the sediments due to strong weathering of bedrock rich in feldspars and this can be related to strong monsoon and humid conditions. High Fe/Al ratio in the lake sediments is due to disturbances of ferricrete surface that would increase the ratio. Fe/Al ratio seems to compare with CaO/MgO and CWI indicating ferricrete formed under wetter environment but was destabilized and eroded under drier conditions. All these processes could be related to agricultural disturbances. Previous study carried out on peat deposits from Sanjiang Plain reveal dry and warm climate environment from 1,940-1,090 yr BP, a warm and humid climate during 1,090-545 yr BP and a cool and dry climate during 545 yr BP (Yu-mei and Pei-fang, 2001).

The chemical index of alteration (CIA) and the chemical weathering intensity (CWI) have been considered as indices of weathering and erosion (Nesbitt and Young, 1982; Colin et al., 1998; Wang et al., 1992, Sun et al., 2009). However, there exist a few clues on response time of these proxies to the climate variability. In our study, using CWI, major oxides and elements normalized to Al reveal three major wet climatic events that have been archived in the sediment core and these are centered around ~3,900 to ~1,900, ~1,400 to ~760 and ~420 to ~140 yr BP. The increasing trend of CaO/MgO ratio and CWI also indicates warm phase with strong monsoon induced precipitation. The geochemical data seems to unveil Medieval Warm Period (MWP)- a period of dry and warm conditions subsequently followed by a cool and wetter episode of Little Ice Age (LIA) (high variation of CWI and ratios of Fe/Al, P/Al, Mg/Al, Ca/Al. K/al, Mn/Al and Ti/Al) (Figures 6, 7). Elevated Ti/Al ratio probably also relates to clastic sedimentation. The two major global climatic events, e.g., MWP from 1,100 and 600 yr BP and LIA between 400 and 250 years BP (Bluemle, 1999) have left their fingerprint in this sediment core (Figures 5-7).

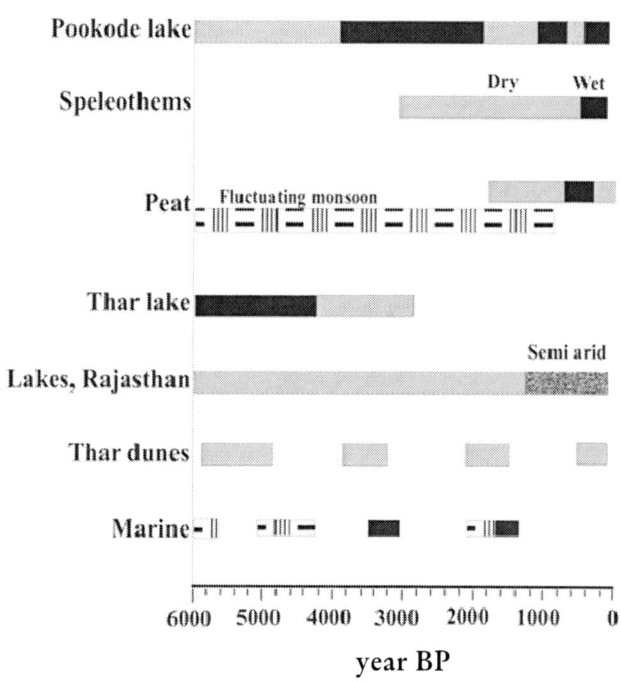

Figure 6. A comparison of Mid–Holocene to Recent paleoclimate data of Pookode lake with paleoclimate variability from different regions following Singhvi and Kale (2008).

In order to obtain a regional trend in palaeoclimate during the past 6,000 yr BP, we have compared our sediment and radiocarbon ages with available marine as well as continental records from the Indian Subcontinent (e.g., von Rad et al., 1999; Anderson et al., 2002; Gupta et al., 2003; Sinha et al., 2006; Achyuthan et al., 2007) (Figure 6). For 2,800 yr BP and past 800 AD (~1,150 y BP), Tiwari et al. (2006), have obtained $\delta^{18}O$ variability in surface dwelling G. *ruber* and G. *sacculifer* from the farther north location of Central Eastern Arabian Sea in an ^{14}C AMS dated sediment core. This region also lies under the high SW monsoon precipitation influenced sea surface salinity (SSS). Gupta et al. (2003) observed wind induced upwelling indices of G. *bulloides* due to the coeval warming off Oman during 1,200-600 yr BP, while an enhancement of monsoon during LIA was established by Anderson et al. (2002), and Chauhan et al. (2006, 2010) and these observations are corroborated by our sediment texture and geochemical studies (Figure 6).

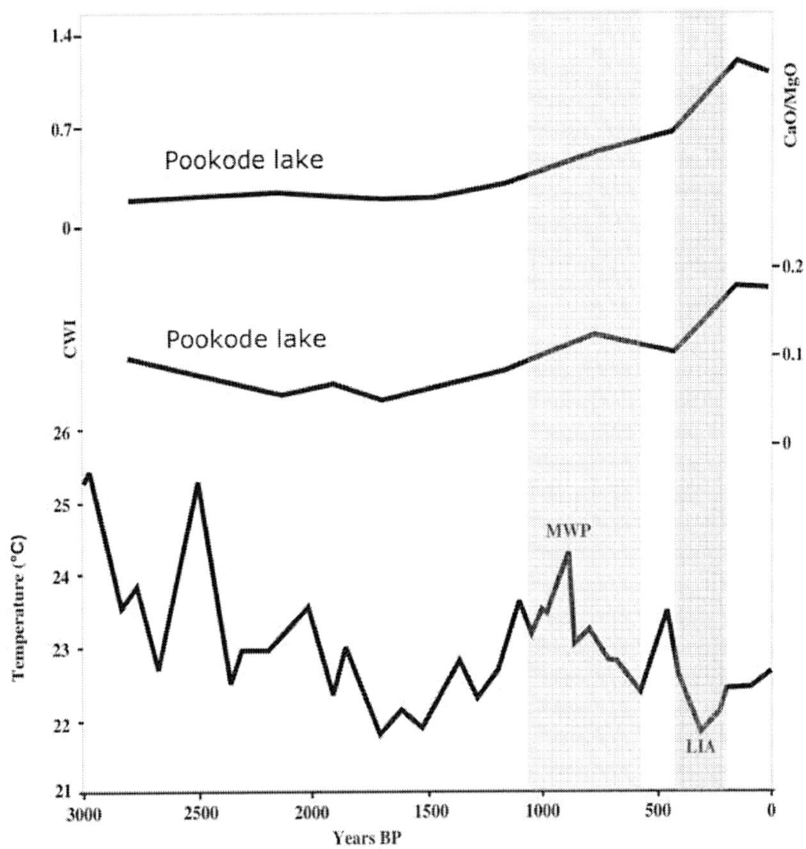

Figure 7. Comparison of MWP and LIA revealed from Pookode lake sediments. This figure is modified after Bluemle, (1999).

LIA has a large temporal and spatial variability across the globe (IPCC report, 2001, Chauhan et al., 2006). Crowley and Lowery, (2000) have described LIA as mild cooling in the Northern Hemisphere. Hu et al. (2001) have found it ~ 1.7°C cooler in Alaska, though its termination is found to vary (in the Sargasso Sea at 400 BP; Keigwin, 1996). In order to estimate a regional linkage between palaeoclimate around the Pookode Lake and the palaeoenvironmental data from surrounding regions, we compared and updated the reconstructions with available climate variability from these regions (see Figure 6; Singhvi and Kale, 2008).

It is surmised, therefore, that since the mid-Holocene to the late Holocene period during the long phases of dry conditions, there were several rapid phases of significant variability in the SW monsoon precipitation (Chauhan et al., 2006), characterized by several wetter events of shorter duration and of greater magnitudes (Singhvi and Kale, 2008) resulting in the expansion of lake margins for short duration (see Figure 7).

Conclusion

A sediment core obtained from the Pookode Lake was analyzed for sediment texture, radiocarbon dating and geochemistry to reconstruct palaeoenvironment around catchment.

Textural data exhibit variation in grain size that reflects the variations in the hydrological regime of the lake basin. The wet and dry conditions are represented by the occurrence of sandy clay and presence of finer sediments such as silty clay, clayey silt. This is also supported by the low sedimentation rate from ~6,200 to ~2,250 yr BP. A decline in CWI and Ti is observed during ~6,200 to ~3,900, ~1,900 to ~1,400 and ~ 760 to ~420 yr BP indicating dry conditions, and indicating predominance of a weaker SW monsoon. From improved concentrations of soluble oxides and CWI values during ~3,900-~1,900, ~1,400-~760, and ~420-~140 yr BP, we infer amelioration in the magnitude of the SW monsoon. During the over all dry phase since mid Holocene 6,240 yr BP to present, there were wetter interludes of short duration and the lake margins expanded for short period and contracted during the long dry phases.

The alternation between warm-dry and wet-humid conditions implies climate change characteristics due to the variations in SW monsoon precipitation.

Acknowledgments

Veena M.P. thanks UGC, New Delhi for the JRF fellowship. The authors thank Dr. N. Karmalkar, Department of Geology, University of Poona, Pune, for his kind help in extending facilities to carry out chemical analyses. The authors thank the reviewers for their constructive comments that helped in the presentation of the paper.

References

Achyuthan, H., Kar, A., Eastoe, C., 2007. Late Quaternary environmental changes and evolution of the Tal Chhapar salt lakes, Rajasthan. *Journal of Palaeolimnology* 38, 493-507.

Anderson, D.M., Overpeck, J.T., Gupta, A.K., 2002. Increase in the Asian Southwest Monsoon during the past four centuries. *Science* 297, 596-599.

Bhattacharyya, A., 1989. Vegetation and climate during the last 30,000 years in Ladakh. *Palaeogeography, Palaeoclimatology, Palaeoecology* 73, 25-38.

Bluemle, J. P., 1999. Global warming: a geological perspective. Arizona Geology, *Arizona Geological Survey* 29 (4), 1-3.

Brown, E.T., Le Callonnec, L., German, C.R., 2000. Geochemical cycling of redox sensitive metals in sediments from Lake Malawi: a diagnostic paleotracer for episodic changes in mixing depth. *Geochimica Cosmochimica Acta* 64, 3515–3523.

Chauhan, O.S., Vogelsang, E., Basavaiah, N., Abdul Kader, U.S., 2010. Reconstruction of the variability of SW monsoon during the past 3 Ka from the continental margin of the South Eastern Arabian Sea. *Journal of Quaternary Science* 25(5), 798-807.

Chauhan, O.S., Jayakumar, S., Menezes, A.A., Rajawat, A.S., Nayak, S.R., 2006. Anomalous inland influx of the River Indus, Gulf of Kachchh, India. *Marine Geology* 229, 91-100.

Chauhan, M.S., Mazari, R.K., Rajagopalan, G., 2000. Vegetation and climate in upper Spiti region, Himachal Pradesh during late Holocene. *Current Science* 79, 373-377.

Colin, C., Kissel, C., Blamart, D., Turpin, L., 1998. Magnetic properties of sediments in the Bay of Bengal and the Andaman Sea: impact of rapid North Atlantic Ocean climatic events on the strength of the Indian monsoon. *Earth and Planetary Science Letters* 160, 623–635.

Crowley, T.J., Lowery, T.S., 2000. How warm was the medieval warm period? *Ambio* 29, 51–54.

Engstrom, D.R., Wright, H.E., 1984. Chemical stratigraphy of lake sediments as a record of environmental change. In: Hayworth E.Y, Lund J.W.G. (Eds.) Lake Sediments, Environmental History, Leicester University Press, pp. 11-67.

Enzel, Y., Ely, L.L., Mishra, S., Ramesh, R., Amit, R., Lazar, B., Rajaguru, S.N., Baker, V.R., Sandler, A., 1999. High resolution environmental changes in the Thar Desert, northwestern India. *Science* 284, 125-128.

Farooqui, A., Ray, J.G., Farooqui, S.A., Tiwari, R.K., Khan, Z.A., 2010. Tropical rainforest vegetation, climate and sea level during the Pleistocene in Kerala, India. *Quaternary International* 213, 2–11.

Faure, H., Manguin, E., Nydal, R., 1963. Formations lacustres du Quaternaire superieur du Niger oriental: diatomites et ages absolus. *Bullettin du Bureau de Recherches de Geologie et Minieres* 3, 41-63.

Fleitmann, D., Burns, S., Mudelsee, M., Neff, U., Mangini, K., Matter, A., 2003. Holocene forcing of the Indian monsoon recorded in a stalagmite from southern Oman. *Science* 300, 1737–1739.

Gupta, A.K., Anderson, D.M., Overpeck, J.T., 2003. Abrupt changes in the Asian southwest monsoon during the Holocene and their links to the North Atlantic Ocean. *Nature* 421, 354–357.

Heiri, O., Lotter, A. F., Lemcke, G., 2001. Loss on ignition as a method for estimating organic and carbonate content in sediments: Reproducibility and comparability of results. *Journal of Paleolimnology* 25,101-110.

Hong, Y.T., Hong, B., Lin, Q.H., Zhu, Y.X., Shibata, Y., Hitota, M., Uchida, M., Leng, T., Jiang, H.B., Xu, H., Wang, H., Yi, L,. 2003. Correlation between Indian Ocean summer monsoon and North Atlantic climate during the Holocene. *Earth and Planetary Science Letters* 211, 371-380.

Hu, S.F., Ito, E., Brown, T.A., Curry, B.B., Engstrom, D.R., 2001. Pronounced climatic variations in Alaska during the last two millennia. *Proceedings of the National Academy of Sciences* 98(19), 10552-10556.

IPCC. 2001. Climate Change 2001, In: The Scientific Basis. Contribution of Working Group I in The Third Assessment Report of the Intergovernmental Panel on Climate Change. Houghton, J.T., Ding, Y., Griggs, D.J., Noguer, M., van der Linden, P.J., Dai, X., Maskell, K., Johnson, C.A., (Eds.), Cambridge University Press, Cambridge.

Jarvis, D. I., 1993. Pollen evidence of changing Holocene monsoon climate in Sichuan Province, China. *Quaternary Research* 39, 325–337.

Keigwin, L.D., 1996. The Little Ice Age and Medieval Warm Period in the Sargasso Sea. *Science* 274, 1503-1508.

Krumbein, W.C., Pettijohn, F.J., 1938. Manual of Sedimentary Petrography, Appleton-Century-Crofts: New York.

Kumaran, K.P.N., Nair, K.M., Shindikar, M., Limaye, R.B., Padmalal, D., 2005. Stratigraphical and palynological appraisal of the Late Quaternary mangrove deposits of the west coast of India. *Quaternary Research* 64, 418–431.

Kusumgar, S., Agrawal, D.P., Bhandari, N., Deshpande, R.D., Raina, A., Sharma, C.,Yadava, M.G., 1992. Lake sediments from the Kashmir Himalayas: Inverted ^{14}C chronology and its implications. *Radiocarbon* 34, 561-565.

Loring, D.H., Rantala, R.T.T., 1992. Manual for the geochemical analyses of marine sediments and suspended particulate matter. *Earth Science Reviews* 32, 235-283.

Mackereth, F.J.H., 1966. Some chemical observations on postglacial lake sediments. *Philosophical Transactions of Royal Society in Biological Sciences* 250,165–213.

Maher, B. A., Thompson, R., Zhou, L. P., 1994. Spatial and temporal reconstructions of changes in the Asian paleomonsoon: A new mineral magnetic approach. *Earth and Planetary Science Letters* 125, 461–471.

Managave, S.R., Sheshshayee, M.S., Borgaonkar, H.P., Ramesh, R., 2010. Past-break monsoon conditions detectable by high resolution intra-annual $\delta^{18}O$ analysis of teak rings. *Geophysical Research Letters* 37, L05702.

Miehe, G., 1996. On the connection of vegetation dynamics with climatic changes in High Asia. *Palaeogeography Palaeoclimatology Palaeoecology* 120, 5–24.

Nair, K.N., Padmalal, D., Kumaran K.P.N., Sreeja, R., Limaye R.B. Srinivas, R., 2010. Late Quaternary evolution of Ashtamudi-Sasthamkotta lake systems of Kerala, Southwest India. *Journal of Asian Earth Sciences* 37, 361-372.

Narayana, A.C., Priju, C.P., Rajagopalan, G., 2002. Late Quaternary peat deposits from Vembanad Lake (lagoon), Kerala, SW coast of India. *Current Science* 83(3), 318-320.

Nesbitt, H.W., Young, G.M., 1982. Early Proterozoic climates and plate motions inferred from major element chemistry of lutites. *Nature* 199, 715-717.

Overpeck, J., Anderson, D., Trumbore, S., Prell, W., 1996. The southwest Indian monsoon over the last 18,000 years. *Climate Dynamics* 12, 213–225.

Pachur, H. J., and Wünnemann, B., 1995. Lake evolution in the Teenager Desert, Northwestern China, during the last 40,000 years. *Quaternary Research* 44, 171–180.

Phadtare, N. R., 2000. Sharp decrease in summer monsoon strength 4000–3500 cal yr BP in the central higher Himalayas of India based on pollen evidences from alpine peat. *Quaternary Research* 53, 122–129.

Porter, C. S., An, Z., Zheng, H., 1992. Cyclic Quaternary alluviation and terracing in a nonglaciated drainage basin on the north flank of the Qinling Shen, Central China. *Quaternary Research* 38, 157–169.

Prasad, S., Enzel, Y., 2006. Holocene paleoclimates of India. *Quaternary Research* 66, 442-453.

Prasad, S., Kusumgar,S., Gupta,S.K., 1997. A mid to late Holocene record of paleoclimatic changes from Nal Sarovar: a paleodesert margin lake in western India. *Journal of Quaternary Science12*, 153-159.

Prell, W. L., Kutzbach, J. E., 1987. Monsoon variability over the past 150,000 years. Journal of Geophysical Research 82, 8411–8425.

Prell, W. L., Van Campo, E., 1986. Coherent response of Arabian Sea upwelling and pollen transport to late Quaternary monsoonal winds. *Nature* 323, 526–528.

Rangarajan, G., Sant, D.A., 2000. Paleoclimatic data from 74KL and Guliya cores: New insights. *Geophysical Research Letters* 27 (6), 787-790.

Roberts, N., 1998. The Holocene, 2nd Ed. Oxford: Blackwell-Wiley.

Sandeep, K., Warrier, A.K., Harshavardhana, B.G., Shankar, R., 2012.Rock magnetic investigations of surface and sub-surface soil samples from five Lake catchments in tropical southern India. *International Journal for Environmental Research 6*(1), 1-18.

Sharma, M., Owen, L. A., 1996. Quaternary glacial history of the Garhwal Himalayas, India. *Quaternary Science Reviews* 15, 335–365.

Sharma, S., Joachimski, M.M., Tobschall, H.J., Singh, I.B, Sharma, C., Chauhan, M.S., 2006. Correlative evidence of monsoon variability, vegetation change and human habitation in Senai lake deposit, Ganga plain. *Current Science* 90, 973-978.

Singhvi, A. K. and Kale, V. S., 2008.Palaeoclimate Studies in India: Last Ice Age to the Present, IGBP–WCRP–SCOPE Report Series: 4, Indian National Science Academy, New Delhi.

Sinha, R., Smykatz-kloss, W., Stuben, D., Harrison, S.P., Berner, Z., Kramar, U., 2006. Late Quaternary paleoclimatic reconstruction from the lacustrine sediments of the Sambhar playa core, Thar Desert margin, India. *Paleogeography Paleoclimatology Paleoecology* 233, 252–270.

Sirocko, F., Sarnthein, M., Erlenkeuser, H., Lange, M., Arnold, M., Duplessey, J.C., 1993. Century scale events in monsoon climate over the past 24000 years. *Nature* 364, 22-324.

Srivastava, P., Rajak, M. K., Sinha, R., Pal, D.K., Bhattacharyya, T., 2010. A high resolution micromorphological record of the Late Quaternary paleosols from Ganga-Yamuna Interfluve: Stratigraphic and Paleoclimatic implications. *Quaternary International* 227, 127-142.

Sukumar, R., Ramesh, R., Pant, R.K., Rajagopalan, G., 1993. A ä^{13}C record of late Quaternary climate change from tropical peats in southern India. *Nature* 364, 703-706.

Sun Q., Wang, S., Zhou, J., Chen, Z., Shen, J., Xie, X., Wu, F., Chen, P., 2009. Sediment geochemistry of Lake Daihai, north-central China: implications for catchment weathering and climate change during the Holocene. *Journal of Paleolimnology* 43, 75–87.

Tiwari, M., Ramesh, R., Somayajulu, B.L.K., Jull, A.J.T., Burr, G.S., 2005. Solar control of southwest monsoon on centennial timescales. *Current Science* 89, 1583-1588.

Tiwari, M., Ramesh, R., Bhushan, R., Somayajulu, B.L.K., Jull, A.J.T., Burr, G.S., 2006. Paleoproductivity variations in the equatorial Arabian Sea: Implications for east African and Indian summer rainfalls and the El Niño frequency. *Radiocarbon* 48, 17–29.

Tiwari, M., Managave, S., Yadava, M.G., Ramesh, R., 2009. Spatial and temporal coherence of paleomonsoon records from marine and land proxies in the Indian region during the past 30 ka. *Current Science Jubilee Volume*, 517-535.

Van Campo, E., Cour, P., Hang, S., 1996. Holocene environmental changes in Bangong Co Basin (Western Tibet). Part 2. The pollen record. *Palaeogeography, Palaeoclimatology, Palaeoecology* 120, 49–63.

Van Campo, E., Gasse, F., 1993. Pollen- and diatom-inferred climatic and hydrological change in Sumxi Co Basin (Western Tibet) since 13,000 yr B.P. *Quaternary Research* 39, 300–313.

von Rad U, Schaaf, M., Michels, K.H., Schulz, H., Berger, W.H., Sirocko, F., 1999. A 5000–yr Record of Climate Change in Varved Sediments from the Oxygen Minimum Zone off Pakistan, Northeastern Arabian Sea. *Quaternary Research* 51, 39–53.

Wang, S.M., 1992. Lake records of the Holocene climatic changes in China. In: Shi YF (ed.) The Holocene climatic optimum and associated environment in China. Science Press, Beijing, pp. 146–152 (in Chinese).

Wang, S.M., Yu, Y.S., Wu, R.J., Feng, M., 1990. The Daihai Lake: environment evolution and climate change. University of Science and Technology of China Press, Hefei, pp 1–191(in Chinese).

Yu-mei, X., Pei-fang, W., 2001. Peat record of climate change for the last 3000 Years in Yangmu, Mishan region of Sanjiang plain. *Journal of Geographical Sciences* ll (4), 455-460.

Xiao, J.L., Wu, J.T., Si, B., Liang, W.D., Nakamura, T., Liu, B.L., Inouchi, Y., 2006. Holocene climate changes in the monsoon/arid transition reflected by carbon concentration in Daihai Lake of Inner Mongolia. *The Holocene* 16, 551–560.

Yadava, M. G., Ramesh, R., 2005. Monsoon reconstruction from radiocarbon dated tropical Indian speleothems. *The Holocene* 15(1), 48–59.

Yadava, M. G., Ramesh, R., 2006. Stable oxygen and carbon isotope variations in speleothems as monsoon proxies: A comparative study of four different locations in India. *Journal of Geological Society of India* 68, 461–475.

Yadava, M. G., Ramesh, R., Pant, G. B., 2004. Past monsoon rainfall variations in peninsular India, recorded in a 331 year old speleothem. *The Holocene* 14(4), 517–524.

In: Holocene
Editor: Bahadur Singh Kotlia

ISBN: 978-1-62257-722-4

Chapter 7

A GLIMPSE OF PALYNOLOGICAL SUCCESSION FROM INDIA: VEGETATION, CLIMATE AND ANTHROPOGENY SINCE MIDDLE HOLOCENE

*Anjum Farooqui**
Birbal Sahni Institute of Palaeobotany, Lucknow, India

ABSTRACT

The Indian sub-continent expresses multi-faceted complexity in the vegetational pattern exhibited by spatial and temporal variability in climatic conditions.

Analysis of the vegetational changes in the past from well acclaimed papers on palaeopalynology have been attributed largely to climatic variability with less focus on comparative localised factors.

This chapter brings together an analysis of vegetational patterns since the middle Holocene with emphasis on depositional environment in varied geomorphological setups and anthropogenic impact interalia vegetational changes in western, central and south-eastern parts of India.

Keywords: Precipitation, Monsoon, Pollen, Climate, India

INTRODUCTION

The most important natural factors which largely influence the vegetation are edaphic, temperature and precipitation. The type of spatial and temporal climate governing the annual duration of dry, arid, moist, humid, cold and frost conditions provide the structure of vegetation in an area. The recent/ present day climate [Banerji, 1952] and vegetation type has been very well classified by Bagnouls and Gaussen [1953]. Later, the detailed classification of Indian forests was outlined by Champion and Seth [1968] in which 16 major types of

* Birbal Sahni Institute of Palaeobotany, 53, University Road, Lucknow, India, E-mail: *afarooqui_2000@yahoo.com.*

forests were recognized. These are primarily grouped into 5 major categories, viz., moist tropical, dry tropical, montane sub-tropical, temperate and alpine forests. Further details with respect to climate and vegetation in the Indian sub-continent have been analysed by Meher-Homji and Gupta [1999] Most of the rainfall in India takes place under the influence of South West monsoons between June to September except in Tamil Nadu where it is under the influence of North-East monsoon during October and November. The rainfall in India shows great variations, unequal seasonal distribution, still more unequal geographical distribution and frequent departures from the normal. It generally exceeds 1000 mm in areas to the East of Longitude 78^0 E and extends to 2500 mm along almost the entire West Coast and Western Ghats and over most of Assam and Sub-Himalayan West Bengal. On the West of the line joining Porbandar to Delhi and then to Ferozpur, the rainfall diminishes rapidly from 500 mm to less than 150 mm in the extreme west. The Indian Peninsula has large areas of rainfall less than 600 mm with pockets of even 500 mm. The complex vegetation of the Indian subcontinent has been altered due to the climatic variability enhanced by human activities in the last few thoUSnd years [Meher-Homji and Gupta, 1999; Farooqui, 2008; Farooqui and Sekar, 2011].

The terrestrial ecosystems are known to be sensitive to monsoon variability and the modeling reveals that the large scale, sustained modification of ecosystems and vegetation cover has significant effects on the monsoon [Wei and Fu, 1998]. Though the palaeo-records of monsoons over India and other parts of Asia are not systematically synthesized, some studies based on different proxy sources, scales and resolutions have been able to bring out a few prominent features of long-term climate change [Parthasarathy et al., 198]. It is well known that the 'Flora' of any region is the result of phylogenetic evolution, but the 'vegetation' reflects the influence of the climate. Thus, the study of pollen/spores preserved in the sediment act like fingerprints in analyzing the past climate on the spatial and temporal basis. The present work puts together the signatures of palynological succession buried in natural archives such as coastal wetland sediments, lake sediments and archaeological sites in India in order to analyze the different aspects of localized factors inducing changes in vegetational pattern at the time of burial since the middle Holocene.

Materials and Method

The pollen/spores assemblage was recorded from 17 natural archives covering 2 archaeological sites in the Saurashtra coast, 1 from Gujarat mainland, 1 in Kerala coast, 3 in south-western Ghat, 4 in south-east coast (Tamil Nadu), 3 in Andhra Pradesh coast, 2 from central India and 1 from the Gangetic plain (Figure 1, Table 1). The pollen recovered in the sediments was used for reconstruction of the vegetation type in these study areas (Figure 2) following Champion and Seth (1968). For the palynological processing of the sediment, the standard method was followed [Erdtman, 1943]. The palynological results have been expressed as a ratio between Evergreen/semi-evergreen/moist deciduous Trees (A) including Mangroves and Dry Deciduous Trees as (B). The chronology of the studied sediment is given in Table 2.

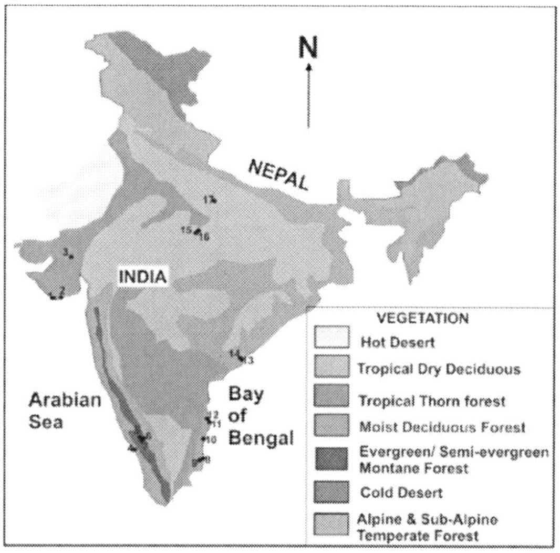

Figure 1. Present Day Vegetation Map and location of the studied sites (Numbers as given in Table-1.

Table 1. Location, Climate and Present Vegetation Type in the studied sites, *Sites at an altitude between 1500-2000m Altitude in South-Western Ghats

No.	Name of the Sites in India	Latitude	Longitude	Average Temp. (°C)	Rainfall (cm)	Present Vegetation type
1	Kanjetar, Saurashtra.	20°43′N	70°38′E	28-30	40-60	Dry deciduous scrub forest and Dry savannah forests.
2	Kaj, Saurashtra.	20°45′N	70°47′E	28-30	40-60	Same as above.
3	Wadhawana, Gujarat.	22° 43' N,	71° 43' E	28-30	76-100	Same as above.
4	Mangalavanam, Kochi, Kerala.	9°59' N	76°11' E	20-35	250-300	Tropical evergreen, semi-evergreen and moist vegetation type.
5	Konalar*,Palni Hills TN.	10°12′ 48″N	77°22′E	14-20	200-400	Tropical Evergreen and Montane Forest.
6	Kuntan*Palni Hills TN.	10°13′N	77° 34′E	14-16	200-400	Tropical Evergreen and Montane Forest.
7	Marian Shola lake* Palni Hills.	10° 15′N	77° 26′E	14-20	200-400	Tropical Evergreen and Montane Forest.
8	Pichavaram, Tamil Nadu (TN).	11°25'N	79°47′E	24-27	80-100	Mangroves and Tropical moist deciduous.
9	Annamalainagar, TN.	11.4N	79.73' E	24-27	80-100	Tropical moist deciduous.
10	Cuddalore, Tamil Nadu.	13°26'–13°43'N	80°03'-80°18'E	25-27	80-100	Back mangroves and Mixed vegetation.
11	Adyar estuary, Chennai, TN.	20° 56' 0" N,	79° 43' E	18-20	80-100	Back mangroves and Mixed vegetation.
12	Sulurpet, Playa Pulicat lagoon, Andhra Pradesh (AP).	13.11N	80°31′ E	20-25	70-100	Exotic plants, wetland ecosystem.
13	Godavari river section AP.	16°43'275" N	82°10' 595" E		100-120	Mixed type of plants, exotics, avenue trees.

Table 1. (Continued)

No.	Name of the Sites in India	Latitude	Longitude	Average Temp. (°C)	Rainfall (cm)	Present Vegetation type
14	Yanam, AP.	16°43'275" N	82°10' 595"	24-28	100-120	Mixed type, wetland and riverine dry and moist deciduous trees.
15	Dhaura Sagar lake, Lalitpur, Uttar Pradesh U.P.	24° 22' 35" N	78° 48' 9" E	20-25	100-120	Dry Deciduous Forest.
16	Mansarovar lake, Talbehat, U.P.	25.038' N	78°431'E	20-25	100-120	Dry Deciduous Forest.
17	Sadatal, Hardoi, U.P.	13.11' N	80°31' E	15-20	140-160	Moist and Dry Deciduous Forest.

RESULTS

Kanjetar and Kaj (North-West Coast)

The Kodinar district in Saurashtra (Table 1) is rich in archaeological remains and the two Harappan period sites are located at Kanjetar and Kaj within 2-7 km range from the present day shoreline. A 60cm trench sample was collected from Kanjetar. The bottom sediment (45-50 cm) dates back to 3,810±100 yr BP and at 15-20cm the date is 780 ±150 yr BP (Table 2). The sediment is loose soil containing small gravels and light to dark brown in colour. The major findings include various types of pottery similar to the Late Harappan Phase. Palynological results from ~4,000-3,000 yr BP reveal high percentage of evergreen moist deciduous tree pollen as compared to the deciduous tree pollen. The ratio is presented in Figure 2. The arboreals include average percentage of evergreen taxa such as *Mangifera* (4%), *Baringtonia* (6%), *Capparis* (4%), *Elaeocarpus* (10%), Garuga (2.5%), *Michelia* (1%), *Dipterocarpus* (3.5%), Moraceae (5.1%), *Shorea* (7%), *Sterculia* (2%) and dry deciduous taxa include *Bombax* (3.5%), *Chorisia* (1.5%), Myrtaceae (2.5%) and *Hardwickia*, (4.2%). The non-arboreal pollen comprise of Asteraceae (4.1%), *Tinospora* (4%), Chenopodiaceae 10% and Poaceae/ Cyperaceae (10.5%). During the last millennium, the arboreals decreased over non-arboreals coupled with sharp decline in pollen assemblage of evergreen taxa. These included *Mangifera* (2%), *Elaeocarpus* ((1.2%), Dipterocarpaceae (2%), *Garuga* (1.8%), Moraceae (1.0%) and Myrtaceae (3.5%). The dry deciduous pollen comprise of the plant taxa such as *Bombax* (6.2%), *Casuarina* (10%), *Eucalyptus* (5.9%), *Holoptelia* (3.5%). The herbs and shrubs comprise of high percentage of *Hibiscus* (8.2%), *Acanthus* (2%), Cucurbitaceae 3.2%) Papavaraceae (3.4%), Umbelliferae (3.6%), Chenopodiaceae ((15.2%) and Poaceae/Cyperaceae (11.5%). The presence of fresh water thecamoebians show high percentage and comprise of *Arcella vulgaris, A. artocrea, A. arenaria, Centropyxis arcelloides, Centropyxis aerophila, Trigonopyxis arcula, Trigonopyxis* spp. (unidentified), *Cucurbitella tricuspis* and *Difflugia* spp. indicating that the study site was a fresh water ecosystem despite its nearness to the shoreline.

Table 2. Depth and Chronology (Radiocarbon Dates: ^{14}C) of the studied sites, India

North-West		North-West Coast				South-West Coast	
Wadhawana, Gujarat		Kaj		Kanjetar		Mangalavanam (Kochi)	
Depth (cm)	Age	Depth	Age	Depth	Age	Depth	Age
102	~6887 (AMS)	60-75	930 ± 90 (BS-3157)	10 -20	780 ± 150 (BS-3187)	180-200	3140 +/- 100 (BS -1713)
70-75	~4824 (AMS)	130-135	3600 ± 120 (BS-3151)	40-50	3810 ± 100 (BS-3158)		

South-Western Ghats					
Konalar		Kuntan		Marian Shola	
160-162	3450+/- 110	100-104	470 +/-90	45-50	5000+/-100
180-200	4000+/- 110	120-150	990+/-110	165-170	30,000+/-100

Tamil Nadu East Coast						Andhra Pradesh East Coast							
Pichavaram		Annamalainagar		Cuddalore		Adyar estuary		Sulurpet (Pulicat lagoon playa)		Gautami-Godavari River Section (Yanam)		Yanam	
40-50	1150 / + - 130 (BS-2179)	640-680	5670 +/- 120 (BS-1860)	120-126	3400 +/- 80 (BS-3188)	800-850	5470 +/- 200	390-400	4608+/-120 (1665)	150-180	650+/-200	800-850	5470 +/- 200 (BS-1666)
185-190	2760 +/- 200 (BS-2138)			186-190 -	6520 +/- 270 (BS-2622)			480-500	6650+/-110	260-270	2080+/-150 (BS-2545)		

Central/Northern India					
Dhaura sagar lake		Mansarovar lake, Talbehat		Sadatal Lake	
75-80	2745+/-100 (BS-1994)	35-40	1410+/-120 (BS-2491)	170-190	4865+/- 110 (BS-2179)
115-120	3865+/-100 (2492)	95-100	4090+/-200 (BS-1992)		

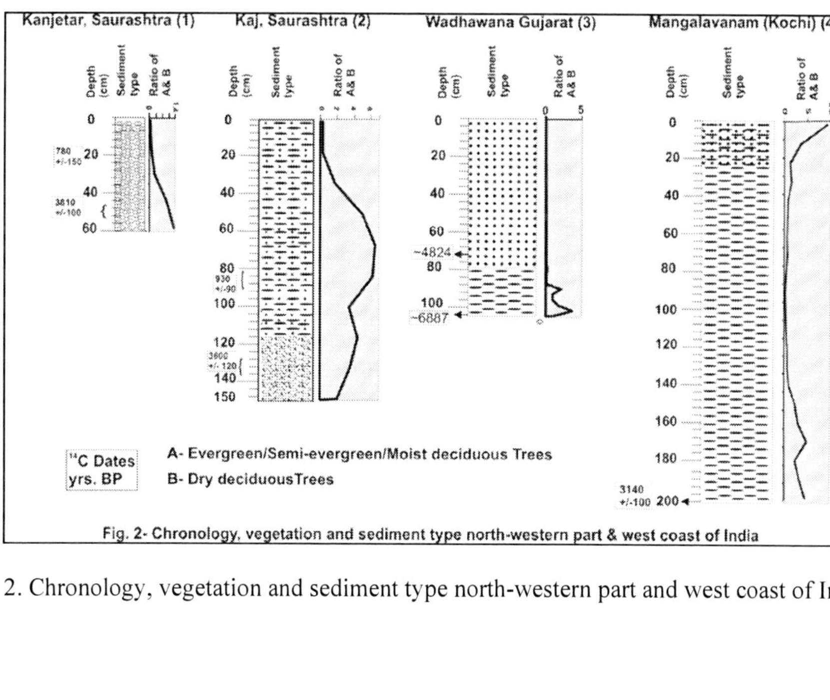

Figure 2. Chronology, vegetation and sediment type north-western part and west coast of India.

A 150 cm trench from Kaj site reveals similar results. The palynological results reveal high ratio of evergreen/moist deciduous trees (Figure 2) until ~800 yr BP.

The average percentages of the evergreen and semi-evergreen pollen are, Anacardiaceae (3.3%), Bignoniaceae (4%), *Canarium* (0.9%), *Chrysophyllum* (3.5%), *Dillenia* (0.9%), *Dipterocarpus* (4.5%), *Elaeocarpus* (6.8%), *Ficus* (8.2%), *Garuga* (1.2%), Moraceae (9.1%), Myrtaceae (2.1%), *Nothopegia travancorica* (1.2%), Oleaceae (1.1%), *Palaquium* (0.9%), Sapotaceae (3.7%), *Schleichera* (4.5%), *Shorea* (3.5%), *Tilia* (2.1%) and *Trewia* (0.5%).

The average percentages of deciduous pollen grains are of *Balanites* (1.8%), *Bombax* (5.5%), *Casaeria* (1.2%), *Chorisia* (1.1%), *Lagerstroemia* (2.5%), *Tectona* (1.2%) and *Oldenlandia* (0.9%). The non-arboreal pollen comprise of *Acanthus* (1.5%), *Argemone* (0.9%), Brassicaceae (0.2%), *Capsicum* (0.2 %), *Clematis* (0.9%), Euphorbiaceae (2.8%), *Hibiscus* (2.8%), *Lawsonia* (0.9%), *Plantago* (2.5%), *Tinospora* (1%), *Cerealea* ((9.2%) and Poaceae/Cyperaceae (3.5%). The aquatic pollen comprises of *Lemna* (4.5%), *Ludwigia* (1.2%), *Myriophyllum* (6.2%), *Trapa* (1.2%) and *Typha* (3.5%). The evergreen to semi-evergreen pollen reduce both qualitatively and quantitatively since the last millennium as shown in Figure 2.

Wadhawana (Gujarat)

Phase-I (~6,000-4,000 Yr BP) Moist Deciduous Forest

The palynological study was carried out in a 102 cm core sample. The sediment up to 80 cm from bottom is blackish clay while the upper sediment is fine to coarse sand.

The bottom sediment dates back to about 6,887 yr BP (AMS dates) and at 65-70 cm, the age is ~4,824 yr BP. During this period, the ratio of evergreen moist deciduous trees is high as compared to the sediments in upper surface as is shown in Figure 2.

The arboreal pollen represents 20-35 % of evergreen and moist semi-deciduous tree taxa like *Canarium (2%)*, *Manilkara/Sapodilla (4.7%)*, *Madhuca (3.7%)*, Meliaceae (3%), *Mimusops (3.5%)*, *Palaquium* (1.2%) and *Sideroxylon (1.2%).*

Phase- II (Since ~4,000 Yr BP)

This period shows increase in 12.4% of dry deciduous tree taxa and pollen assemblage comprises of *Acacia*, *Mimusops*, *Madhuca*, *Terminalia*, *Mangifera indica*, *Holoptelea*, *Sesbania* and Anacardiaceae.

Mangalavanam (Kochi, West Coast)

A 200 cm core sample dates back to ~3,000yr BP. The sediment is blackish to dark brown clay with abundant roots and semi-decayed plant material in upper 100 cm sediment.

The ratio of evergreen and moist deciduous tree pollen is high in bottom sediments which however reduces until 40 cm from below and shows a sharp increase again in surface/sub-surface samples.

These constitute about 14.8% and belong to Bignoniaceae, *Melia*, *Murraya*, *Caraya*, *Hopea*, Sapotaceae, *Strombosia*, *Shorea*, *Schefflera*, *Syzygium*, Rosaceae, *Hibiscus*, *Croton*, Myrtaceae, Euphorbiaceae, *Cocos*, palmae and Araceae.

The mangroves show about 15% of the total pollen count in the beginning of the Late Holocene but show a sharp decline in the middle part of the Late Holocene. A high ratio of mangroves (27%) is evident in the surface and sub-surface sediment (Figure 2). The mangroves comprise of *Excoecaria, Sonneratia, Aegiceras, Xylocarpus, Derris, Achrostichum* and *Baringtonia*.

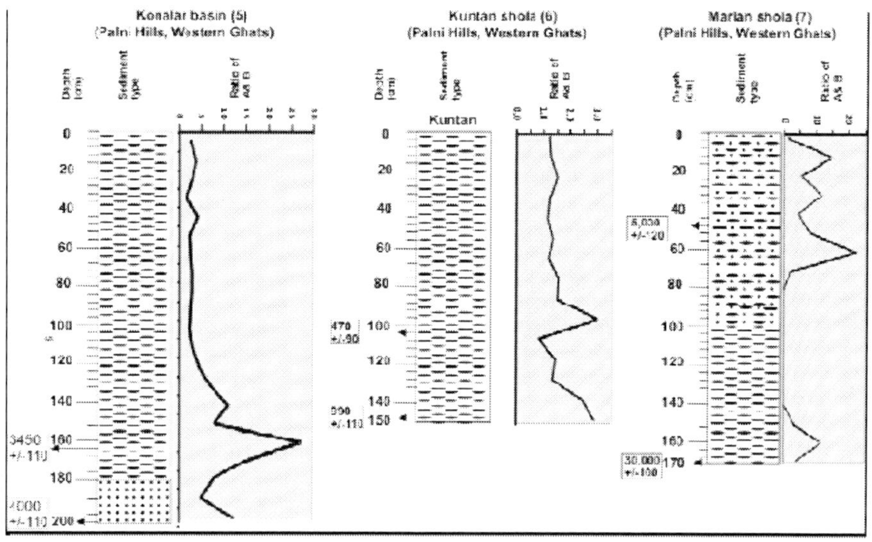

Figure 2. Contd.-Chronology,Vegetation and sediment type from South-western Ghats, India.

Konalar*,Kuntan* and Marian Shola Lake* (South-Western Ghats; Altitude,1,500-2,200 M)

Phase-I (~6,000-~3,000 Yr BP) Evergreen, Moist Deciduous and Montane Forest

The sediment during this period is blackish clay in all the three sites in Palni Hills. the palynological study shows a high ratio of moist and evergreen tree pollen along with montane forest community (Figure 2) until ~ 3,450± 110 yr BP. These mainly constitute 32% of the total pollen count which include *Celastrus, Dodonaea, Elaeocarpus ferrugineous, Euonymous , Eurya japonica, Glochidion, Gordonia, Ilex, Rhododendron, Sideroxylon, Symplocos, Syzygium, Tabernaemontana, Ligustrum, Mallotus, Myristica, Osbeckia* and *Palaquium*.

Phase-II (Since ~3,000 Yr BP) Evergreen and Subtropical Montane Forest

A sharp decline in the ratio is indicated by increase in exotics like *Pinus, Eucalyptus* and *Acacia* trees, constituting 12.3% of the total count in the surface sediment. The lake sediments from Kuntan shola and Marian shola show similar results whereby, the ratio of evergreen and moist deciduous trees is quite low during the latter phase of the Late Holocene. The Poaceae/ Cyperaceae and algal/fungal spores dominate along with the cryptogamic spores.

Pichavaram, Annamalainagar, Cuddalore, Adyar Estuary, Sulurpet, Nilarevu River Section and Yanam (South-East Coast)

Phase-1 (~6,500-3,500 Yr BP) Evergreen to Semi-Evergreen and Moist Deciduous Forest

The depth, chronology and the lithology of the respective sedimentary profiles from the south-east coast is shown in Figure 2. In general, the sediment is sandy clay or clayey sand dominated by either thin blackish peaty layer in between or peaty lenses. The palynological study reveals high percentage (40%) of mangroves during this period. These include *Rhizophora apiculata, R. mucronata, Aegiceras corniculatum, Bruguiera, Ceriops, Excoecaria, Lumnitzera racemosa, Sonneratia alba, S. caseolaris, S. apetala, Nypa fruiticans.*

Evergreen tree pollen constitute 12% of the total count. These include pollen of *Baringtonia, Cullenia, Derris, Dipterocarpus, Shorea, Hopea, Ilex, Eurya, Elaeocarpus, Symplocos, Syzygium* and *Osbeckia.*

Phase-II (~3,500-2,000 Yr BP) Mixed Forest Type (Moist and Dry Deciduous)

During this period, mangroves comprise about 20% of total pollen count. In this, *Avicennia* pollen out number Rhizophoraceae and other true mangrove taxa. No evergreen terrestrial pollen was recorded. About 40% of the pollen constitute mixed type of forest. These are *Acacia, Cassia, Combretum, Diospyros, Flacourtia, Zizyphus. Anogeissus, Bombax,* Fabaceae, *Ipomoea, Grewia, Lagerstroemia, Madhuca, Mytragyna, Nyctanthus,* Lythraceae, Myrtaceae, *Schleichera, Sterculia, Terminalia* and *Xeromphis*. The rest of the total pollen count constitute poaceae/cyperaceae, chenopodiaceae/ amaranthaceae and cryptogamic/algal/fungal spores.

Phase-III (since ~2,000 Yr BP)

A sharp decline in semi-evergreen tree pollen along with the local extinction of many mangrove species from most part of the studied area is evident. The ratio between moist and dry deciduous trees decline sharply but an instant increase in the ratio is seen in the surface sediments in some studied sites (Figure 2). Species of *Sonneratia* have become extinct from Tamil Nadu coastal wetland.

Nypa fruticans that was present during the middle Holocene all along the Indian coastal wetland became extinct except in Andaman and Nicobar islands. Several species of Rhizophoraceae, *Xylocarpus, Aegiceras* etc. show stray occurrence and were at the verge of extinction. Most of the present day coastal wetlands show pure strands of *Avicennia officinalis, A. marina* and fringes are occupied by *Rhizophora apiculata, R. mucronata* or the hybrid form of *Rhizophora*. Other mangrove pollen during this period are rare. The mixed vegetation show about 14.7% of the total count. These comprise *Combretum, Dendrocalamus, Flacourtia, Terminalia, Murraya, Putranjiva, Madhuca, Syzygium, Melia, Ficus,* Moraceae *Baringtonia, Thespesia, Schefflera, Sesbania and* Brassicaceae.

DHAURA SAGAR LAKE, TALBEHAT, SADATAL LAKE

Phase I (4,090 to 3,865 Yr BP) Moist Deciduous Forest

The palynological spectrum during this period mainly comprises elements typical to moist deciduous forest and includes ~70% of pollen taxa dominated by *Anogeissus,* Asteraceae, *Bombax, Buchanania, Casearia, Cassia, Croton, Diospyros, Hopea, Lagerstroemia, Randia, Shorea, Terminalia.* Out of this, *Shorea* (Sal) and *Hopea* constitute 8 and 6% respectively. The remaining 30% comprises the elements of mixed dry deciduous forest and shows dominance of pollen taxa like *Boswellia, Dalbergia, Lannea, Mitragyna* and *Schleichera.*

Phase Iia (3,865 to 2,350 Yr BP) Mixed Moist Deciduous Forest

This period is dominated by about 80% of mixed moist deciduous forest elements and include pollen taxa of Anogeissus, Bombax, Boswellia, Buchanania, Butea, Diospyros, Fabaceae,Gardenia, Garuga, Madhuca, Mitragyna, Schleichera, Sterculia and Terminalia. The remaining 20% comprises pollen taxa of Anogeissus, Butea, Casearia, Cassia, Croton, Diospyros, Lagerstroemia and Shorea.

Phase Iib (2,350 to 1,410 Yr BP) Mixed Dry Deciduous Forest

This period exhibits dominance of 60% of dry deciduous forest elements and include pollen taxa like Acacia, Boswellia, Buchanania, Dalbergia, Diospyros, Elaeodendron, Lannea, Madhuca, Mitragyna, Myrtaceae, Schleichera, Shorea,Terminalia. The remaining 40%t included pollen taxa like Anogeissus, Casearia, Cassia, Croton, Hopea and Lagerstroemia.

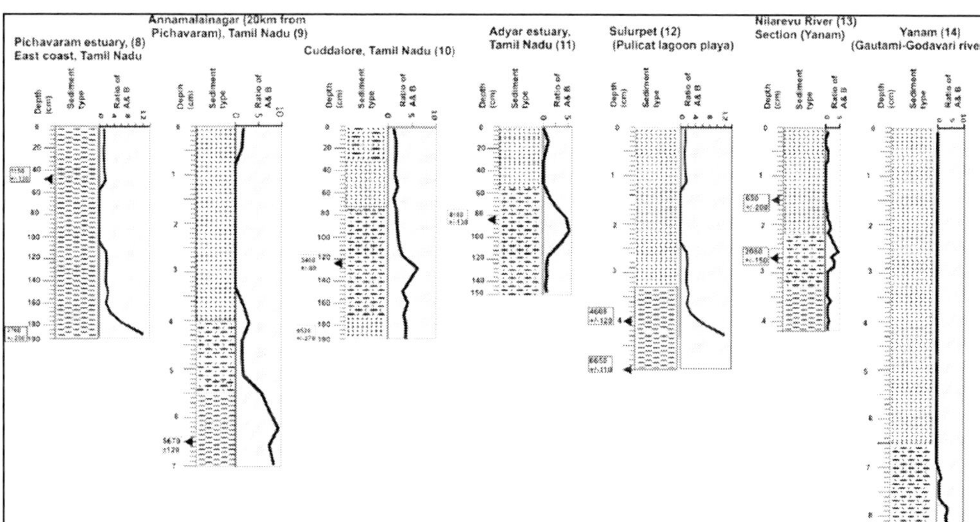

Fig 2. Contd.- Chronology, Vegetation and sediment type in South-eastern coast of India

Figure 2. Contd.-Chronology,Vegetation and sediment type from South-western Ghats, India.

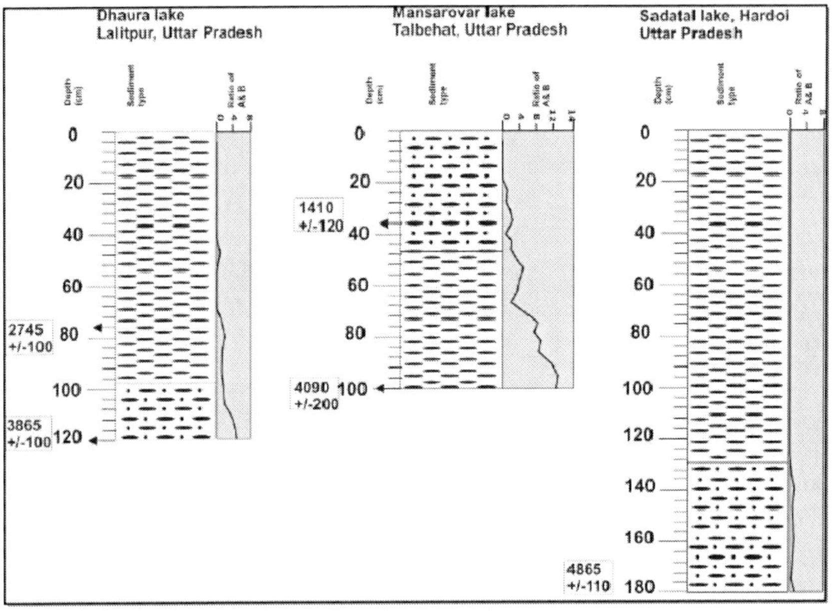

Figure 2. Contd.-Chronology,Vegetation and sediment type in Central India.

Phase III (since 1,410 Yr BP) Dry Deciduous Forest

The palynological spectrum in this phase comprises 45% of dry deciduous forest taxa like *Acacia, Butea, Carissa, Cassia, Combretum, Diospyros, Flacourtia, Gymnosporea* and *Zizyphus*. Another 35% comprises pollen taxa like *Anogeissus, Boswellia, Grewia, Helectris, Lagerstroemia, Madhuca, Mytragyna, Nyctanthus, Tectona* (Teak) and *Woodfordia*. The remaining 20% includes *Annona, Bauhinia, Bombax, Bridelia, Butea, Cocculus, Elaeodendron*, Fabaceae, *Ipomoea,* Lythraceae, Myrtaceae, *Schleichera*, *Terminalia* and *Xeromphis*.

DISCUSSION

Forest ecosystems have long been subjected to many human-induced pressures such as land-use change, over-harvesting, overgrazing by livestock, fire, and introduction of exotic species, agricultural expanse, aquaculture and so on. Climate change constitutes an additional pressure that could change or endanger the natural ecosystems. Among the climatic factors determining the vegetation and its zonation, rainfall and temperature are the only source for which data are easily available and from these the length of dry season can be calculated [Pascal, 1982]. The potential impacts of climate change on forest ecosystem have been highlighted recently [Ravindranath et al., 2011]. There are many publications on the bioclimates of the Indian subcontinent [Mooley and Parthasarathy, 1984; Kumar et al., 1992; Parthasarthy et al., 1993, 1994; Pant and Kumar, 1997; Meher-Homji and Gupta, 1999]. Except in much protected areas, dense forest is rare and the cause is attributed to the intensity of the anthropogenic pressure. Scattered shrubs stage is obtained by the degradation of moist deciduous forest and gradual replacement by dry deciduous forest/ savanna grassland. The original moist deciduous climax formation is obtained when rainfall exceeds 1200 mm and edaphic conditions are favourable. On the contrary, under less than 1200 mm of rainfall, the climax is probably not a dense forest but an open forest or a savanna-woodland. The duration and importance of anthropogenic activities in these regions in disturbing the natural vegetation in their respective biomes in the Indian sub-continent is fast and alarming.

Palynological signatures recorded here from varied natural archives comprise Northern Gangetic plain (Sadatal lake), rugged terrain of Central India (Dhaura lake, Talbehat lake), endemic areas of South-western Ghats (Konlar, Kuntan and Marian Shola), coastal wetlands along the north-west (Kaj, Kanjeter), south-west (Mangalavanam) and east coast (Pichavaram, Annamalainagar, Cuddalore, Adyar estuary, Sulurpet, Nilarevu River section and Yanam) of India. The results show a general trend in the reduction and evolution of the vegetation (evergreen-moist to dry deciduous) in all the sites except in rugged terrains of central India where although the average annual rainfall (high seasonality) is low but the anthropogenic pressure in terms of agricultural expanse is also low.

The coastal zone is a dynamic area with many cyclic processes owing to a variety of resources and habitats. In the historical past, the coastal areas remained an important site for trading between neighbouring countries and therefore, a lot of focus in human settlement took place along the Indian coastline as well [Rao, 1979; Shinde, 1992; Sankalia, 1974; Gaur et al., 2009]. Thereby, a disbalance in the vegetation was prominent which is evident from the palynological changes observed in the core or trench samples from the coastal wetlands.

In the Indian peninsular region, the western Ghats montane forests ($\geq$ 1800 m above sea level) are locally called as 'Sholas', that naturally occur as discrete patches restricted to the folds in the hillside, less subject to winter frost. Palynological analysis [Vishnu-Mittre and Gupta, 196] since 30 ka BP in peat deposits in swamps of Nilgiri and Palni hills (> 2000 m a.s.l.) provides support for historical reconstruction of the shola–grassland vegetation during climatic shifts in the Pleistocene and Holocene [Vasanthy 1988; Sukumar et al. 1993; Blasco, 1974; Mohandass and Davidar, 2010]. If succession is the major process regulating shola expansion, it can be used as a process to restore shola forests where they have been damaged and destroyed. 'Sholas' are now restricted to a very limited area (comprising of endemic plant species and is regarded as one of the hot spot of biodiversity) in the upper montane regions of the Western Ghats of India, and need stringent protection and restoration in order to survive in the climatic region where insignificant changes may not hamper the vegetation. However, iintroduction of exotic trees for commercial exploitation has significantly affected the natural regeneration of the endemic species. The shallow root system and high rate of transpiration of *Pinus, Eucalyptus* is likely to draw more of the surface soil moisture inhibiting the growth of understorey plants which somehow favour the regeneration and protect the seedlings of endemic top storey trees in the region.. The grasslands found in the region are not natural plant formations but have originated secondarily due to destruction of natural forests in some places. Exposed land lead to excess soil erosion affecting the edaphic conditions which lead to modifications in the floristic composition.

The monsoonal and other weather conditions in the north and central part of India are variable. The dry deciduous forests and scrubs grow in areas where the rainfall is between 50 cm and 100 cm. These are found in Central Deccan plateau, south-east of Rajasthan, Punjab, Haryana and parts of Uttar Pradesh and Madhya Pradesh which experience seven to eight months of dry period in the annual cycle. Our palynological study reveals the existence of moist deciduous forest in the beginning of late Holocene but later these species declined and the forest was replaced by dry deciduous Teak (*Tectona*) forest since the last millennium.

This suggests that the climate ~4000to 1400 Yr BP, in the study area, was relatively moist having fewer dry months in the annual cycle than now. The length of the dry season (month/year) according to the bioclimatic maps prepared from climate data at the 71 pollen sampling sites have been studied from southern India [Pascal, 1982]. A gradual colonization by Teak (*Tectona grandis*) forest during the last millennium indicates dry and semi-arid climatic conditions characterised by relatively low rainfall and increased number of dry months [Champion and Seth, 1968; Pal et al., 2009]. Thus, the climate of the Lalitpur district as evident through palynological data [Farooqui and Sekhar, 2011] since ~4,000 yr BP was sub-humid moist in Phase-I (4 - 6 months yr–1 dry period) to semi-arid/sub-humid dry in Phase IIa and IIb (5-7 months yr–1 dry period) and finally became semi-arid in Phase III (7-9 months yr–1 of dry period). Edaphic conditions in the undulating rocky nonarable terrain with shallow soil cover and less agricultural expanse allowed water and nutrient retention in phase - III which supported the diversification of the dry deciduous vegetation despite the gradual increase in aridity and weakening of monsoon during the latter part of Late Holocene. As a result, about 35 pollen taxa were recorded during the last millennium as compared to about 17 taxa recorded during ~4,000 to 2,340 Yr BP in these areas. Archaeological investigations along many important sites in the southern, central and western parts of the Indo-Gangetic Plain suggest considerable progress from incipient agricultural activities to well developed agricultural practices over a span of the last 10,000 years [Sharma et al., 1980; Williams et al., 2006]. Increased anthropogenic activity in the highly fertile Gangetic plain led to enormous deteriorating changes in the landscape, vegetation and hydrological conditions [Singh 2005; Chauhan et al., 2004; Sharma et al., 2004; Saxena et al., 2006]. In the contemporary eastern side of central India [Madhya Pradesh, M. P], moist and dry tropical deciduous Sal forests occur today [Champion and Seth, 1968] but palynological study reveals the dominance of moist deciduous forest during the beginning of late Holocene (4600-1800 Yr BP) that gradually changed to mixed dry deciduous Sal forest [Chauhan, 2002, 2005]. This transition is attributed to increased aridity [Chauhan, 2005; Chauhan et al., 2004]. In the eastern margin of the Thar Desert, Rajasthan, a saline lake existed between 6,000 and 1,300 yr BP and finally the present day semiarid conditions set in since 1,200 Yr BP [Achyuthan et al., 2007]. Similar palynological and climatic analysis from the western part of India is evident [Bryson and Swain, 1981; Singh et al., 1990; Caratini et al., 1991; Enzel, et al., 1999; Madella and Fuller, 2006]. The lake sediment profiles and the vegetational changes recorded from southern India [Bera et al., 1996, 1997; Bera and Farooqui, 2000; Farooqui, 2008; Farooqui et al., 2010] also correlate well with records from other parts of the Indian subcontinent indicating a climate shift towards increased aridity and reduced months of monsoon. In view of the above and other records from India [Meher-Homji and Gupta 1999; Singh et al., 2007], it is concluded that throughout the Indian subcontinent and in most parts of the world [Brooks, 2006] the climate (increasing aridity) played an important role in vegetation succession enhanced by anthropogenic activity during Late Holocene. The vegetational diversification is likely to take place depending on edaphic conditions and less anthropogenic pressure (agricultural practices) as recorded in the Dhaura land Mansarovar lake sediments profiles, Lalitpur, Uttar Pradesh. Therefore, the present day forest cover is more prominent in the rugged terrains of southern parts of Uttar Pradesh, confined to Lalitpur, Mirzapur and Sonbhadra districts, and in the north, the 'Tarai region' (foothills of Siwaliks) unlike the forest cover in the more conducive climate and fertile Gangetic plain where the anthropogeny has played a major role in the local extinction of the climax vegetation of the region. More high resolution palaeo-vegetational data are required from contemporary areas in order to establish the Holocene climate fluctuation versus anthropogenic pressure responsible for regional vegetation changes through time and space.

Conclusion

Areas which receive comparatively high average annual rainfall is likely to show insignificant short-term changes in the vegetation type and less effort would be required for natural regeneration and climax vegetation stage provided the anthropogenic activity is minimum. Areas that receive low average annual rainfall as in most part of the central India show a significant shift in vegetation type from moist tropical forest to dry tropical forest or scrub land where anthropogenic impact is very low or negligible due to rugged terrain not conducive for an agricultural expanse. However, high anthropogenic pressure as is observed in the Gangetic plains have reduced the forest cover drastically where the climax vegetation is not in the equilibrium with the present climatic conditions. Human interference has greatly influenced the climax vegetation. The ppalynological study in areas such as rugged terrains with thin soil cover and undulating topography accompanied by low anthropogenic activity (in terms of agricultural expanse) are best suited for palaeoclimate and paleoenvironmental comparison and interpretations. It is predicted that future increases in aridity and may give way to dry deciduous forest or scrub land replacing the economically more important evergreen and moist deciduous forest from most parts of the Indian subcontinent.

Acknowledgments

I thank the Director, Birbal Sahni Institute of Palaeobotany for providing necessary facilities and Radiocarbon dates to accomplish this work. Thanks are due to Dr. A. S. Gaur, NIO, Goa for providing samples from Kaj and Kanjetar, and to Dr. V Prasad, B.S.I.P., Lucknow for providing the samples from Wadhawana, Gujarat for palynological study.

References

Achyuthan, H., Kar, A., Eastoe C., 2007. Late quaternary-Holocene lake-level changes in the eastern margin of the Thar Desert, India. *Journal of Paleolimnology* 38, 493-507.

Bagnouls, F., Gaussen, H., 1953. Saison sèche et indice xérothermique. *Docum. pour les Cartes des Prod. Végét.* Série: Généralité 1, 1-49.

Banerji, S. K., 1952. The climate of India. *Journal of Bombay National History Society* 50(4), 718-733.

Bera, S. K., Gupta, H. P., Farooqui, A., 1996. Berijam Lake: 20.000 Years Sequence of Pala co floristics and palaeoenvironment in Palni Hills, South India. *Geophytology* 26(1), 99-104.

Bera, S. K, Farooqui, A., Gupta, H. P., 1997. Late Pleistocene/Holocene vegetation and environment in and around Marian Shola, Palni Hills, Tamil Nadu. *The Palaeobotanist* 190-195.

Bera, S. K., Farooqui, A., 2000. Mid-Holocene vegetation and climate of South Indian Montane. *Journal of Palaeontological Society of India* 45, 49-56.

Blasco, F., Thanikaimoni, G., 1974. Late Quaternary vegetational history of southern region. In: Surange, K.R. (Eds.), Aspects and Appraisal of Indian Palaeobotany. *Birbal Sahni Institute of Palaeobotany*, Lucknow, pp. 632–643.

Brooks, N., 2006. Cultural responses to aridity in the middle Holocene and increased social complexity. *Quaternary International* 151, 29-49.

Bryson, R. A., Swain, A. M., 1981. Holocene variations of the Monsoon rainfall in Rajasthan, *Quaternary Research* 16, 135-145.

Caratini, C., Fontugne, M., Pascal, J. P., Tissot, C., Bentaleb, I., 1991. A major change at ca. 3500 Years B. P. in the vegetation of the Western Ghats in North Kanara. Karnataka. *Current Science* 61, 669-672.

Champion, H. G., Seth, S. K., 1968. A revised survey of the forest types of India. *Manager of Publications*, Delhi, India, pp. 404.

Chauhan, M. S., 2002. Holocene vegetation and climatic changes in southeastern Madhya Pradesh, India, *Current Science* 83, 1444-1445.

Chauhan, M. S., 2005. Pollen record of vegetation and climatic changes in north eastern Madhya Pradesh during last 1,600 years, *Tropical Ecology* 46, 265-271.

Chauhan, M. S., Sharma, C., Singh, I. B., Sharma, S., 2004. Proxy records on late holocene vegetation and climate changes from Basaha Jheel, Central Ganga Plain. *Journal of Palaeontological Society of India* 49, 27-34.

Enzel, Y., Ely, L. L., Mishra, S., Ramesh, R., Amit, R., Lazar, B., Rajaguru, S., Baker, V. R., Sandler, A., 1999. High-resolution Holocene environmental changes in the Thar desert, Northwestern India. *Quaternary Science* 284, 125-128.

Erdtman, G., 1943. An Introduction to Pollen Analysis (a new series of Plant Science Books, vol. 12, Waltham Mass: *Chronica Botany Company*, pp. 239.

Farooqui, A., Ray, J. G., Farooqui, S. A., Tiwari, R. K., Khan, Z. A., 2010. Tropical rainforest vegetation, climate and sea level during the Pleistocene in Kerala, India. *Quaternary International* 213, 2–11.

Farooqui, A., Sekhar, B., 2011. Climate change and vegetation succession in Lalitpur area, Uttar Pradesh (India) during late Holocene, *Tropical Ecology* 52(1), 69-77.

Farooqui, A., 2008. The evolution of Pichavaram mangrove and its environment since middle Holocene. *Journal of Geological Society of India - IGCP- 464,* Special Volume 96, 40-54.

Gaur, A. S., Sundaresh., Tripati, S., 2009. New evidence on Maritime archaeology around Mul Dwarka (Kodinar), Gujarat Coast, India. *Man Environment* 34(2), 72-76.

Kumar, R. K, Pant, G. B., Parthasarathy, B., Sontakke, N. A., 1992. Spatial and sub-seasonal patterns of the long-term trends of Indian summer monsoon rainfall. *International Journal of Climatology* 12, 257-268.

Madella, M., Fuller, D. Q., 2006. Paleoecology and the Harappan civilisation of South Asia: A Reconsideration. *Quaternary Science Reviews* 25(11-12), 1283-1301.

Meher-Homji, V. M., Gupta, H. P. 1999. A critical appraisal of vegetation and climate changes during Quaternary in the Indian Region. *Proceedings of Indian National Science Academy* B65 (5), 205–244.

Mohandass, D., Davidar, P., 2010. The relationship between area, and vegetation structure and diversity in montane forest (Shola) patches in southern India. *Plant Ecology and Diversity* 3(1), 67–76.

Mooley, D. A., Parthasarathy, B., 1984. Fluctuations in All-India summer monsoon rainfall during 1871-1978. *Climatic Change* 6, 287-301.

Pal, D. K., Bhattacharya, T., Srivastava, P., Chandran, P., Ray, S. K., 2009. Soils of the Indo-Gangetic plains: their historical perspective and management. *Current Science* 96, 1193-1202.

Pant, G. B., Kumar, R. K., 1997. Climates of South Asia. *John Wiley and Sons*, Chichester pp. 320.

Parthasarathy, B., Munot, A., Kothawale, D. R., 1994. All-India monthly and seasonal rainfall series 1887-1993. *Theoretical and Applied Climatology* 49, 217-224.

Parthasarathy, B., Kumar, R. K., Munot, A., 1993. Homogeneous Indian monsoon rainfall: variability and prediction. *Proceedings of Indian Academy Sciences (Earth and Planatery Science)* 102, 121-155.

Parthasarathy, B., 1984. Inter-annual and long term variability of Indian summer monsoon rainfall', *Proceedings of Indian Academy Sciences (Earth Planet Science)* 93, 371-385.

Pascal, J. P., 1982. Bioclimates of the Western Ghats at1/250000 (2 sheets) *Inst. fr. Pondichéry, Trav. Sec.Sci. Tech. Hors Série n* 17.

Rao, S. R. 1979. Lothal–A Harappan Port Town (1955-1962), MASI-78, Part-1. New Delhi: Archaeological Survey of India.

Ravindranath, N. H., Chaturvedi, R. K., Joshi, N. V., Sukumar, R., Sathaye, J., 2011. Implications of climate change on mitigation potential estimates for forest sector in India. *Mitigation Adaptation Strategy Global Change* 16, 211–227.

Sankalia, H. D. 1974. Prehistory and Protohistory of India and Pakistan. Pune: Post Graduate Research Institute, Deccan College, India.

Saxena, A., Prasad, V., Singh, I. B., Chauhan, M. S., Hasan, R., 2006. On the holocene record of phytoliths of wild and cultivated rice from Ganga plain: evidence for rice-based agriculture. *Current Science* 90, 1547-1552.

Sharma, G. R., Misra, V. D., Mandal, D., Misra, B. B., Pal. J. N., 1980. Beginning of agriculture (Epipalaeolithic to Palaeolithic), Excavations at Chopani-Mando, Mahadahan and Mahagara. Abinash Prakashan, Allahabad.

Sharma, S., Joachimski, M., Sharma, M., Tobschall, H. J., Singh, I. B., Sharma, C., Chauhan, M. S ., Morgenroth, G., 2004. Late glacial and Holocene environmental changes in Ganga plain, Northern India. *Quaternary Science Reviews* 23, 145-159.

Shinde, V., 1992. Padri and the Indus Civilization. *South Asian Studies* 8, 55-66.

Singh, G., Wasson, R. J., Agrawal, D. P., 1990. Vegetational and seasonal climatic changes since the last full Glacial in the Thar Desert. Northwestern India. *Review of Palaeobotany and Palynology* 64, 351-358.

Singh, I. B., 2005. Quaternary palaeoenvironments of the Ganga plain and anthropogenic activity. *Man and Environment* 30, 1-35.

Singh, V., Prasad, V., Chakraborty, S., 2007. Phytoliths as indicators of monsoonal variability during mid–late Holocene in mainland Gujarat, western India. *Current Science* 92 (12), 1754- 1759.

Sukumar, R., Ramesh, R., Pant, R. K., Rajagopalan, G., 1993. A record of Late Quaternary change from tropical peats in Southern India. *Nature* 364, 703-705.

Vasanthy, G., 1988. Pollen Analysis of Late Quaternary Sediments: Evolution of upland savannah in Sandynallah (Nilgiris, South India). *Review of Palaeobotany and Palynology* 55, 175-192.

Vishnu-Mittre., Gupta, H. P., 1968. A living fossil plant community in south Indian hill. *Current Science* 37, 671-672.

Wei, H. L., Fu, C. B., 1998. Study of the sensitivity of a regional model in land cover change over northern China. *Hydrological Processes* 12, 2249-2285.

Williams, M. A. J., Pal, J. N., Jaiswal, M., Singhvi, A. K., 2006. River response to quaternary climatic fluctuations: evidence from the Son and Belan valleys, north-central India. *Quaternary Science Review* 25, 2619-2631.

In: Holocene
Editor: Bahadur Singh Kotlia

ISBN: 978-1-62257-722-4

Chapter 8

THE BLACK SEA AND WORLD OCEAN LEVELS CHANGE DURING THE HOLOCENE

N. V. Esin[1]*, *A. G. Zacepin*[2] *and N. I. Esin*[1]

[1]Southern Branch of P. P. Shirshov Institute of Oceanology of Russian Academy of sciences (I. O. R. A. S.), Gelendzik, Russia

[2]P. P. Shirshov Institute of Oceanology of Russian Academy of Sciences (I. O. R. A. S.), Moscow, Russia

ABSTRACT

The climate change on the Earth is periodical and the global warming replaces glacial periods and vice versa. For about last 20 thousand years, the process of global warming has been continuing on our planet. As a result, the rise of the World Ocean level and deglaciation of ice which was formed during the last glacial period have been occurring for this time period. This chapter focuses on the problem of determining the eustatic change of the World Ocean and Black Sea levels during the Holocene. At present, there are differences among opinions about this level change. The reason for the large number of descriptions of the same process is the transfer of the marks of the sea level change (in form of vertical elevations of sediment) to other horizons through the earth's crust movements and other processes. We calculate the absolute (eustatic) sea level change using the known local curves (each curve was obtained for an individual region) by means of excluding the effect of vertical (smooth) crustal movements. This makes it possible to draw together the known local curves, bringing them in a certain corridor defined by the effect of random processes (we imply uneven movement of the earth crust, earthquakes, landslides, compression of sediments, determination of age etc.). The significant weakening of the effect of random processes and compressing corridor can be reached by vertically averaging the obtained corrected curves.

We used 7 local curves of the Mediterranean Sea level change and one local curve of the Black Sea change, provided by different authors, and as result the absolute curve of the World Ocean level change with a certain degree of accuracy was obtained. We suggest that the process of global warming continues on earth and the World Ocean level rise is 0.5 mm/year. We believe that more number of the local curves may improve the

* E-mail: *ovos_oos@mail.ru.*

accuracy of our method result. Our curve is almost identical to the World Ocean level curve which has been obtained earlier by analyzing date sets of the ocean level position from various parts of the globe, mainly from the tectonically stable areas. Thus, the two different methods provide almost identical results.

INTRODUCTION

The late Pleistocene and Holocene encompass the period of most intensive development of human society, the creation of first civilizations, the settlement in Europe and America. During this process, there was conversion firstly from primitive man to present man, global warming, melting glaciers, and secondly, the cultural migration or disappearance. These processes are controlled by World Ocean level and climate. The transgression of the World Ocean level stimulated the permanent migration of the people tribes towards the land on the hills. Perhaps, this historical process is reflected in the myths about the "World Flood".

So far, the courses of Ocean level as well as the levels of geographically isolated seas, e.g., the Black Sea are not defined in a clear approach. Moreover, the Ocean level course is unidentified even during historical times (at least for the last 2000 years). However, there are numerous curves "course of level" of the Ocean and the Black Sea level, proposed by various researchers. Since there are several curves available, yet there may be one curve of eustatic ocean level changes that should be followed universally.

The reason for the large number of various descriptions of the same process is in the fact that traces of the absolute (eustatic) changes in ocean level in form of vertical marks of sediment after its formation are transferred to other horizons by movements of the earth crust and other processes. Identification of the modern vertical position of sediment (and other factors) with its ancient position in the geological past has led to a variety of the level curves. Consequently, the curves are presented as "course of ocean levels" which are local (relative) because they include the vertical movement of the earth crust, as well as the result of random processes.

This also applies to the Black Sea level. At present, there are dozens of curves which allegedly describe the change in its level (Figure 1). However, as shown by Brückner [2010], these are local (relative) curves, representing the set of points obtained from sediment samples taken from various parts of the Black Sea coast. Therefore we propose a method of calculating the absolute sea-level curve with help of a few local curves.

To check the accuracy of the obtained (or accepted) curve of the absolute sea-level changes, we describe the method of reverse calculation, i.e., calculation of the local curves for coasts with different tectonics by using the absolute curve.

Another problem is the study of the influence of the World Ocean level at the level of the Black Sea. The course of the Black sea level is considered to be independent from the ocean level course as described by earlier workers. However, Esin et al. [2010, 2010b, 2011] have shown that the level of Black Sea repeated the ocean level changes and exceeded by 20-40 cm during the Holocene. In this chapter, our aim is to confirm this. In general we showed, the curve of the ocean level changes [Rohde, 2007] and the curves of the Mediterranean and Black Sea level changes obtained by us have an accuracy of 1-2 m, and they describe the process of level increasing with one minor regressive-transgressive cycle.

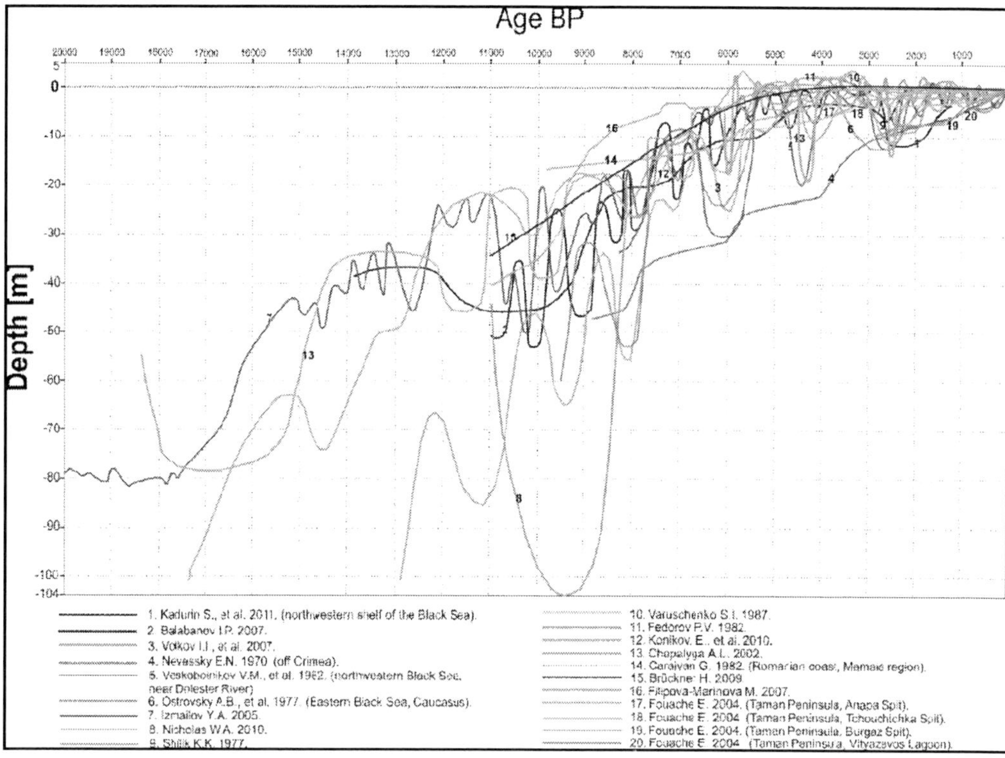

Figure 1. Curves of the Black Sea level change by various authors.

Conversion of Absolute Sea Level Change to Relative (Local) Curves by Vertical Movements of the Earth Crust

Marine sediments are formed in the shallow sea waters and have fixed position of the level at the time of formation, after which the vertical marks of the sediment are changed as a result of vertical movements of the earth crust, caused by tectonics, erosion, accumulation, sediment compaction as well as other random processes. As a result, the coeval sediments in sea water areas, currently located at different levels show the marks of sediment levels of ancient sea at time of their formation. Therefore, the curves of the level change, constructed through the modern marks of the sediments do not show absolute but only relative course of the level that counted from moving Earth's surface. Such curves are called local since the local vertical movement of the earth surface have influenced their shape. According to the mathematical formula, if the absolute sea level course is denoted as $H = F(t)$ (where H_{-} is vertical marks of level), the following expression can explain the relative level change as $H_{\text{OTH}} = F(t) \pm U_0 t + k(t)$ where U_{0}- is average velocity of the vertical movement of Earth crust and $k(t)_{-}$ is contribution of random factors in the shape of local curve (for random factors, we attribute the uneven vertical movement of the Earth crust, sediment compaction, displacement of the Earth crust during the earthquakes and other processes) and t_{-} is the age of the sediment that determines the position of sea level in the past.

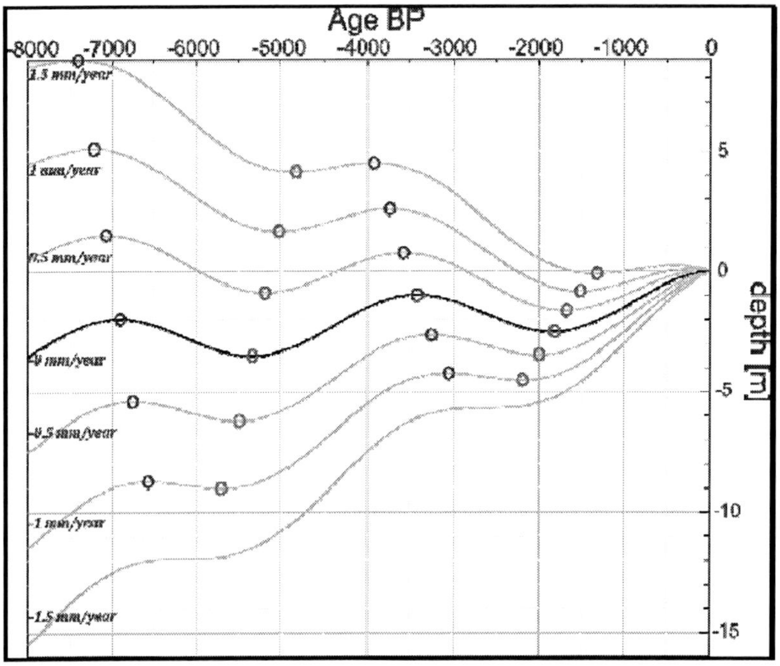

Figure 2. The curve of the absolute sea-level changes as an example (U = 0) and corresponding local curves for uplifting coasts (+) and dipping coasts (-). - extremum values of fluctuations.

An example on how the tectonics contributes in transformation of the eustatic sea-level movement will be discussed here. The eustatic (absolute) rise of the sea level with two fluctuations as well as local curves corresponding to the given absolute sea level course for uplift coasts with velocities 1.5, 1, 0.5 mm /year and dipping coasts with velocities -0.5, -1, -1.5 mm /yr. are shown in Figure 2.

The analyses of the obtained local curves have shown that the vertical movements of the Earth crust can distort the absolute course of the sea level to uncertain amount. Thus, two regressive-transgressive cycles, with the value of scope fluctuation are about 2 m for the dipping coast with velocity 1.5 mm/yr. transform into the continuous transgression of the sea with variable velocity. In such a case, velocity values of the eustatic regression are decreased in the local level curves and the values of the eustatic transgression are increased with increasing velocity of the coast dip. If the coast is uplifted, the reverse occurs.

In general, the eustatic transgression with two regressive-transgressive cycles convert into almost continuous regression in the local curve with variable velocity of regression provided the velocity of the uplift of the Earth crust is about 2 mm/yr. Thus, the values of velocities of the eustatic regression increase in the local level curves, and the values of the eustatic transgressions velocities decrease in the local-level curves provided the velocity of the coast uplift increases. Thus, the general eustatic transgression of the level with two regressive-transgressive cycles can be represented in the local curves by reverse movements of the level on the uplifting and dipping coasts.

In addition, the following factors are very important. If the velocity of tectonic movement increases from -1.5 to 1.5 mm/yr., the start of each regression is shifted to the right (Figure 2). In this case, the regression in each overlying local curve starts earlier than in the underlying curve. The transgression on the each overlying curve starts later than in the underlying. As a result, the duration of the regression increases from the bottom to up, and transgression reduces. Hence, the almost symmetric fluctuations of the absolute sea level course are converted to degenerate asymmetric fluctuations by the vertical movements of the Earth crust.

The vertical crustal movements may very significantly distort the eustatic level changes, up to establish the opposite results. In this case, the time of the beginning of the transgressions and regressions change depend upon the velocity and value of vertical movement of the Earth crust. Consequently, the local curve not show any sign of eustatic level change in itself. It gives some information about the change of the level only if we know the velocity of the coast tectonic movements.

Calculation of the Mediterranean Sea Level Course during the Holocene

The local curves of the sea level, as mentioned above, are obtained by adding the value to the ordinate of absolute sea level change. In order for obtain the absolute sea level curve from the local curves, we need to have the reverse action. We show a method which allows to obtain only one absolute sea level course from known local Mediterranean Sea level curves.

The most representative local curves of the Mediterranean Sea level are described by Brückner [2010] (see Figure3).

The relative curves are quite different from each other because the vertical movements of the Earth crust were different in the different regions. These can be bring closer, if the value of sediment displacement which was caused by the tectonic movements of the Earth crust will be subtracted. Based on the World Ocean level changes [Rohde, 2007] (which correspond to the change of the Mediterranean Sea level), it is widely accepted that the level of Mediterranean Sea was on the mark of -1.5 m about 5 ka BP which correspond to the age of the sediment formation. This can be chosen as the identification level point. Thereafter, the sediments were raised or lowered on the modern mark by the tectonic movements. The average velocity of the Earth crust movement can be determined as quotient of the difference between the mark -1.5m and the modern mark of sediment at the age of sediment, i.e., 5 ka BP. If we multiply the velocity of the vertical movement of the Earth crust with the age of sediments which were formed in diverse times, we obtain the value of sediment displacement for each curve in different times from 7 ka BP to present (displacement of modern sediment is considered as 0). Then, the ordinates of each curve are raised for dipping coast at the value of displacement and lowered for uplifting coast and thus tectonic displacements are subtracted. If the velocity of the tectonic movements does not change in time, other processes of the vertical sediment displacement are unaffected, and calculations and measurements are carried out with perfection, all relative curves can be brought together in one curve which may describe the absolute sea level change. However, as the processes and measurements were not ideal (uneven movements of the Earth crust, inaccurate determination of the mark and the age of sediment etc.), the corrected relative curves were not brought together but much closer to each other (Figure 4).

The phenomena, caused by tectonics in the corrected curves have reduced but there were the imprecisions were caused by the action of random processes. Their action can be significantly reduced by vertical averaging of the corrected curves. The obtained curve describes the absolute course of the Mediterranean Sea level and of course the World Ocean with some degree of accuracy. After completing the process of averaging, we obtained a curve (Figure 5). The precise solution at present seems very difficult due to the action of the random processes. The only accessible method allows to bring together the known local curves by subtracting the average vertical velocity of the Earth crust and vertical averaging. The Rohde's curve of the Ocean level change and the curve obtained by us are shown in Figure 5.

As we can see, Rohde's curve and our results are almost identical in the time interval from 7 to 5 ka BP and differ only by less than 1 m between 5 ka BP to present. At the same time, our curve shows a slight regression of the ocean in the time interval from 4.5 to 3 ka BP, while the Rohde's curve shows unaffected position of the ocean level.

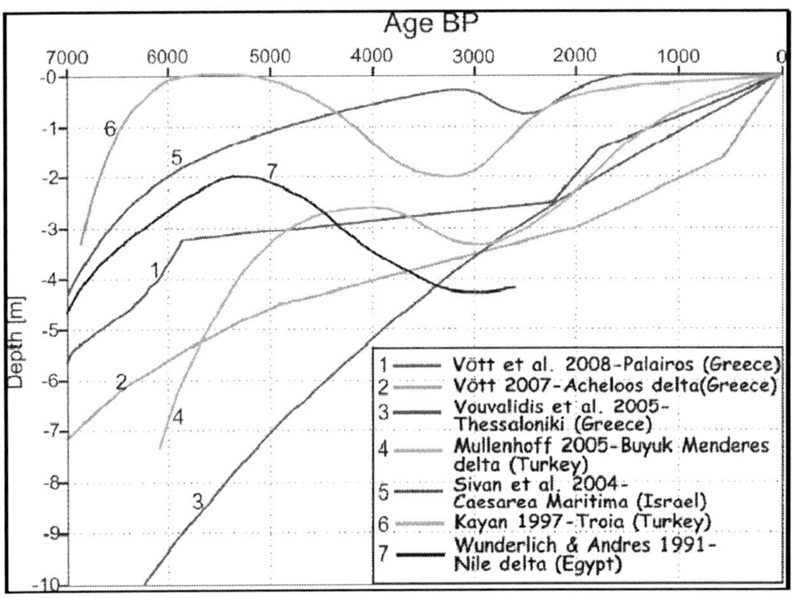

Figure 3. Relative (local) curves of the Mediterranean Sea level change during the Holocene [modified after Brückner, 2010].

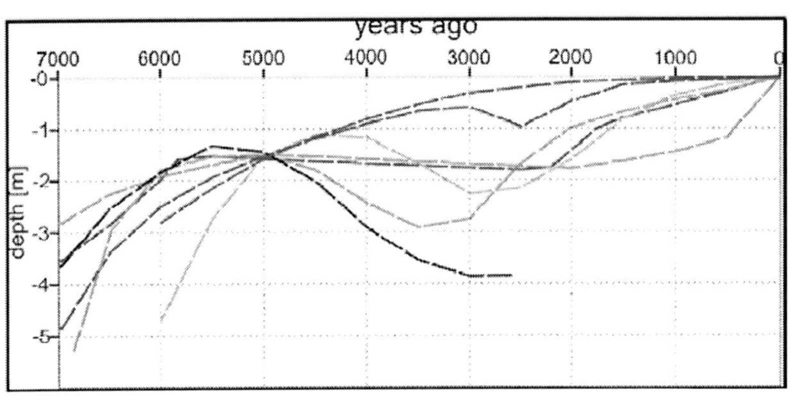

Figure 4. Corrected curves of the Mediterranean Sea level change relative to the vertical movements of the Earth crust.

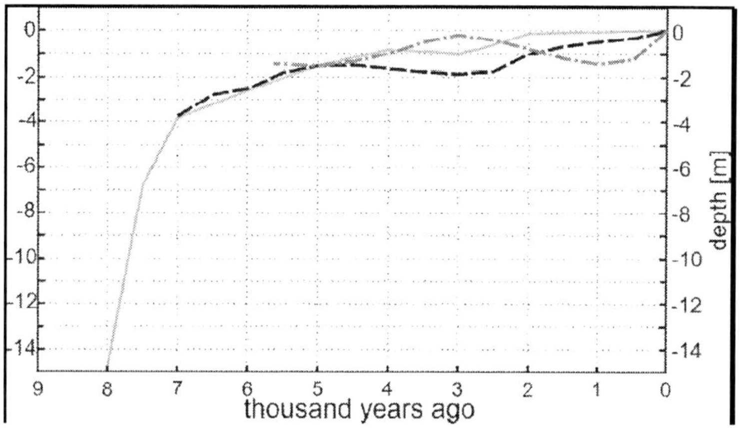

Figure 5. Curves of the World Ocean level change (solid line is Rohde's curve; dashed line is present study; dash-dotted line is obtained by us for the Black Sea level change).

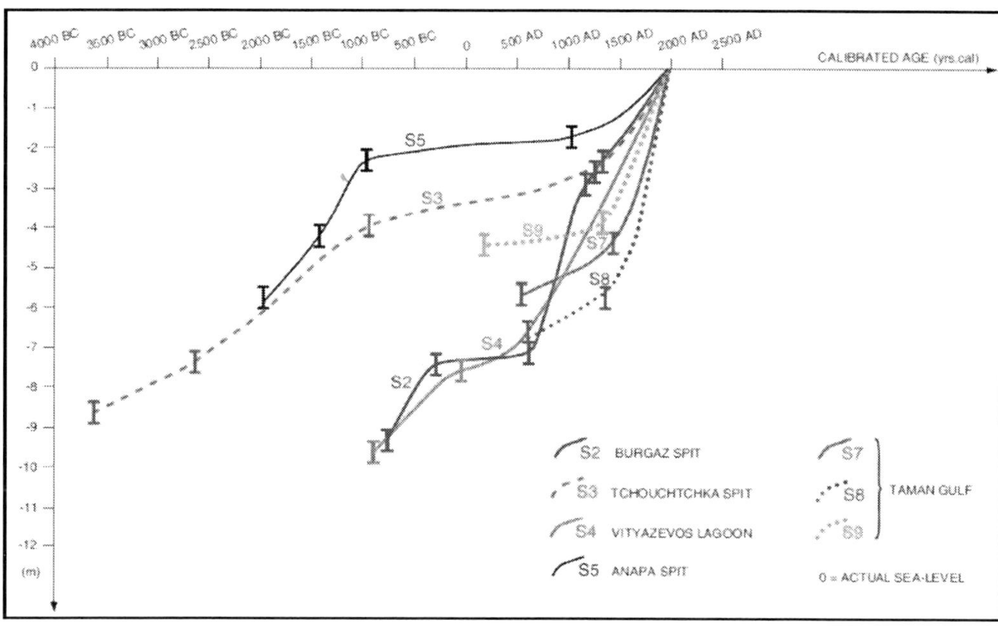

Figure 6. Local curves of the Black Sea level change for diverse water areas [modified after Fouache, 2004].

Fouache [2004] described the set of local curves of the Black Sea level change (see Figure 6). We calculated the absolute course of the Black Sea level (Figure 5) with the method described in the text. In this case, the Rohde's curve and our curves for the Mediterranean and Black Seas are very close to each other as they have signatures of the regressive-transgressive cycle. In our opinion, the same process is described with a theoretical position. According to Esin et al. [2010], the Black Sea level repeated the course of the ocean level during the Holocene. This also is proved the theoretical conclusion that there may not be a large number of secondary fluctuations of the Black Sea level during the Holocene.

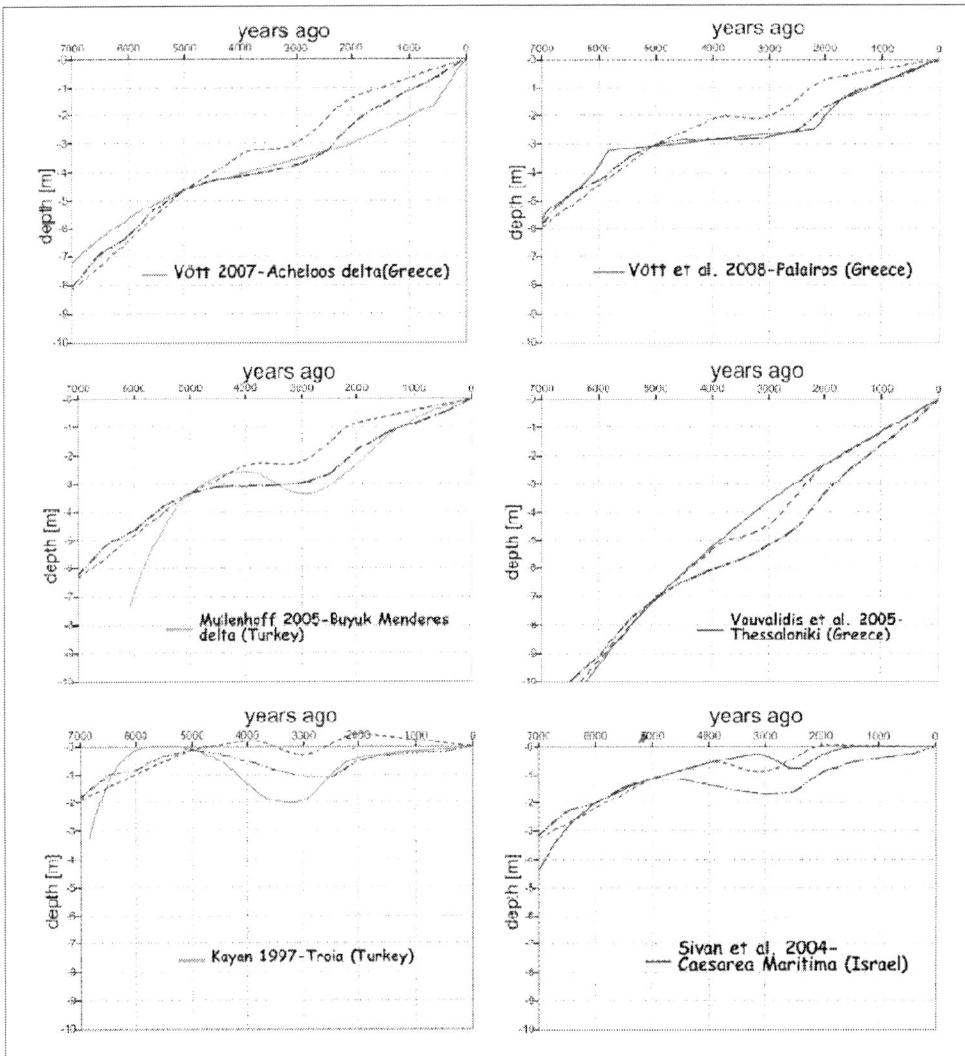

Figure 6. Comparison of local curves of Mediterranean Sea level change (solid line) with their theoretical images, calculated for the Rohde's curve (dotted line) and our curve (dash dotted line).

It is possible to assess the accuracy of our curve of the Mediterranean Sea level change. For this, we performed inverse operation to calculate the local curves for the same areas of the Mediterranean coast, assuming that the absolute course of level is obtained by us. The theoretical and "geological" local curves of the sea level change are very close to each other (Figure 7). The same calculations were performed for Rohde's curve. The comparison of the real local curves and their theoretical images allows the following conclusions. The curve obtained by us more accurately describes the ocean level change (in two cases we have almost precise coincidence of theoretical and "geological" local curves) than the Rohde's curve. In addition, two independent methods of research have given two very similar curves of the ocean and the Mediterranean Sea level change. As can be seen from Figure5, the ocean level made one regressive-transgressive cycle with a range of 0.7 m during the Holocene, which probably caused the Fanagorian regression in the Black Sea.

CONCLUSION

Based on the present study as well as the earlier works [Esin et al., 2010, 2011], we conclude that the Black sea level have been changed during the Holocene like the Mediterranean Sea level. However, the Black Sea level was higher than Mediterranean Sea level at 20-30 cm. Also, there may be some fluctuations with scope less than 0.45 m if the climatic fluctuations of the freshwater balance were in the range of 100-700 km^3/year. The fluctuations of the Black Sea level could not have the scope 0.5 m and more as it was described earlier [Chepalyga, 2002; Izmailov, 2005; Balabanov, 2009]. In case of the deep and wide Bosphorus Strait, it is physically impossible (if only freshwater balance values were not much more than 700 km3/year), as there was always an acting mechanism of leveling of the neighboring seas which was in intense water flow through the strait. Our curve of the World Ocean level change also confirms the current continue of the natural warming, and that the velocity of the Ocean level rising is approximately 0.5 mm/year for last 2 thousand years.

REFERENCES

Balabanov, I. P., 2007. Holocene sea-level changes of the Black Sea. In: Yanko-Hombach, V., Gilbert, A. S., Panin, N., and Dolukhanov, P. (Eds.), *The Black Sea Flood Question: Changes in Coastline, Climate, and Human Settlement*. Dordrecht, The Netherlands, Springer, pp. 711–730.

Balabanov I. P., 2009. Paleogeograficheskie predposylki formirovanija sovremennyh prirodnyh uslovij i dolgosrochnyj prognoz razvitija golocenovyh terras Chernomorskogo poberezh'ja Kavkaza. Moskva-Vladivostok. *Dal'nauka*, 350 p. (in Russian).

Brückner H., Kelterbaum D., Marunchak O. et al., 2010. The Holocene sea level story since 7500 BP – Lessons from the Eastern Mediterranean, the Black and the Azov Seas. *Quaternary International* 225, 159-179.

Caraivan G., 1982. Upper Quaternary evolution of the Mamaia zone. *Pontica* 15, 15-32.

Chepalyga A. L., 2002. Chernoe more, Razvitie landshavtov i klimata Severnoj Evrazii: Pozdnij plejstocen-golocen – aspekty buduwego. *M. GEOS*, pp. 205-285. (in Russian).

Esin N. V., Yanko-Hombach V., Kukleva O. N., and Esin N. I., 2010. Main Regularities of the Late Pleistocene–Holocene Transgression of the Black Sea. *Doklady Earth Sciences* 430 (2), 194–197.

Esin N. V., Yanko-Hombach V., Kukleva O. N., 2010b. Mathematical model of the Late Pleistocene and Holocene transgressions of the Black Sea. *Quaternary International* 225, 180-190.

Esin N. V., Esin N. I., Potetunko E. N., 2011. Estimation of the Double-Amplitude Peak of the Black Sea Level Secondary Fluctuations in Late Pleistocene-Holocene. *Mathematical Modelling*. Chapter 6. Nova Science Publishers, Inc. US.

Fedorov, P. V., 1982. Poslelednikovaia transgressiia Chernogo moria i problema izmeneniia urovnia okeana za poslednie 15,000 let [The post-glacial transgression of the Black Sea and the problem of ocean level change during the last 15,000 years. In: Kaplin, P. A., Klige, R. K., and Chepalyga, A. L. (Eds.), Kolebaniia urovnia morei i okeanov za 15,000 let [Sea and Oceanic Level Fluctuations for 5,000 Years]. *Nauka, Moscow*, pp. 151—156. (In Russian).

Filipova-Marinova M., 2007. Archaeological and Palaeontological evidence of climate dynamics, sea-level change, and coastline migration in the Bulgarian sector of the Circum-Pontic region. In: Yanko-Hombach V., Gilbert A. S., Panin N., and Dolukhanov P. (Eds.), *The Black Sea Flood Question: Changes in Coastline, Climate, and Human Settlement*. Springer, Dordrecht. pp. 453-481.

Fouache, E., Porotov, A., Muller, C., Gorlov, Y., 2004. The role of neo-tectonics in the variaton of the relative mean sea level throughout the last 6000 years on the Taman Peninsula (Black Sea, Azov Sea, Russia). In: Colloque Rapid Transgressions into Semi-enclosed Basins. PICG 464. *Polish Geological Institute, Special Papers*, vol. 11. Polish Geological Institute, Gdansk, pp. 47–58. 8–9–10 Mai 2003.

Izmailov Y. A., 2005. Evolutional Geography of the Black Sea and the Sea of Azov seas. Vol. 1. *Anapskaya barrier*. Sochi, 174 p. (In Russian).

Kadurin S., Larchenkov E., 2011. C14 dating and facies control of paleo-shorelines location on NW Black Sea shelf in Holocene. *INQUA 501 Seventh Plenary Meeting and Field Trip*, Odessa. Ukraine, 21-28 August 2011. pp. 107-110.

Konikov E., Ivanova S., Vinogradova E., Kiosak D., and Govedarica B., 2010. The role of the liman-baymouth complexes in human settlement in the Eneolithic to Bronze Age. *1NQUA 501- IGCP 521 Sixth Plenary Meeting and Field Trip, Rhodes, Greece, 27 September- 5 October 2010*. pp. 101-103.

Nevessky, E. N., 1970. Holocene history of the coastal shelf zone of the U. S. S. R. in relation with processes of sedimentation and condition of concentration of useful minerals. *Quaternaria* 12, 78–88.

Nicholas W. A., Chivas A. R., Murray-Wallace C. V., and Fink D., 2010. Radiocarbon and amino acid racemisation dating of the post-glacial Black Sea: rapid transgression and gradual salinisation of the palaeo-lake by Mediterranean-sourced water. *INQUA 501-1GCP 521 Sixth Plenary Meeting and Field Trip, Rhodes, Greece, 27 September 5-October 2010*. pp. 145-147.

Ostrovsky, A. B., Izmaylov, Y. A., Balabanov, I. P., Skiba, S. I., Skryabina, N. G., Arslanov, S. A., Gey, N. A., and Suprunova, N. I., 1977. Novie dannie o palegidrologicheskom rezhime Chernogo moria v verkhnem pleistotsene i golotsene [New data on the paleohydrological regime of the Black Sea in the Upper Pleistocene and Holocene]. In: Kaplin, P. A., and Shcherbakov, F. A. (Eds.), Paleogeografi ia i otlozheniia pleistotsena iuzhnykh morei S. S. S. R. [Pleistocene Paleogeography and Sediments of the Southern Seas of the U. S. S. R.]. *Moscow, Nauka*, p. 131–140 (in Russian).

Rohde R.A. image: Post-Glacial Sea Level. 2007. *http//www.globalwarmingart.com/*.

Shilik K. K., 1977. Izmenenia urovnia Chernogo moria v pozdnem golotsene i paleotopografia arkheologicheskikb pamianikov Sevemogo Prichernomor'ia antichnogo vremeni [Changes of the Black Sea level in the Late I lolocene and paleotopography of archaeological finds of the Not hem Black Sea]. In: Kaplin, P. A., and Shcherbakov P. A. (Eds.), Paleogeografia i otlozhenia pleistocene juzhnykh morei S. S. S. R. [Paleogeography and Deposits of the Pleistocene of the Southern Seas of the U. S. S. R.]. *Nauka, Moscow*. pp. 158-163. (In Russian).

Varuschenko S. I., Varuschenko A. N., and Klige R. K., 1987. Izmenenija rejima Kaspijskogo morja i besstochnih vodoemov v paleovremeni [Variations of the Caspian Sea regime and of closed lakes in palaeotimes] Moscow. Nauka Publishers, (In Russian).

Volkov I. I., Neretin L. N., 2007. *Hydrogen Sulfide in the Black Sea: The Handbook of Environmental Chemistry*. Springer. pp. 309-333.

Voskoboinikov, V. M., Rotar', M. F., and Konikov, E. G., 1982. Sviaz' ritmichnosti stroeniia tolshch golotsenovykh otlozhenii Prichernomorskikh limanov s kolebatel'nym rezhimom urovnia Chernogo moria [Relationship between the rhythmic composition of thick Holocene layers of the Black Sea region lagoons and the oscillatory level regime of the Black Sea]. In: Kaplin, P. A., Klige, R. K., and Chepalyga, A. L. (Eds.), *Izmeneniia urovnia moria* [Sea-Level Fluctuations]. Moscow, Moscow State University, p. 264–274 (in Russian).

In: Holocene
Editor: Bahadur Singh Kotlia
ISBN: 978-1-62257-722-4

Chapter 9

GEOLOGICAL EVIDENCE OF PRE-HISTORICAL AND HISTORICAL TSUNAMIS IN A TIDAL CHANNEL OF SOUTHWESTERN SPAIN

M. L. González-Regalado[1], J. Borrego[2], M. Abad[1], F. Ruiz[1,*] B. Carro[2], J. Rodríguez Vidal[1], L. M. Cáceres[1], M. I. Carretero[3], M. Pozo[4], A. Toscano[1], F. Gómez Toscano[5] and E. X. García[6]

[1]Departamento de Geodinámica y Paleontología,
Universidad de Huelva, Huelva, Spain
[2]Departamento de Geología, Universidad de Huelva, Huelva, Spain
[3]Departamento de Cristalografía,
Mineralogía y Química Agrícola, Universidad de Sevilla, Spain
[4]Departamento de Geología y Geoquímica,
Universidad Autónoma de Madrid, Spain
[5]Departamento de Historia I,
Universidad de Huelva, Huelva, Spain
[6]Departamento de Botánica y Zoología,
Universidad de Guadalajara, México

ABSTRACT

The sedimentological, palaeontological and radiometric analyses of two cores collected in the Domingo Rubio tidal channel (Huelva estuary, SW Spain) permit us to recognize the main features of several tsunamigenic layers with different ages. Two tsunamigenic facies were differentiated with a foraminiferal record, largely dependent on the palaeogeography of the estuary. These bioclastic beds were deposited during a prehistorical tsunami (3.6-3.0 cal ka BP) and a historical tsunami (1531 CE), the latter is the first geological record of this high-energy event in Spain.

* E-Mail: ruizmu@dgyp.uhu.es, ruizmu@uhu.es.

Keywords: Tsunamigenic facies, Foraminiferal record, Facies, Spain

INTRODUCTION

In the last decade, numerous investigations focused on geological records of Holocene tsunamis. The main features of tsunamigenic layers have been studied in several ways, including: general remarks (Dawson et al., 1991), sedimentology (Dawson and Smith, 2000), geomorphology (Rodríguez-Vidal et al., 2011), geochemistry (Vött et al., 2007), palaeontology (Mamo et al., 2009), magnetic susceptibility (Font et al., 2011) and recurrence periods (Pinegina et al., 2003).

An important part of the Holocene tsunamigenic record was recorded in estuaries. These coastal environments act as sedimentary traps during these high-energy events and the foraminiferal record permits their detection (e.g., Williams and Hutchinson, 2000; Haslett et al., 2010)

The aim of this chapter is to delimitate the main facies and the foraminiferal record of tsunamigenic layers observed in a small tidal channel located in southwestern Spain. The calibrated ages of these layers were compared with the pre-historical and historical catalogs of tsunamis in this area.

STUDY AREA

Huelva Estuary: The Domingo Rubio Tidal Channel

The Huelva estuary is located on the southwestern Spanish Atlantic coast, at the junction of the Tinto and Odiel rivers (Figure 1). Near the mouth, these rivers constitute an incised-valley estuary filled by the Holocene sediments overlying the Miocene clays, Pliocene bioclastic sands and Pleistocene fluvial conglomerates. The present-day geomorphological structure is formed by wide salt-marsh bodies separated by two main channels (Padre Santo Channel and Punta Umbria Channel), drained by numerous ebb-tide channels and protected from the sea by two main spits (Punta Umbria spit and Punta Arenillas spit). The Punta Arenillas spit protects the inner Domingo Rubio Channel, a tributary of the Tinto River. This channel draws a wide valley bounded by Pleistocene, fluvial conglomerates and coarse sands. Its watercourse is sinusoidal, with a narrow width (< 50 m) and a depth depending on tidal floods. This tidal regime is mesotidal and semi-diurnal, with a low diurnal inequality (Borrego et al., 1993).

Holocene Tsunamis in Southwestern Spain

The southwestern Spanish coast is a low-probability tsunamigenic area compared with other areas of the world. The Holocene record of tsunamis has been extensively studied in the last decade and includes at least 20 tsunamis during the last 10,000 years (Ruiz et al., 2005; Gràcia et al., 2010; Lario et al., 2011).

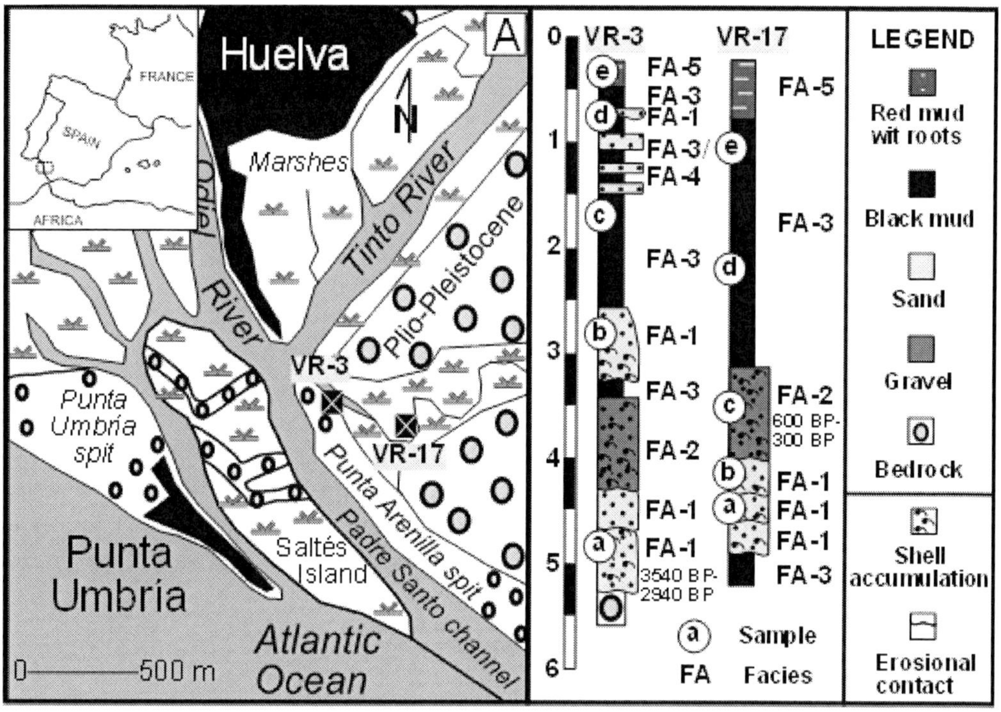

Figure 1. Location of study area and cores.

Sixteen of them have a historical record (Campos, 1991), although the most important imprints were derived from the 218 BCE-210 BCE tsunamis (Luque et al., 2001; Ruiz et al., 2004; Rodríguez-Vidal et al., 2011) and 1755 CE tsunami (Luque et al., 2001; Whelan and Kelletat, 2005). Morales et al. (2008) review the tsunami record of the Huelva Estuary. They indicate the presence of five layers during the last 5,000 years.

Methodology

Two cores (Figure 1) were recovered in the Domingo Rubio tidal channel (Figure 1: VR-3 and VR-17) using the method described by Laneski et al. (1979), with a length comprised between 4.5 and 5.1 m. A preliminary study was developed in order to distinguish the main facies and their main visual features. In a second step, ten samples were taken and divided into sub-samples for the different analyses, according to the facies distribution.

Grain Size

Owing to the predominance of detrital facies, the lithological description has been complemented with a grain-size analysis. The grain-size distribution was determined by wet sieving the coarser fractions (>100 μm).

Micropalaeontology

In addition, the palaeontological record was studied, including both macro- and microfauna. In each core, five samples (50 g) were selected for micropalaeontological analysis, mainly according to the different sedimentary facies observed in each core. About 300 individual specimens were picked, counted and determined in each sample, or the total foraminiferal population if the number of individuals was less than 300. In a final step, percentages of each species were calculated (Table 1).

Radiocarbon Dating

Two samples were dated at the University of Barcelona, Spain (Table 2: UBAR-404 and UBAR-405) by radiocarbon methods applied to mollusc shells. The data was calibrated using CALIB version 6.0.1, with the regional marine reservoir effect (ΔR) suggested by Soares and Martins (2009) for this area. The final results correspond to calibrated ages (cal.) using 1σ and 2σ intervals.

Results and Discussion

Facies: Description, Macrofauna and Interpretation

Five Facies Can Be Distinguished (Figure 1).

Facies 1 (FA-1): Fining upward sand. These layers present a shell accumulation disposed on an erosional base, followed by bioturbated muddy sands. The mollusc record constitutes carapaces and valves showing evidence of transport and reworking (rupture, disarticulation, erosion of the edges, loss of the hinge). The marine species of bivalves (*Dosinia exoleta*, *Chamelea gallina*) and gastropods (*Hinia incrassata*, *Natica catena*, *Calliostoma zizyphinus*) are very abundant. All these features indicate a tsunamigenic origin, as pointed by Morales et al. (2008).

Facies 2 (FA-2): Bioclastic gravel and sand. These coarse-grained deposits consist of gravels with numerous shells embedded in a sandy matrix. The faunal remains include gastropods (*Bittium reticulatum*, *Rissoa* sp) and disarticulated and/or fragmented valves of bivalves (mainly *Glycymeris glycymeris*). These tsunamigenic layers (Morales et al., 2008) could be the result of tsunamis with a run-up exceeding that caused FA-1.

Facies 3 (FA-3): Black mud. These fine sediments are strongly bioturbated and present high organic matter contents. The macrofauna is very scarce. In other estuaries of SW Spain, these features are observed in the bottom of some ebb-tide channels or the channel margins of tributary and wider channels (Morales, 1997).

Facies 4 (FA-4): Red mud with roots. Facies FA-3 is covered by the reddish sediments that constitute the recent salt marsh bodies in this area. The macrofauna is absent, whereas numerous roots of *Salicornia* and *Sarcocornia* are present.

Facies 5 (FA-5): Yellow sand. This facies was only found in the upper part of core VR-3, interbedded with FA-3. Its periodical apparition in the geological record is present. The marine faunas collected (mainly *Chamelea gallina*) and a similar lithology to the adjacent Punta Arenillas spit (Borrego, 1992) indicate an origin from wash over avalanches derived from this spit.

In a general overview, the tsunamigenic facies are interbedded with low-energy sequences of FA-3. The transition from FA-3 to FA-5 represents the progressive infilling of this tidal channel, linked to the growing external spits (Punta Umbría, Punta Arenillas) and the partial closing of the marine connection.

Facies and Foraminiferal Record

Autoecology of the Main Species

Seventeen species were identified in the taxonomical analysis with almost 1,200 individuals. The most abundant species are *Ammonia tepida*, *Astrononion stelligerum*, *Elphidium crispum*, *Elphidium williamsoni*, *Quinqueloculina seminulum* and *Triloculina oblonga.*

Ammonia tepida is the most abundant species in the southwestern Spanish estuaries. This species is abundant in the channel margins or the bottom sediments of distributary channels, whereas *Astrononion stelligerum* and *Elphidium williamnsoni* prefer the sandy sediments of distributary channels located near the river mouths (Ruiz et al., 2005). These species are widely represented in other Spanish estuaries (e.g., Cearreta et al., 2008). *Elphidium crispum*, *Quinqueloculina seminulum* and *Triloculina oblonga* are marine species that live in shelf environments, the marine zones of estuaries and outer bays of southwestern Spain (Villanueva-Guimaraes and Canudo, 2008).

Table 1. Percentages of species in the samples selected, diversity, density and facies

SPECIES / SAMPLES	VR-3-a	VR-3-b	VR-3-c	VR-3-d	VR-3-e	VR-17-a	VR-17-b	VR-17-c	VR-17-d	VR-17-e
Ammonia beccarii	13.3					33.3				
Ammonia tepida	33.3	37.8	44.7	45.5		66.7	66.6	43.3	67.3	
Astrononion stelligerum	6.7	21.6	39.5	9.1				37	21.7	
Bulimina subulata	6.7	2.7								
Cornuspira involves		2.7								
Cribroelphidium vadescens		10.8							0.4	
Elphidium crispum		16.2		9.1			33.3			
Elphidium macellum	13.3		7.9							
Elphidium oceanensis								1.7		
Elphidium williamsoni			2.6					10.8	9.4	
Haynesina germanica								7		
Jadammina macrescens					28.6					
Quinqueloculina jugosa	13.3									
Quinqueloculina seminulum	13.3	2.7	2.6	18.2						
Rosalina globularis		5.4							1.1	
Triloculina oblonga			2.6	18.2				0.3		
Trochammina inflata					71.4					
SPECIES	7	8	6	5	2	2	3	6	5	0
INDIVIDUALS / 10 g	3	7.4	7.6	2.2	4.2	3	< 1	15.5	5.5	0
FACIES	FA-1	FA-2	FA-3	FA-2	FA-5	FA-1	FA-1	FA-2	FA-3	FA-3

Foraminiferal Record of Tsunamigenic Facies

Tsunamigenic facies of external (e.g., core VR-3) and internal (e.g., core VR-17) areas of the Domingo Rubio channel show important differences:

a) *External areas.* The foraminiferal record is more diverse, with a wider presence of marine species. Nevertheless, the density is very low, indicating a possible transport from the bottom sediments of the adjacent main channels, where these micro-organisms are poorly represented (González-Regalado et al., 2001).
b) *Internal areas.* The foraminiferal assemblage is composed of rich populations of a few estuarine species (*A. tepida*, *A. stelligerum*). Consequently, an important part of the tsunamigenic sediments are derived from the erosion of shallow channels and channel margins where these species are abundant.

These variations may be due to a difference in the palaeogeographical situation. The lower tsunamigenic layers of core VR-3 were deposited in an open estuary scenario, with a moderate development of sandy spits and a more intense exposition to the waves. On the contrary, the Domingo Rubio Channel was a more protected area during the basal sedimentation of core VR-17. A palaeogeographical evolution of this area can be found in Pendón et al. (1998) and Borrego et al. (1999).

Foraminiferal Record of Other Facies

Although there has been no systematic study, a general review of sandy sediments belonging to FA-4 permits to detect an absence of foraminifera. This feature has been observed in the aeolian sediments of other littoral spits (Pozo et al., 2010).

The salt marsh deposits of FA-5 are clearly dominated by agglutinated species (*Trochammina inflata*, *Jadammina macrescens*). The low salt marsh usually contains frequent specimens of *T. inflata*, with *J. Macrescens* as additional species (Hawkes et al., 2010). This assemblage has been found in sample VR-3-e.

Dating and Holocene Tsunamis in SW Spain

In the core VR-3, a radiometric dating (Table 2) was obtained at the basal tsunamigenic layer, indicating a calibrated age of 3,540 yr BP - 2,940 yr BP. This age coincides with those reported from other tsunamigenic beds of southwestern Spain (Ruiz et al., 2005; Pozo et al., 2010; Gutiérrez-Mas, 2011) and a turbidite deposited in the abyssal areas of the SW Iberian Margin (Vizcaino et al., 2006). Consequently, a regional tsunami occurred in this region, probably between 3,600-3,000 cal. yr BP.

Table 2. Radiocarbon dating results

Core	Lab. Ref.	^{14}C age	cal BP		cal AD/BC	
		(BP)	1σ	2σ	1σ	2σ
A	UBAR-404	3490±60	3100-3390	2940-3540	1150 BC-1440 BC	990 BC-1590 BC
B	UBAR-405	700±80	380-530	300-600	1420 AD-1570 AD	1350 AD-1650 AD

Another sample was dated in the core VR-17 (600-300 cal. yr BP). This age includes a historical tsunami (1531 CE; Campos, 1991); the latter record would be the first evidence of this high-energy event in Spain. In Portugal, a possible tsunamite, derived from this event, was located near Lisbon (Andrade et al., 2003).

Conclusion

The multidisciplinary analyses of two cores collected from the Domingo Rubio Channel (SW Spain) have allowed to recognize five sedimentary facies, two of them with a tsunamigenic origin. Both facies present important percentages of gravels to medium sands with numerous fragmented bioclasts of marine and estuarine species. Some tsunamigenic layers were deposited by two tsunamis that occurred between 3,600-3,000 cal. yr BP and 1531 CE. The foraminiferal record depends on the estuarine palaeogeography, with higher percentages of marine species in an open marine scenario and dominant estuarine assemblages in a more restricted estuary. These high-energy beds are interbedded in a regressive sequence, with a progressive infilling from ebb-tide channels to salt marsh environments.

Acknowledgments

This work was financed by the DGYCIT Project CGL2010-15810 (funded by FEDER), an international Project funded by the Fundacao para a Ciencia e a Tecnología (PTDC/CTE-GIX/110205/2009), an Andalusia Excellence Project funded by FEDER (SEJ-4770) and four Research Groups of the Andalousia Board (RNM-238, RNM-276, RNM-293 and RNM-349). This chapter is a contribution to the IGCP 588.

References

Andrade, C., Freitas, M.C., Miranda, J.M., Baptista, M.A., Cachão, M., Silva, P., Munhá, M., 2003. Recognizing Possible Tsunami Sediments In The Ultradissipative Environment Of The Tagus Estuary (Portugal). Coastal Sediments' 03, Florida, USA.

Borrego, J., 1992. Sedimentología del estuario del río Odiel (Huelva, SO España). Ph.D. Thesis, Universidad de Sevilla, Spain.

Borrego, J., Morales, J. A., Pendón, J. G., 1993. Holocene filling of an estuarine lagoon along the mesotidal coast of Huelva: the Piedras River mouth, southwestern Spain. *Journal of Coastal Research* 9, 242-254.

Borrego, J., Ruiz, F., González-Regalado, M. L., Pendón, J. G., Morales, J. A., 1999. The Holocene transgression into the estuarine central basin of the Odiel River mouth (Cádiz Gulf, SW Spain): lithology and faunal assemblages. *Quaternary Science Reviews*, 18, 769-788.

Campos, M.I., 1991. Tsunami hazard on the Spanish coast of the Iberian Peninsula. *Science of Tsunami Hazards* 9, 83-90.

Cearreta, A., Alday, M., Irabien, M. J., Etxebarria, N., Soto, J., 2008. Modern conditions and recent environmental development of the Muskiz estuary: historical disturbance by the largest oil refinery of Spain. *Journal of Iberian Geology* 34, 191-213.

Dawson, A. G., Smith, D.E., 2000. The sedimentology of mid-Holocene tsunami facies in northern Scotland. *Marine Geology* 170, 69-79.

Dawson, A. G., Foster, I. D. L., Shi, S., Smith, D. E., Long, D., 1991. The identification of tsunami deposits in coastal sediment sequences. *Science of Tsunami Hazards* 9, 73-82.

Font, E., Nascimento, C., Omira, R., Baptista, M. A., Silva, P. F., 2011. Identification of tsunami-induced deposits using numerical modeling and rock magnetism techniques: A study case of the 1755 Lisbon tsunami in Algarve, Portugal. *Physics Earth Planetary International* 182, 187-198.

González-Regalado, M .L., Ruiz, F., Baceta, J. I., González-Regalado, E., Muñoz, J.M., 2001. Total benthic foraminifera assemblages in the southwestern Spanish estuaries. *Geobios*, 34, 39-51.

Gràcia, E., Vizcaino, A., Escutia, C., Asioli, A., Rodés, A., Pallàs, R., García-Orellana, J., Lebreiro, S., Goldfinger, C., 2010. Holocene earthquake record offshore Portugal (SW Iberia): testing turbidite paleosismology in a slow-convergence margin. *Quaternary Science Reviews* 29, 1156-1172.

Gutiérrez-Mas, J. M., 2011. Glycymeris shell accumulations as indicators of recent sea-level changes and high-energy events in Cádiz Bay (SW Spain). *Estuarine, Coastal and Shelf Science* 92, 546-554.

Haslett, S.K., Davies-Burrows, K., Panayotou, K., Jones, B.G., Woodroff, C. D., 2010. Holocene evolution of the Minnamurra River estuary, southeast Australia: foraminiferal evidence. *Zeitschut für Geomorphologie* 54, 79-98.

Hawkes, A.D., Horton, B.O., Nelson, A.R., Hill, D.F., 2010. The application of intertidal foraminifera to reconstruct coastal subsidence during the giant Cascadia earthquake of AD 1700 in Oregon, USA. *Quaternary International* 221, 116-140.

Lanesky, D.E.B., Logan, B.W., Brown, R.H., Hine, A.C., 1979. A new approacj to portable vibracoring underwater and on land. *Journal of Sedimentary Patrology* 654-657.

Lario, J., Zazo, C., Goy, J.L., Silva, P.G., Bardají, T., Cabero, A.M., Dabrio, C.J., 2011. Holocene palaeotsunami catalogue of SW Iberia. *Quaternary International* 242, 196-209.

Luque, L., Goy, J.L., Zazo, C., Dabrio, C.J., Silva, P.F., Lario, J., 2001. Tsunami deposits as paleoseismic indicators: examples from the Spanish coast. *Acta Geologica Hispanica* 36, 197-212.

Mamo, B., Strotz, L., Dominey-Howes, D., 2009. Tsunami sediments and their foraminiferal assemblages. *Earth-Science Reviews* 96, 263-278.

Morales, J. A., Borrego, J., San Miguel, E. G., López-González, N., Carro, B., 2008. Sedimentary record of recent tsunamis in the Huelva Estuary (Southwestern Spain). *Quaternary Science Reviews* 27, 734-746.

Pendón, J. G., Morales, J. A., Borrego, J., Jiménez, I., López, I., 1998. Evolution of estuarine facies in a tidal channel environment, SW Spain: evidence for a change from tide- to wave-domination. *Marine Geology* 147, 43-62.

Pinegina, T.K., Bourgeois, J., Bazanova, L. I., Melekestsev, I. V., Braitseva, O.A., 2003. A millenial-record of Holocene tsunamis on the Kronotskiy Bay coast, Kamchatka, Russia. *Quaternary Research* 59, 36-47.

Pozo, M., Ruiz, F., Carretero, M. I., Rodríguez Vidal, J., Cáceres, L. M., Abad, M., González-Regalado, M. L, 2010. Mineralogical assemblages, geochemistry and fossil associations of Pleistocene-Holocene complex siliciclastic deposits from the Southwestern Doñana National Park (SW Spain): A palaeoenvironmental approach. *Sedimenatary Geology* 225, 1-18.

Rodríguez-Vidal, J., Ruiz, F., Cáceres, L.M., Abad, M., González-Regalado, M.L., Pozo, M., Carretero, M.I., Monge Soares, A.M., Gómez Toscano, F., 2011. Geomarkers of the 218-209 BC Atlantic tsunami in the Roman *Lacus Ligustinus* (SW Spain): A palaeogeographical approach. *Quaternary International* 242, 201-212.

Ruiz, F., Rodríguez-Ramírez, A., Cáceres, L.M., Rodríguez Vidal, J., Carretero, M.I., Clemente, L., Muñoz, J.M., Yañez, C., Abad, M., 2004. Late Holocene evolution of the southwestern Doñana Nacional Park (Guadalquivir Estuary, SW Spain): a multivariate approach. *Palaeogeography Palaeoclimatology Palaeoecology* 204, 47-64.

Ruiz, F., Rodríguez-Ramírez, A., Cáceres, L.M., Rodríguez Vidal, J., Carretero, M.I., Abad, M., Olías, M., Pozo, M., 2005. Evidence of high-energy events in the geological record: Mid-Holocene evolution of the southwestern Doñana National park (SW Spain). *Palaeogeography Palaeoclimatology Palaeoecology* 229, 212-229.

Soares, A.M.M., Martins, J.M.M., 2009. Radiocarbon dating of marine samples from Gulf of Cadiz: The reservoir effect. *Quaternary International* 221, 9-12.

Villanueva-Guimaraes, P., Canudo, J.I., 2008. Assemblages of recent benthic foraminifera from the northeastern gulf of Cadiz. *Geogaceta* 44,139-142.

Vizcaino, A., Gràcia, E., Pallàs, R., García-Orellana, J., Escutia, C., Casas, D., Willmott, V., Diez, S., Asioli, A. Dañobeitia, J.J., 2006. Sedimentology, physical properties and ages of mass-transport deposits associated to the Marquês de Pombal Fault, Southwestern Portuguese Margin. *Norwegian Journal of Geology* 86, 177-186.

Vött, A., Brückner, H., May, M., Lang, F., Brockmüller, S., 2007. Late Holocene tsunami imprint at the entrance of the Ambrakian Gulf (NW Greece). *Méditerranée* 108, 43-57.

Whelan, F., Kelletat, D., 2005. Boulder deposits on the southern Spanish Atlantic coast: possible evidence for the 1755 AD Lisbon tsunami?. *Science of Tsunami Hazard*, 23, 25–38.

Williams, H., Hutchinson, I., 2000. Stratigraphic and Microfossil Evidence for Late Holocene Tsunamis at Swantown Marsh, Whidbey Island, Washington. *Quaternary Research* 54, 218 –227.

In: Holocene
Editor: Bahadur Singh Kotlia

ISBN: 978-1-62257-722-4

Chapter 10

Multi-Proxy Reconstruction of Past Environmental Changes and Climate Conditions over the Last Millennium in Southern Iberia (Sierra Nevada, Spain)

Marc Oliva[1•,3] **and** ***Antonio Gómez Ortiz***[2#,3†]
[1]University of Lisbon Alameda da Universidade, Lisboa, Portugal
[2]University of Barcelona Montalegre, Barcelona, Spain
[3]University of Barcelona, Barcelona, Spain

Abstract

In the Iberian Peninsula, there are a lack of studies focusing on the reconstruction of past climate conditions in the high mountain environments over the recent millennia. In this chapter, the reconstruction of palaeoenvironments is used as the base for palaeoclimatic interpretations. Understanding the past climatic evolution in the southern fringe of the Iberian Peninsula is crucial for better framing the present climate trend, since it is one of the most sensitive parts of the Mediterranean region concerning the future climate scenario. With this objective, we discuss in this chapter the data of sedimentary and documentary sources in Sierra Nevada, the highest mountain range in Iberia. Concerning sedimentary archives, we have analyzed two of them located in the modern-day periglacial environment; solifluction lobes and mountain lake sediments. Both the archives provide data about the landscape evolution in the high lands of this massif for the last millennium. On the other hand, historical sources are written documents left by travelers who travelled around the massif during the past centuries and described various features of the landscape of Sierra Nevada. This information is used to

• Institute of Geography and Territorial Planning, University of Lisbon Alameda da Universidade, 1600-214, Lisboa, Portugal.
Department for Physical and Regional Geography, University of Barcelona Montalegre, 6-8, 08001, Barcelona, Spain.
† Landscape Research and Mediterranean Mountain Palaeoenvironments Research Group, University of Barcelona, Spain. E-mail: *oliva_marc@yahoo.com.*

infer the geomorphological processes, and therefore the associated climate conditions, prevailing in the highest cirques of this range during the last eight centuries. The sedimentary records suggest that the Little Ice Age (LIA) has been the coldest period of the last millennium in Sierra Nevada and historical sources enable to quantify the temperature increase since the mid 19th century in 0.93°C with respect to the present-day thermal conditions.

Keywords: Sierra Nevada, solifluction, lake sediments, landscape changes, climate conditions

INTRODUCTION

The high mountain areas are very appropriate environments to study climate variability since the altitude reinforces the variations in temperature and moisture conditions [Wanner et al., 2000]. Over the last decade, Europe has experienced a greater warming than the global mean, which has been even more pronounced in mountain regions [Böhm et al., 2001]. Considering the high population living in mountainous areas in Europe where climate hazards are numerous and diverse, the study of past climate variability in these environments through the analysis of sedimentary records is crucial in order to better understanding of the future climate trends.

Within this context, the Iberian Peninsula is a particularly vulnerable area regarding future climate scenario. Some forecasts estimate a diminution of total rainfall in about 20-30% accompanied by a temperature increase of 2 to 6°C [IPCC, 2007]. Therefore, it is necessary to quantify the ongoing warming trend within the climate variability driven by natural factors that has occurred in the Iberian Peninsula during the last millennium.

Until the last decade, a few researchers studied the interaction among different geomorphological processes in mountain areas and their operation in the past with regard to climate conditions. However, recently there have been enlarged studies attempting to reconstruct the past activity of cold-climate geomorphological processes from a palaeoclimatic point of view [Giraudi, 2005; Schwamborn et al., 2008; Knight and Harrison, 2009]. But, no studies on these topics exist on the Iberian mountains. The aim of this paper is to present research, focused on the modern-day periglacial environment of the Sierra Nevada, where we have reconstructed the past environmental evolution and climate conditions using an innovative multi-approach for the last millennium. We have analyzed two different sedimentary records (mountain lake sediments and solifluction records) as well as the historical documents, described earlier by the travellers.

REGIONAL SETTING

Sierra Nevada forms a part of the Betique range and constitutes the highest massif in the Iberian Peninsula with several peaks exceeding 3,000 m asl [Mulhacén, 3,478 m; Veleta, 3,398 m]. The mean annual temperature at 2,507 m is 4,4°C and precipitation reaches 710 mm, mostly concentrated between October and April as snow [Oliva et al., 2009].

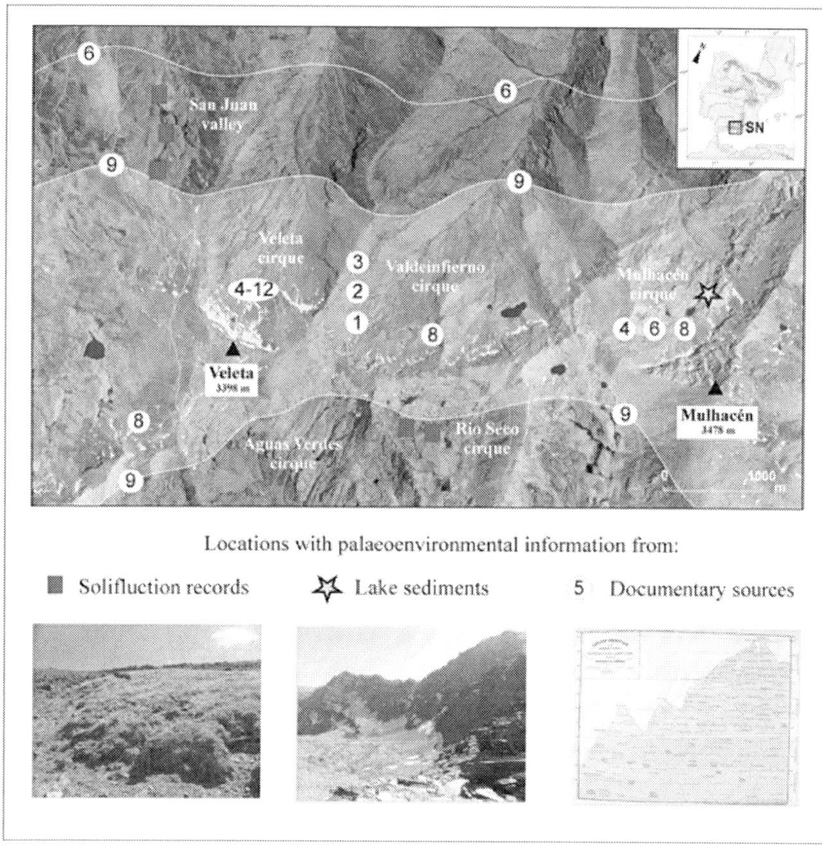

Figure 1. Sierra Nevada and sites cited in the text. Documentary sources are listed in Table 2.

Sierra Nevada contains all the characters of a high semiarid Mediterranean massif. The massif has a very scarce vegetation cover above 2,500 m, but contains a large number of endemic species [Molero and Raya, 1987]. The landscape of the summit area (focus of this paper) is composed of highly weathered mica schists and was strongly affected by glacial processes during the Pleistocene glaciation.

Here, we discuss data of the sedimentary records and documentary sources, centered between 2500 and 3100 m (Figure 1). The studied sedimentary archives correspond to solifluction landforms and mountain lake sediments. The solifluction landforms are widespread features in the present-day periglacial belt, and we have studied those existing in the headwaters of Rio Seco and San Juan valleys. We have also analyzed sediments from La Mosca lake, one of the highest lakes in Europe. Finally, the use of historical sources constitutes a complementary tool for palaeoenvironmental reconstruction for the last centuries in the highest parts of this massif to infer geomorphological and environmental implications.

Materials and Methods

The data generated during the last few years of the palaeoenvironmental evolution in Sierra Nevada are discussed although a few data have already generated by earlier workers. However, we attempt to reconstruct the environment and climate conditions in Sierra Nevada for the last thoUSnd years. Therefore, we only use the data on solifluction landforms and mountain lake sediments that provide information of this time of period. We have used up to 12 Accelerator Mass Spectrometer (AMS)-^{14}C datings, 9 of them correspond to solifluction landforms and 3 to lake sediments [Oliva, 2009; Oliva and Gómez Ortiz, 2012]. The dating was performed at the Angström Laboratory of Uppsala (Sweden) and the dates have been calibrated with the CALIB program [Reimer et al., 2004].

Historical Documents as a Source of Palaeoenvironmental Information

Over the last eight centuries, travelers and others interested in the landscape and resources of Sierra Nevada have described its landscape. We have used these descriptions to unveil the evolution of the landscape in this massif to infer the geomorphological processes prevailing in the summits of Sierra Nevada since the middle of the 12th century.

Geomorphic Activity and Glacier Development Inferred from Lake Sediments

Several lakes located both on the northern and southern slopes of Sierra Nevada have been sampled for palaeoecological purposes. A part of the data has been discussed by earlier workers [Oliva et al., 2010; Oliva and Gómez Ortiz, 2012]. We analyzed the organic matter content and the grain size of the sediments, which are indicative of the past environmental dynamics in the high lands of the massif. Here, we have used only the data of La Mosca lake, the lake that provides a higher accuracy for the last millennium.

Table 1. Dating on samples from solifluction landforms and lake sediments [* Esteban, 1995]

Sample	Laboratory code	Dated material	Conventional age (BP)	Calibrated age
Solifluction landforms				
SJB3.6a	Ua-34876	Pollen	1035 ± 40	1020 ± 126
SJB3.4eL	Ua-36082	Pollen	1030 ± 35	957 ± 51
SJB38.4	Ua-36081	Pollen	830 ± 35	737 ± 59
SJB38.6	Ua-34882	Pollen	730 ± 40	650 ± 83
SJC69.2	Ua-34883	Pollen	700 ± 35	627 ± 66
RSB1.7*	-	Bulk sediment	-	170 ± 120
RSB1.4	Ua-34880	Pollen	670 ± 40	619 ± 64
RSB1.1*	-	Bulk sediment	-	1120 ± 120
RSA30.5	Ua-34879	Pollen	290 ± 40	376 ± 91
Lake sediments				
LM-12	Ua-40162	Plant fragments	238 ± 30	293 ± 24
LM-25	Ua-40163	Plant fragments	536 ± 30	539 ± 23
LM-61	Ua-40164	Plant fragments	1872 ± 32	1800 ± 73

Geomorphic Processes Determined from Solifluction Records

Hundreds of solifluction landforms are distributed in the present-day periglacial belt of Sierra Nevada [Oliva et al., 2009]. Tens of these landforms have been excavated for characterizing the sedimentary sequences. Samples were collected from various lithostratigraphic units in order to determine the organic matter content and measure the grain size of the sediments. Geochronology of the periglacial activity over the last millennia has been discussed in earlier publications [Oliva et al., 2009, 2011].

Palaeoenvironmental Evolution Inferred from Historical Sources and Sedimentary Records in Sierra Nevada

The interpretation provides an approximate chronology for the landscape changes over the last 870 yr BP, while the sedimentary sequences in solifluction landforms and lake sediments provide an approximate chronology for landscape changes over the last millennia.

Landscape Changes Inferred from Historical Sources

Historical documents provide information about the landscape changes that have taken place in the high lands of this massif over the last centuries. Therefore, the geomorphological processes prevailing in the summit area of Sierra Nevada may be inferred.

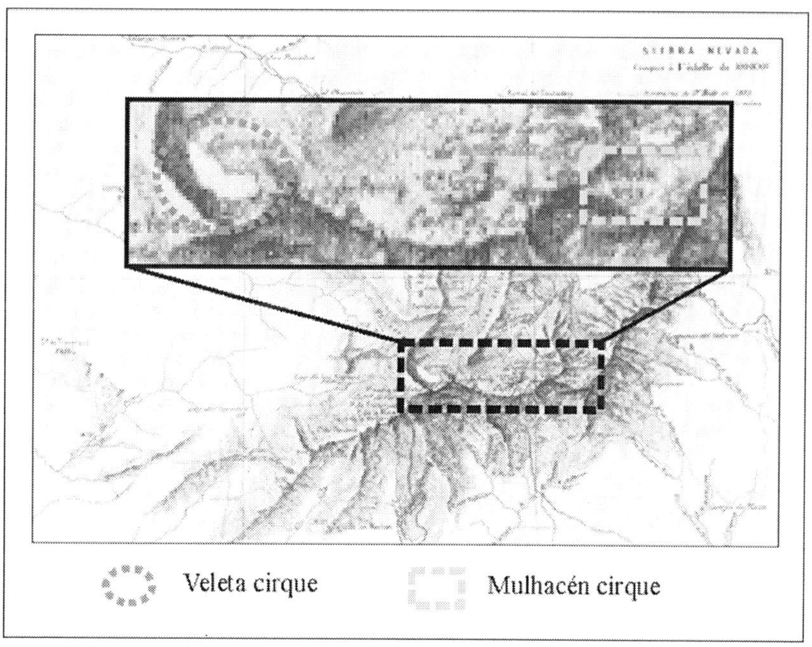

Figure 2. Sketch of summit area in Sierra Nevada from the end of 19th century [Bide, 1893], in which a glacier can be observed in the Veleta cirque but not in the Mulhacén area.

Table 1. Descriptions referring to the past environmental evolution in the Sierra Nevada [Oliva and Gómez Ortiz, 2012]

ID	Autor (year)	Literature description	Geomorphological interpretation
1	Abi Baku al-Zuhri (1137).	"And this mountain is one of the wonders of the world because it is free of snow neither in winter nor in summer. There can be found the snow of past years, which, darkened and solid, looks like black stone, but when it is broken open white snow is to be found on the inside. Plants do not grow at the summit of this mountain, nor can animals live there. Nobody can climb this mountain, nor walk upon it, except in the warm season, when the Sun is in the sign of Scorpio, access then being possible".	GL, SNP.
2	Mahmud al-Qazwini (1263).	"Mount Sulayr of al-Andalus, from which the snow disappears neither in winter nor in summer and from where can be seen most of the territory of al-Andalus due to its height. And the cold there is intense and continuous".	Possible GL, SNP.
3	Bermúdez de Pedraza (1638).	"Sierra Nevada is five leagues from Granada and appears to overhang the houses, such is its height; and the snow extends across eight leagues, from the Ragua Pass, in the marquisate of Zenete, to Órgiva. And there can be no lack of water there, the snow sent to it from heaven after the Flood still being preserved. The snow of this Sierra is a gift to Granada, and it has the three qualities medicine demands for its use to be healthy: it is clean and fresh every year".	GL, SNP.
4	Fernández Navarrete (1732).	"The Sierra wears its beautiful canvas of snow, the summits and upper parts reveal their nakedness in bare rocks and gleaming crags. The places of permanent snow. The main one is Veleta. It has already been said where it is, and its snow-filled cirque. In the shaded parts of these ravines (referring to the stretch between Mulhacén and Veleta) the snow remains perpetually and in very hot years they make tracks on the left and carry it on their backs more than a quarter league. There is also snow all year two leagues above Güejar in the place called Haza de Mesa, and half a league beyond, in the place called Haza de la Escaramuza, at the source of the River Bacares and in the lake of La Caldera".	GL northern cirques, SNP northern and southern cirques.
5	Ponz (1754).	"In the neighbourhood, called Corral del Veleta, a name well suited to its proportions as it is wide and deep, closed off with steep-sided ravines with no entry to them from any side, a gulley filled with snow, considered the first snow that fell after the Flood, turned to stone; as it lies open to the North, here there is ice where in other places there is only snow; and it never melts but for the surface of it, which the Sun discovers".	GL Veleta cirque, SNP.
6	Rojas Clemente(1809).	"I noted here that all the high peaks of the Sierra Nevada are bare, not because the snow does not fall on them, but because the strong winds they are exposed to sweep the snow from them, so that close by, some 100 yards or less below the Mulhacén and Veleta, there can be found enormous perpetual ice fields. At above 2900 varas there are the very cold and glacial areas (what he calls frigidísimo areas)".	GL Mulhacén and Veleta cirques, SNP. Limit of cryonival processes at 2900 varas (2436 m).

ID	Autor (year)	Literature description	Geomorphological interpretation
7	Boissier (1845)	"The glacier has a very steep slope, and a perpendicular height of 200 or 300 feet, and a width of some 600 paces. The peculiarity of it is that this is the only one in all the Sierra and the most southerly in Europe: due to its formation and its position in the depths of a sheltered cirque and dominated in every part by the high peaks where the storms sweep away the snow in the winter. It presents in miniature all the characteristics of the alpine glaciers, clefts, impure ice and muddy moraines. This region (referring to the snowy area) includes all the upper parts of the Sierra Nevada, above 8000 feet. In the Alps at this altitude there may already be perpetual snow, but here nothing similar; however, the snow characterizes our region, although only in the form of patches or drifts that accumulate in the depths and folds of the land. From the end of September, the whole area is covered with new snow, which will only partially disappear from June onwards and covers it completely for eight months".	GL Veleta cirque, SNP.
8	Madoz (1849).	"The regions of these two lofty mountains, Mulhacén and Veleta, and their immediate surroundings are covered with eternal and hardened snows, with layers or stratifications which can be counted easily. The snow, blown by the wind, which can be found collected in the great repository of these cirques, is petrified to such an extent that it has the consistency of the hardest marble".	GL Mulhacén and Veleta cirques.
9	Quelle (1908).	"The Veleta glacier owes its existence solely and exclusively to the fact that it faces north, protected by high walls. The lower edge of the glacier is at an altitude of 2850 m; according to Hellmann, 2845 m; according to Boissier, 2850 m; while I myself set it at 2835 m. It is clear that we cannot use the Veleta glacier to determine the climatic limit of perpetual snow, because it owes its existence only to its extremely favourable location. Only ice fields can provide us with a starting point. On the south face of the Sierra the average altitude of the lower limit of these ice fields is approximately 3000 m, but on the north face 2850 m".	GL Veleta cirque, SNP above 2850 in northern valleys and 3000 m in southern cirques.
10	Enciclopedia Univ. (1912-1919).	"This ice field is about 580 m long by 250 m wide and within it is found the blue ice, cracks and moraine characteristic of other ice fields".	Description of the Veleta glacier.
11	Obermaier (1916).	"The ice field of Veleta is a mass of dead ice, without movements or fluctuations".	Degradation of the Veleta glacier.
12	García Sainz (1947).	"It presents the same stratigraphic superimposition of layers of snow and bluish ice as can be observed in the glaciers of the Pyrenees".	Last reference to the Veleta glacier.

GL: existence of glaciers; SNP: abundant snow patches in summer.

Arab records document cold conditions during the Muslim occupation in southern Iberian Peninsula, stating that the cold, snow, ice and wind were widespread during the 12th and 13th

centuries [Torres Palomo, 1968]. Therefore, snow must have been a more abundant element in the landscape of the massif than what it is today.

No documented records are available about the landscape of the massif from the 13th century to the beginning of the 17th century. In view of the lack of reports, we assume a decreasing presence of snow in summer. Most of the historical descriptions of the Sierra Nevada coincide with the summer season, when the snow melting made easier access to the travelers. Even during warmer periods (than present), snow would have covered the slopes of the Sierra Nevada during the cold season as it used to happen in other ranges of the regional setting, but with less frequency. It was perhaps the presence of snow during the extremely high summer temperatures in Andalusia when it led the visitors abandon. In turn, one of the causes that there are no written works corresponding to this period might be related to the low abundance of snow patches during the summer season, so that the snow-free semiarid landscape of the summits was not so surprising for the travelers.

Site locations of the references are mapped in Figure 1 [first three references cited in Torres Palomo,1968].

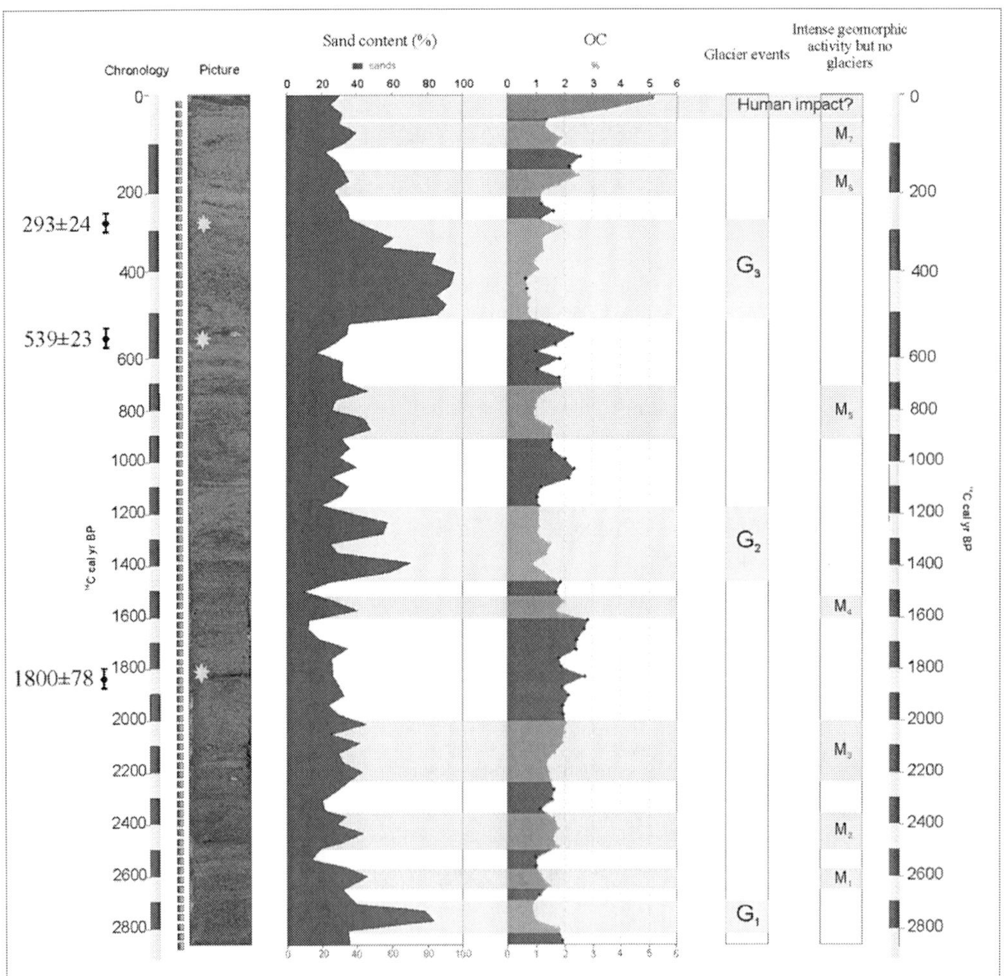

Figure 3. Organic carbon and sand content in the sediments of core LM-1.

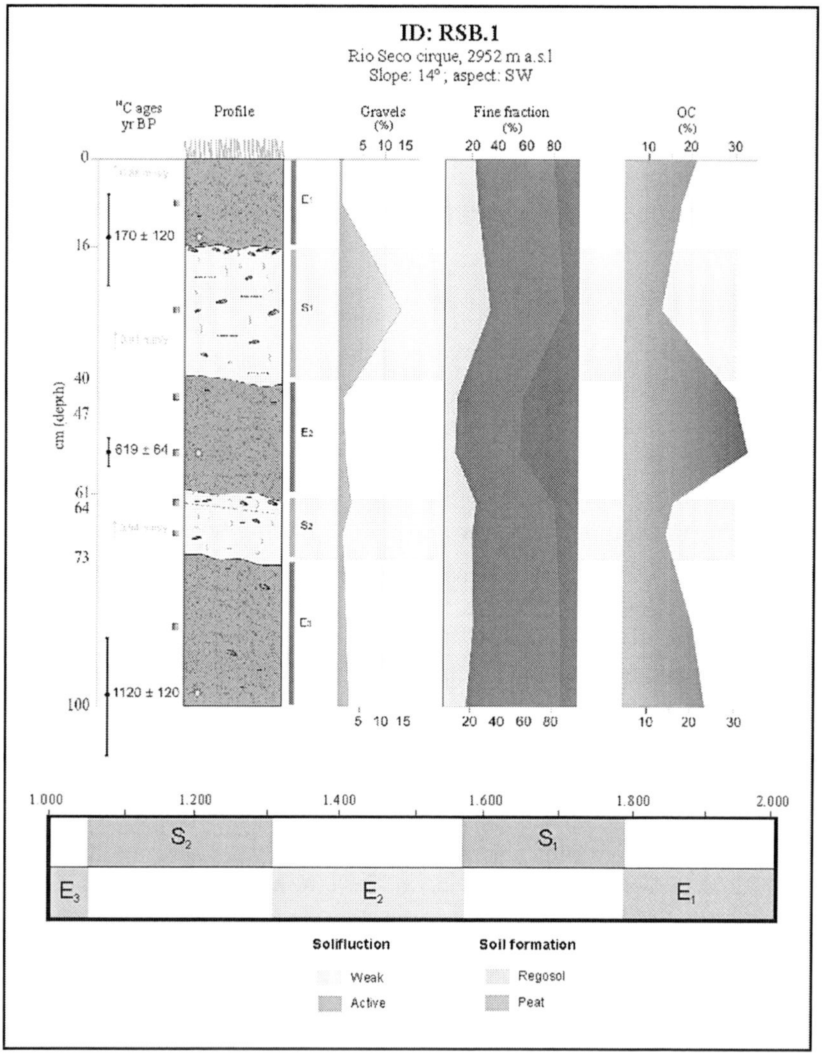

Figure 4. Lithostratigraphy and sedimentological properties of a solifluction landform in Sierra Nevada [modified after Oliva et al., 2011].

From the middle of the 17th century until the 20th century there are tens of related, sketches and maps that give evidence of the existence of a glacier in the Veleta and Mulhacén cirques until the end of the XIX century and beginning of the 20th. Documents also point to the presence of long-lying snow patches in the massif much more abundant than today in the present-day periglacial environment, lasting for more than eight months at altitudes of 2436-2438 m [Rojas Clemente, 1809; Boissier, 1845].

Active Geomorphological Processes in the Mulhacen Catchment

La Mosca lake is located in a glacial depression in the bottom of the Mulhacén cirque (Figure 1), closed by a moraine ridge. Depending upon the climatic conditions, snow only persists as small s patches at the foot of the Mulhacén, the highest mountain in Iberia. Several moraine arches are distributed in the bottom of this cirque, although the chronology of their formation is not yet resolved. A 90-cm long core was collected manually from the centre of this lake.

Varrious lithostratigraphic units contain large variations in the grain size of the sediments (Figure 3). The increase in sands is always accompanied by decrease in the organic carbon (OC) content. Three units are very sandy, with sands exceeding 60% of the mineral fraction in the sediments (G_1-G_3). Other units (M_1-M_7) contain a moderate percentage of sands (40-50%), while the rest has between 25-30%. The content of OC is generally low, fluctuating between 1 and 5% (Figure 3). The very low OC content along the core is indicative of a very low vegetation cover in the Mulhacén cirque during the Late Holocene. However, variations in grain size suggest periods with enhanced activity of geomorphic processes in this cirque, especially during phases G_1-G_3 and M_1-M_7 (Figure 3).

According to the three dates in this core (Table 1), the environmental evolution in this catchment for the last three millennia can be described. During G1-G3 phases, the deposition of sandy sediments may have been influenced by higher erosion and runoff in the catchment, probably derived from a small glacier.

During the summer season, the melting of the snow may have provided the water, capable of transporting sand particles to the lake. Thus, much colder and wetter climate conditions must have prevailed during these three periods: 2,800-2,700 (G1), 1,400-1,200 (G2) and 510-240 cal. yr BP (G3) [Oliva and Gómez Ortiz, 2012]. In the units M1-M7, a similar pattern is observed though the magnitude of changes in grain size and OC is lower than that in phases G1-G3. The geomorphic processes may have also been active during these stages with high runoff and erosion, probably driven by cold and moist conditions although not promoting the development of a glacier in the cirque. During those phases, a higher proportion of silts and OC values, the geomorphological stability may have been prevailed with a slight increase in the extension of sparse grass cover in the catchment. Probably, the warmer temperatures may have prevailed and induced slope stabilization in the highest cirques.

Palaeoecological Evolution in the Modern-Day Periglacial Environment, Inferred from Solifluction Landforms

At present, the periglacial environment of Sierra Nevada has numerous solifluction landforms. Their contemporary activity has been analyzed by earlier workers [Oliva et al., 2008, 2009]. Similarly, we have inferred the past activity of solifluction processes during the Late Holocene based on the sedimentary sequences which the landforms contain [Oliva et al., 2009, 2011].

We focus on the solifluction landforms from San Juan and Rio Seco valleys (Figure 1). These show an alternation between slope deposits and organic layers. The slope deposits have high gravel content (10-50%), low organic proportion (OC < 10%) and a sandy matrix, while organic layers contain high organic matter concentrations (OC > 15%), scarce presence of gravels and a silty matrix (Figure 4).

The study of these sedimentological sequences suggests a significant landscape change in these valleys during the last millennium, with two periods with solifluction processes and three phases with prominent soil formation. The sedimentological characteristics of the slope deposits suggest that these were originated by solifluction processes [Oliva and Gómez Ortiz, 2012]. The periods of active solifluction did not favour vegetation growth, probably due to colder and wetter climatic conditions. During these phases, more prone to periglacial processes at elevations above 2500 m, the slopes remained covered by debris and solifluction evacuated the material down into the valley.

The warm conditions were more conducive to soil formation in the valley bottoms. As per the moisture regime and duration of the warm stage, soils were more or less developed. In turn, soil formation was more intense in the southern face of the Sierra Nevada, where insolation is higher. The periods with little precipitation tended to diminish the activity of periglacial processes, generating a relative stability of slopes, increasing the vegetation cover slightly and promoting an incipient soil development in the valley floors (regosols). The warm and wet phases induced the development of organic-rich soils (histosols) in water-rich and flat areas of the valley floors and headwaters of the highest ravines of the massif, with poorly structured soils (regosols) in steeper hillslopes. The more organic-rich soils are related to the denser vegetation cover. Even in the phases with warmer temperatures, the soil formation would not be possible in the steepest slopes above 2,500 m and in areas surrounding the highest cirques, where a scarce or inexistent vegetation cover would exist.

Environmental Evolution and Climatic Conditions Inferred from Documentary Sources and Sedimentary Records for the Last Millennium

The interpretation of the data generated from the sedimentary records (La Mosca lake sediments and solifluction landforms) and documentary sources points to a succession of palaeoecological changes in the high lands of the massif over the last millennium. This evolution may have been framed within a natural long-term aridification trend that has been inferred from lake sediments from water bodies located in the southern face of the massif and also from solifluction records [Oliva et al., 2011]. However, this climate trend has not been observed in the sediments of La Mosca lake.

Environmental and Climatic Interpretation of the Data

During the last millennium, there have been colder and warmer periods with fluctuating moisture conditions. The colder and wetter regimes implied that the snow melted more gradually at the end of spring and beginning of summer, therefore the ground remained frozen for longer. A shorter warm season may have limited the growing of vegetation and reduced the vegetation cover [Oliva et al., 2009]. The periglacial processes were more intense, supplemented by more effective weathering and erosion. A greater amount of surface runoff along with the longer-lasting seasonal frozen ground may have created the right conditions for saturation of topsoil during the melting season and its slippage by solifluction. There was also an increase in transporting and depositing of sediments in the lakes, where more thickly-textured particles were settled. Parallel to this, the longer duration of ice cover on the lakes meant that their capacity for biological production was restricted. With regards to the historical sources, these records point to the development of glaciers during the coldest, wettest phases (i.e., Little Ice Age) in the highest north-facing cirques of the massif and the existence of late-melting snow patches in greater abundance and much longer-lasting than today and this extended to levels of 2,500 m [Gómez Ortiz et al., 2009]. However, the magnitude of this cooling trend may not have been enough to expand significantly the permanent frozen ground temperatures to other areas in the massif. Further, the permafrost conditions probably may have been only restricted to the highest cirques, close to the small glaciers.

The phases with less extreme cold and wet conditions would have brought about more isolated slope instability with scattered vegetation cover but without widespread solifluction processes. In the high mountain lake basins, these periods may have been involved in greater mineral mobilization enhanced by the lack of vegetation to act as a brake, although the size of mobilized grains was smaller than in those phases with appreciably lower temperatures and higher humidity levels.

On the other hand, warmer phases favoured the development of soil in what today is periglacial ground. The edaphogenesis was more intense in gentle areas with abundant available liquid, giving rise to peaty soils during wet periods (histosols) and weakly structured edaphic horizons during the phases of reduced liquid supply (regosols). In both the cases, the herbaceous cover extended to a greater or lesser degree along the valley bottoms and in isolated patches on the hillsides of the massif, in a similar way to what is presently observed. The higher temperatures may have brought a decrease in solifluction efficiency and periglacial process activity with greater slope movement stability. The rapid thawing of late-melting snow patches and the reduced duration of deep-frozen grounds were unfavorable for activating solifluction. The geomorphic stability would have prevailed, influenced by the extension of vegetation which would slow down the slope dynamics. On the edges of the lakes, the higher temperatures associated with high humidity inherent to the lacustrine micro-environment would have favoured peaty development, with a thick blanket of plant life surrounding the lake. The less periglacial process activity would also result in less mineral mobilization and deposition in the bottom of the lakes, with enhancement in the organic matter content of the lake sediments. The longer duration of the ice-free period on the lakes would bring an increase in the lacustrine biological productivity. Even during warm periods, the travelers have described that the Sierra Nevada landscapes had existence of snow in summer as a common feature in this massif, but unusual in its regional context (e.g., early 20th century, Obermaier, 1916), although during these phases are characterized by higher temperatures and there is no mention of the presence of glaciers deep in the highest cirques.

Synthesis of Climate Variability over the Last Millennium in Sierra Nevada

The sedimentary records in the Sierra Nevada shows an increase in climate variability in Southern Iberian Peninsula over the last 4 ka BP, with an enhanced geomorphological activity manifested in shorter and more frequently alternating cycles of slope stability-instability [Oliva et al., 2011] However, the phases of slope instability are appreciably shorter than those in which the slope stability prevailed.

In the Sierra Nevada, the last millennium has seen an alternation of cold and warm regimes with fluctuating moisture conditions (Figure 5). The solifluction records suggest the prevalence of geomorphic stability in the valley bottoms of the present-day cryonival environment until 850 yr BP, a warm phase that corresponds to the Mediaeval Warm Period (MWP, 1.6-0.85 ka BP) (Figure 4). These conditions are also confirmed by the lake sediments (Figure 3). Martínez Cortizas et al. [1999] have estimated temperatures higher by 1.9°C with respect to present-day values in the NW Iberian corner. The MWP was a warm phase with lower rainfall in southern Iberia [Gil García et al., 2007; Martín-Puertas et al., 2008] and North Africa [Marquer et al., 2008].

The three sources point to the occurrence of two cold and wet periods after the MWP, the first took place around the 12th and 13th centuries (Late Mediaeval Warm Period). The second phase, more intensive, has the coldest and wettest conditions between the 16th to the first half of the 19th century (LIA). In between these periods, a phase with slope stability and weak soil formation occurred, paralleling to a decrease of the water level in the close Zoñar lake [Martín-Puertas et al., 2008].

Both the cold phases suggest the existence of glaciers in the northern cirques and highlight the role of snow as a defining element in the landscape of the high lands of the massif. The solifluction landforms with periglacial slope activity and lake sediments point to an increase in the geomorphic activity in both the southern and northern catchments, with glacier formation in the Mulhacén cirque. Therefore, the sedimentary records suggest that the LIA has been the coldest period in the Late Holocene in the Sierra Nevada, as also is recorded from the Northern Hemisphere [Grove, 2004)] The historical sources document the temperature increase by 0.93°C in this massif from the last decades of the 19th century until present [Oliva and Gómez Ortiz, 2012]. This temperature increase is very similar to that reconstructed for the Pyrenees, which ranges from 0.9 to 0.95°C, and which also promoted glacier expansion [Chueca et al. 1998; González Trueba et al., 2008].

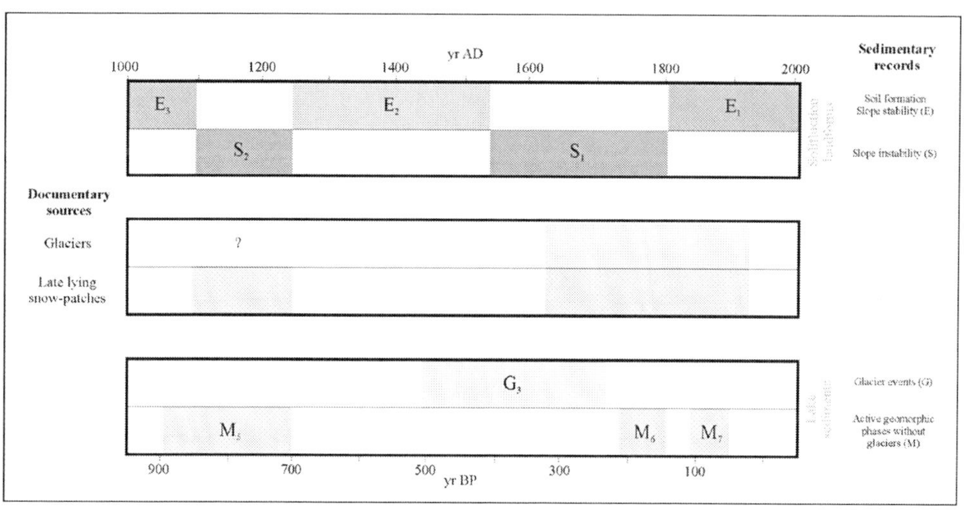

Figure 5. Geochronological framework of the geomorphic processes and environmental conditions from different sources of information.

Since the second half of the 19th century, the sedimentary records in the Iberian Peninsula show a trend towards warming with fluctuating levels of humidity. In the Sierra Nevada, this temperature increase has brought about the disappearance of glaciers, a migration of periglacial processes towards higher altitudes and the formation of organic soils in the highest valley bottoms.

CONCLUSION

Three sources have been used in this chapter for reconstructing the past environmental evolution in Sierra Nevada as well as the climatic conditions that have influenced the landscape transformations. In general, there is a wide agreement between the information collected from historical documents and the data inferred from the sedimentary archives.

The sedimentary records reveal phases with enhanced geomorphic processes (i.e., solifluction, formation of glaciers) and others with prevailing geomorphic stability (i.e., slope stability, soil formation). These changes have been induced by the varying climatic conditions affecting the massif during the last millennia. They were also decisive in promoting geomorphic activity (cold and wet periods) and stabilizing slopes (warm phases with varying moisture conditions). The historical documents also confirm alternating colder and warmer phases over the last few centuries, relating to the existence of glaciers in the highest northern cirques and abundant snow-patches in the massif during the cold and wet stages. Both the sedimentary and historical sources suggest that the Little Ice Age has been the coldest period of the last millennium, with temperatures being ~0.93°C below current values.

ACKNOWLEDGMENTS

The first author thanks the Agency for the Management of University and Research Grants (AGAUR) of Catalonia for funding a postdoctoral fellowship. We would like to thank the National Park of Sierra Nevada for the facilities offered during the field work campaigns. This research was financially supported by the Spanish research project "El interés científico de la documentación de época para el estudio del glaciarismo histórico (PEH) de Sierra Nevada (CS02009-06961)" and the Consolidated Research Group "Landscape Research and Mediterranean Mountain Palaeoenvironments".

REFERENCES

Bermúdez de Pedraza, F., 1638. Historia eclesiástica de Granada. Granada: Universidad de Granada, new edition from 1989.

Boissier, C. E., 1845. Voyage botanique dans le midi de l´Espagne pendant l´anné 1837. Granada: Fundación Caja de Granada and Universidad de Málaga, new edition from 1995.

Böhm, R., Auer, I., Brunetti, M., Maugeri, M., Nanni, T., Schöner, W., 2001. Regional temperature variability in the European Alps: 1760-1998 from homogenized instrumental time series. *International Journal of Climatology* 21, 1779-1801.

Chueca, J., Peña Monné, J. L., Lampre, F., Julián, A., 1998. La Pequeña Edad de Hielo en el Pirineo central y meridional. Inferencias paleoambientales a partir de datos geomorfológicos. In: Gómez Ortiz, A., Pérez Alberti, A. (Eds.), *Las huellas glaciares de las montañas españolas Santiago de Compostela: Servicio de Publicaciones de la Universidad de Santiago de Compostela,* pp. 307-328.

Enciclopedia Universal Ilustrada Europeo Americana, 1964. Sierra Nevada. Madrid-Barcelona: Espasa Calpe, 5, 425-426.

Esteban, A., 1995. Evolución del paisaje durante los últimos 10.000 años en las montañas del Mediterráneo Occidental: ejemplos del Pirineo Oriental y Sierra Nevada. Ph. D. Thesis, University of Barcelona.

Fernández Navarrete, F., 1732. Cielo y suelo granadino. Almeria-Barcelona: G. B. G. Editora, new edition from 1997.

García Sainz, L., 1947. El clima de la España cuaternaria y los factores de su formación. Valencia: Secretariado de Publicaciones de la Universidad de Valencia.

Gil García, M. J., Ruiz Zapata, M. B., Santisteban, J. I., Mediavilla, R., López Pamo, E., Dabrio, C. J., 2007. Late Holocene environments in Las Tablas de Daimiel (South central Iberian Peninsula, Spain). *Vegetation History and Archaeobotany* 16, 241-250.

Giraudi, C., 2005. Middle to Late Holocene glacial variations, periglacial processes and alluvial sedimentation on the higher Apennine massifs (Italy). *Quaternary Research* 64 (2), 176-184.

Gómez Ortiz, A., Palacios, D., Schulte, L., Salvador, F., Plana, A., 2009. Evidences from historical documents of landscape evolution after Little Ice Age of a Mediterranean high mountain area, Sierra Nevada, Spain (eighteenth to twentieth centuries). *Geografiska Annaler, Series A, Physical Geography* 91, 279-289.

González Trueba, J. J., Martín Moreno, R., Martínez de Pisón, E., Serrano, E., 2008. Little Ice Age glaciation and current glaciers in the Iberian Peninsula. *The Holocene* 18 (4), 551-568.

Grove, M., 2004. Little Ice Age. Ancient and modern. London: Routledge.

I. P. C. C., 2007. Intergovernamental Panel on Climate Change. World Meteorological Organization.

Knight, J., Harrison, S., 2009. *Periglacial and paraglacial processes and environments.* London: Special Publication of the Geological Society, 320.

Madoz, P. (1849*). Diccionario Geográfico-Estadístico-Histórico de España y sus posesiones de ultramar*. Madrid 14, 383-386.

Marquer, L., Pomel, S., Abichou, A., Schulz, E., Kaniewski, D., Van Campo, E., 2008. Late Holocene high resolution palaeoclimatic reconstruction inferred from Sebkha Mhabeul, southeast Tunisia. *Quaternary Research*, 70 (2), 240-250.

Martínez-Cortizas, A., Pontevedra-Pombal, X., García-Rodeja, E., Nóvoa-Muñoz, J. C., Shotyk, W., 1999. Mercury in a Spanish peat bog: archive of climate change and atmospheric metal deposition. *Science* 284, 939-942.

Martín-Puertas, C., Valero-Garcés, B., Mata, M. P., González-Sampériz, P., Bao, R., Moreno, A., Stefanova, V. 2008. Arid and humid phases in southern Spain during the last 4000 years: the Zoñar Lake record. *The Holocene* 18 (6), 907-921.

Molero, J., Pérez, F., 1987. La flora de Sierra Nevada. Granada: Universidad de Granada.

Obermaier, H., 1916. Los glaciares cuaternarios de Sierra Nevada. Trabajos del Museo Nacional de Ciencias Naturales, *Geología*, 17, 1-68.

Oliva, M., 2009. Holocene alpine environments in Sierra Nevada (Southern Spain). Ph. D. Thesis, University of Barcelona.

Oliva, M., Gómez Ortiz, A., 2012. Late Holocene landscape evolution and climate variability in a Mediterranean high mountain environment (Sierra Nevada, Spain) inferred from lake sediments and historical sources. *The Holocene*, doi: 10.1177/0959683611434235.

Oliva, M., Gómez Ortiz, A., 2011. Holocene slope dynamics in Sierra Nevada (Southern Spain). Sedimentological analysis of solifluction landforms and lake deposits. In: Martini, I. P., French, H. M., Pérez Alberti, A. (Eds.), Ice-Marginal and Periglacial Processes and Sediments, *London: Special Publication of the Geological Society* 354, pp. 227-239.

Oliva, M., Schulte, L., Gómez Ortiz, A., 2011. The role of aridification in constraining the elevation range of Holocene solifluction processes and associated landforms in the periglacial belt of the Sierra Nevada (Southern Spain). *Earth Surface Processes and Landforms*, 36, 1279-1291.

Oliva, M., Gómez Ortiz, A., Schulte, L., 2010. Tendencia a la aridez en Sierra Nevada desde el Holoceno Medio inferida a partir de sedimentos lacustres. *Boletín de la Asociación de Geógrafos Españoles*, 52, 27-42.

Oliva, M., Schulte, L., Gómez Ortiz, A., 2009. Morphometry and Late Holocene activity of solifluction landforms in the Sierra Nevada (Southern Spain). *Permafrost and Periglacial Processes* 20 (4), 369-382.

Oliva, M.; Schulte, L., Gómez Ortiz, A., 2008. Solifluction lobes in Sierra Nevada (Southern Spain): morphometry, process monitoring and palaeoenvironmental changes. *Proceedings of the Ninth International Conference on Permafrost (N. I. C. O. P.)* 2, 1321-1326.

Ponz, A., 1754. Relación del viaje que hizo desde Granada a Sierra Nevada D. Antonio Ponz a influxo del Excmo. Sr. Marqués de la Ensenada. *Granada: Mensajero Económico y Erudito de Granada*, 25-30.

Quelle, O.,1908. Beiträge zur Kenntnis der spanischen Sierra Nevada. Ph. D. Thesis, University of Friedrich-Wilhelm, Germany.

Reimer, P. J., Baillie, M., Bard, E., Bayliss, A., Beck, J. W., Blackwell, P. G., Bronk, C., Buck, C. E., Burr, G. S., Edwards, R. L., Friedrich, M., Grootes, P. M., Guilderson, T. P., Hajdas, I., Heaton, T. J., Hogg, A. G., Hughen, K. A., Kaiser, K. F., Kromer, B., Mc Cormac, F. G., Manning, S. W., Reimer, R. W., Richards, D. A., Southon, J. R., Talamo, S., Turney, C. S., van der Plicht, J., Wyhenmeyer, C. E., 2009. IntCal09 and Marine09 radiocarbon age calibration curves, 0-50,000 years cal B. P. *Radiocarbon* 51, 1111-1150.

Rojas Clemente, S., 1809. Viaje a Andalucía. Historia Natural del Reino de Granada (1804-1809). Barcelona: G. B. G. Editora, new edition from 2002.

Schwamborn, G., Fedorov, G., Schirrmeister, L., Meyer, H., Hubberten, H. W., 2008. Periglacial sediment variations controlled by Late Quaternary climate and lake level change at Elgygytgyn Crater, Arctic Siberia. *Boreas*, 37, 55-65.

Torres Palomo, M. P., 1968. Sierra Nevada en los escritores árabes. *Miscelánea de Estudios Árabes y Hebraicos* 16-17, 57-88.

Wanner, H., Holzhauser, H., Pfister, C., Zumbühl, H., 2000. Interannual to century scale climate variability in the European Alps. *Erdkunde* 54, 62-69.

In: Holocene
Editor: Bahadur Singh Kotlia

ISBN: 978-1-62257-722-4

Chapter 11

A FIVE-CENTURY LONG LIMNOLOGICAL AND ENVIRONMENTAL RECORD FROM NORTHEASTERN INDIA

Hema Achyuthan[1*], Anjum Farooqui[2**], Christopher Eastoe[3#], R. Ramesh[4•], M. Devi[5♣] and P. Prabhu Ganesh[1*]

[1] Anna University, Chennai, India
[2] Birbal Sahni Institute of Palaeobotany, Lucknow, India
[3] University of Arizona, Tucson, US.
[4] Physical Research aboratory, Navrangpura, Ahmedabad, India
[5] Wadia Institute of Himalayan Geology, Itanagar, India

ABSTRACT

Lakes are the major geomorphic landforms that hold signatures of the past climate and environment. The Ganga Lake near Itanagar, Arunachal Pradesh is a freshwater lake and structurally controlled basin. The sediment cores were collected from this lake towork out the paleoenvironmental shifts. The radiocarbon ages indicate high rates of sedimentation and higher terrestrial influx into the lake system in the last ~500 years. The high content of Total Organic Carbon (TOC) in the lake sediments indicate decaying of organic material in the form of plant debris, lake organic biota, and anaerobic conditions. The higher Si/Al values signify intense weathering of the surrounding bedrock and soil sediments around the lake. The high percentage of fruiting bodies of Epiphyllous Lichen (EL) in the sediments recorded in association with the pollen/spores during the past ~500 yrs BP indicate cool and moist climatic conditions of longer duration punctuated by the

* Department of Geology, Anna University, Chennai, 600 025, India.
** Birbal Sahni Institute of Palaeobotany, Lucknow, 226 007, India.
# Department of Geosciences, University of Arizona, Tucson, 85721,US.
• Physical Research Laboratory, Navrangpura, Ahmedabad, 380 009, India.
♣ Wadia Institute of Himalayan Geology, Itanagar, 791 111, India.
* Department of Geology, Anna University, Chennai, 600 025, India. E-mail: *hachyuthan@yahoo.com.*

warmer phases. EL serves as a dependable proxy for understanding cold and moist conditions on a short term scale. This chapter concerns with our results from the Ganga Lake core that has revealed about 5 warmer phases intercepted by cold moist phases which occurred during the Little Ice Age (LIA). The warmer events were of shorter duration during LIA.

Keywords: Lake sediment core; Palynofacies; Paleoenvironment; Weak monsoon; Little Ice Age; North eastern India

INTRODUCTION

Lakes are excellent repositories of air-borne and especially stream-borne materials [Mackereth, 1966; Thompson et al., 1975; Oldfield, 1977; Engstrom and Wright, 1984; Burden et al., 1986; Edwards and Whittington, 1993], including catchment soils and sediments [Pearsall et al., 1960; Mackereth, 1965, 1966]. Lake sediment furnishes a record of the past terrestrial events in the lake catchments. The radiocarbon dating provides a chronology for understanding the events of palaeoenvironmental shifts and their variation in terms of the temporal pattern of sedimentation (including inverted sequences of the dates).

Lakes in the northeastern Himalaya, especially in Arunachal Pradesh, have not been studied for their palaeoenvironmental potential barring a few preliminary observations [Bhattacharyya et al., 2007].

The Ganga Lake near Itanagar (Arunachal Pradesh) was chosen, as this site offered a variety of geological inputs from a wide catchment area. In particular, the rainfall is abundant between March and May at a time when the northern India has its rainfall minimum. The area is tectonically active [Singh, 2007]. Through this chapter, we trace the palaeoenvironmental changes of Ganga Lake.

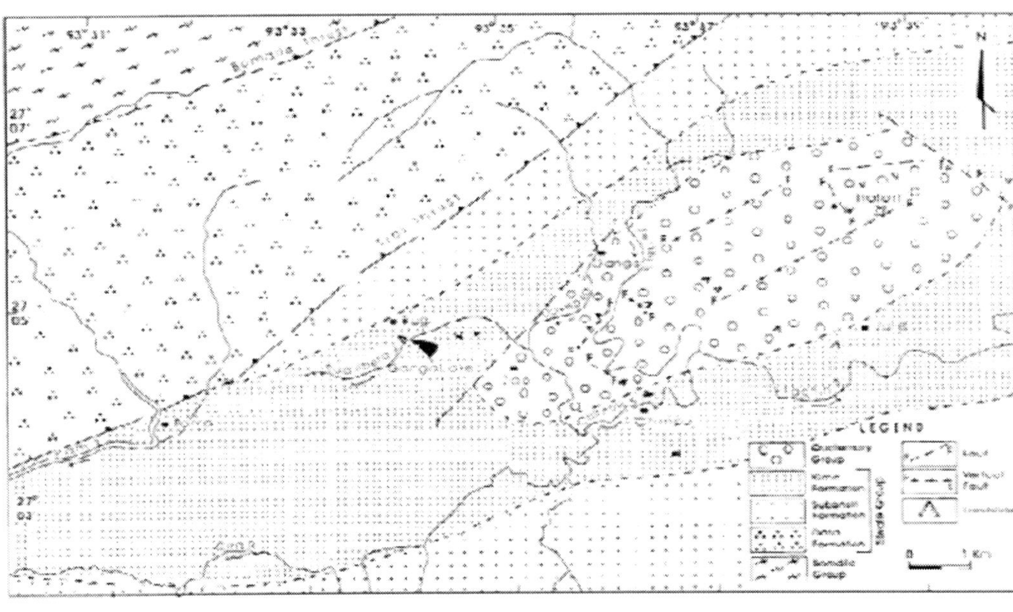

Figure 1. Geological map of the study area [after Acharya, 1980; Devi and Singh, 2006].

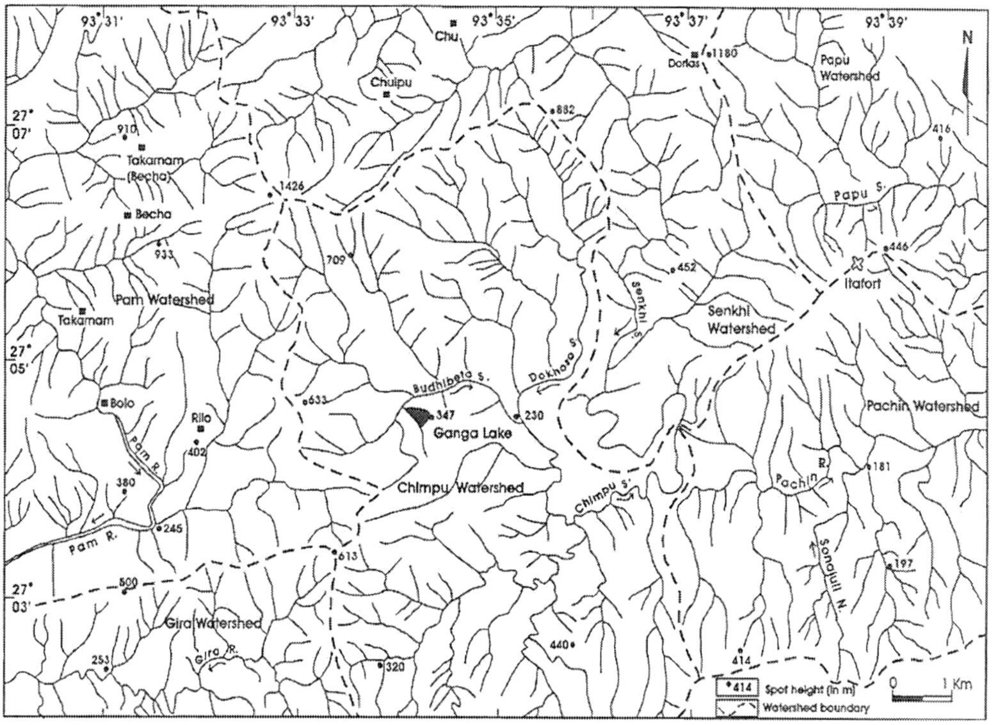

Figure 2. Geomorphological map of the study area [after Devi and Singh, 2006].

STUDY AREA

Ganga lake (Figure 1) is a triangular fresh water basin that lies in a rugged hilly terrain in the foothills of the Himalaya in the Papum Pare District, Arunachal Pradesh (India) in an area referred to as the Itanagar capital complex. This lake is located nearly 7.6 km WSW of Ita fort, and lies between 27°04'28"; 27°04'36"N and: 93°34'04";93°34'22"E, covering an areal extent of about 35 kms^2. The Ganga Lake is about 24 to 25 m deep, and is surrounded by low relief hills and corrugated landforms at altitudes between 347 and 630 m.a.s.l (Figure 2). The tributary streams have a gentle gradient and form narrow V-shaped valleys. To the north, the hills rise to >1,000-1,500 m and locally more than 2,000 m.a.s.l. The Ganga Lake has its catchment in two watersheds, namely the Pam (WSW) and Pachin (ENE).

A series of southeast-directed thrust faults have juxtaposed the Proterozoic Bomdilla Group metamorphic rocks with the Neogene Siwalik Group. The Bomdilla group is represented by green phyllites and massive white coarse-grained quartzites, garnetiferous gneiss, sericite, muscovite, biotite, biotite-garnet schist and bands of limestone [Kumar and Singh, 1980]. The gneisses are intruded by amphibolites along the foliation plane and are highly deformed. The Siwalik Group is made up of the Dafla, Subansiri and Kimin Formations in ascending order. The Quaternary fluviatile terrace and basin-fill alluvial sediments fill a fault-bounded basin east of the lake. These deposits comprise poorly sorted boulders and pebbles of gneiss, schist, granitic gneiss, quartzite, etc., set in a loose and unconsolidated matrix of sand, silt and clay. The boulders and pebbles are mainly derived from the Bomdila Group, though sandstone boulders from the Gondwana and Siwalik groups are also present [Kumar and Singh, 1980].

The lake lies between two regional thrust faults, viz., the Himalayan Frontal Fault (HFF) in the south (exposed south of the present study area) and Bomdila Thrust (BT) in the north. The Tipi Thrust is another major structure within the Siwalik deposits. The Neogene and Quaternary sediments form the core of a regional syncline. The area has experienced extensional neotectonic movements represented in the form of multiple terrace sets along the nearby Budhibeta and Poma rivers and their tributaries [Devi and Singh, 2006]. The lake appears to have formed as a result of subsidence and consequent ponding of a tributary of the Budhibeta river. The study area has a moderately hot and humid tropical climate, and receives monsoon rains. The southwest monsoon begins in middle of June and is the heaviest in the country, owing to the orographic effect of the Himalaya. Over a span of 11 years, the rainfall data have varied from 228.9 cm (in 2001) to 392.3 cm (at present).

Vegetation Around the Ganga Lake

Altitudinal and climatic conditions have given rise to different forest types in Arunachal Pradesh. The categories are tropical forests, sub-tropical forests, temperate forests, pine forests and alpine forests and occupy 40% of the floral species of India, many of which are endemic to the region. Of the 8,000 reported species of flowering plants from the Himalaya, 5,000 species are found in the northeastern region. Apart from the large number of timber species, there are innumerable varieties of orchids, medicinal plants, ferns, bamboos, canes, wild relative of large number of cultivated plants, and even plants of biological curiosities, such as parasites, saprophytes, etc. The orchids form a dominant group comprising more than 600 species. About one third of the 105 species of Indian bamboo are found in Arunachal Pradesh [Chowdhery, 1999, 2008].

Materials and Methods

A reconnaissance survey of the lake sediments, geomorphology, structural and drainage pattern was carried out. Three short sediment cores were retrieved using 5 cm diameter PVC pipes from the southwest (SW1 58 cm long), northern (N2 65 cm long) and southeast (SE3 75 cm long) parts of the lake where the water column is about 3 m deep. The samples were collected at every 2 cm and the sediment colour and texture were observed. The samples were oven dried at 50°C and then subjected to sedimentology, textural and mineralogical studies following Carver [1971] and chemical analyses. The total organic carbon percentage was determined following Walker [1947] and Jackson [1958]. Silica and alumina were analysed following the methods described by Shapiro and Brannock [1952]. The TOC rich samples from the SW core were radiocarbon dated (Table 1) at the Radiocarbon Laboratory, Department of Geosciences, University of Arizona, Tucson, US. Pollen and remains of Epiphyllous lichen (EL) fruiting bodies were identified following Erdtman [1943]. See Plate 1.

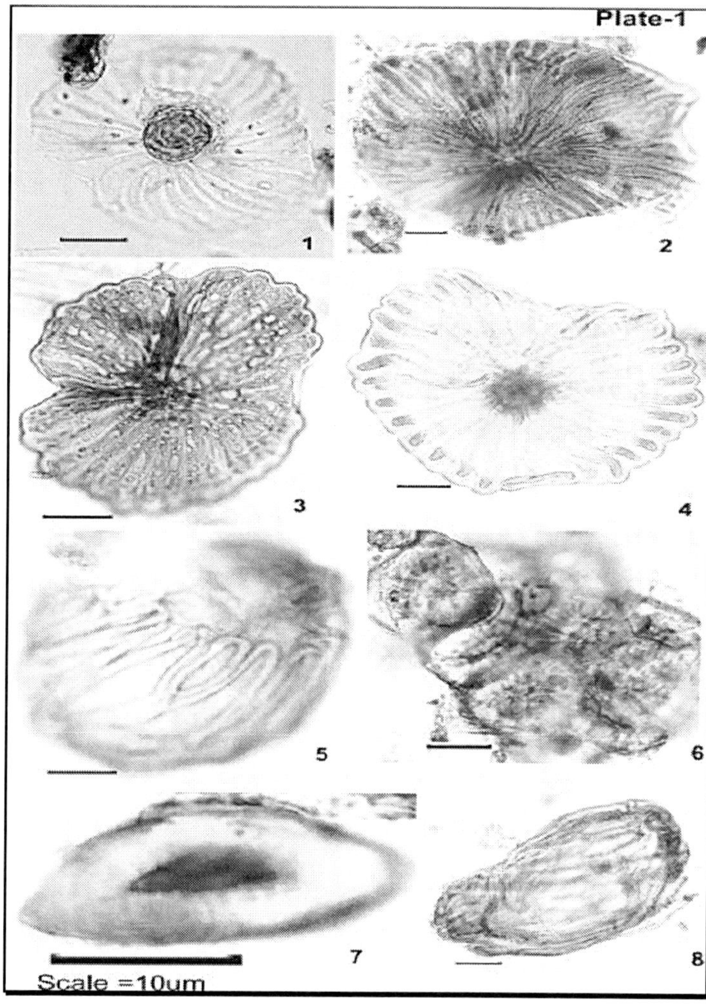

Plate 1. Light microscopic photomicrographs (1. Circular cup- shaped isidia (common in all the three cores), a characteristic reproductive feature of lichens. The radiating light and black colour bands represent algal and the fungal body, respectively; 2-4. Lichen body showing different stages of asci formation (site for mycobiont ascospores); 5. Globose flask-shaped Pycnidia (very common in core N2), differs from isidia in shape and arrangement of the symbiont thallus; 6-7. Group of Lichen germlings having both algal and fungal symbionts; 8. Ear-shaped Campylidia type (common in core N2 and SE),characteristic reproductive feature of lichens belonging to family Ectolechiaceae).

RESULTS

Sediment Core Description

The sediment cores are represented by dark greyish brown organic carbon rich fine silty clay and sand with varying stages of decayed leaf particles (Figures. 3-5). This unit (Phase-I) is interrupted by a thin layer of light grayish brown silty sand and sandy layers. Subsequently, the Phase-II is deposited as a layer of dark brown, organic carbon rich, fine silty clay with decayed leaf and plant material that occur at the top (15 cm thick). This is the sequence of sedimentation in all the three cores.

Table 1. Radiocarbon dates of the Ganga lake sediments

S.No.1	Depth cm	Conventional date (yr BP)	δ13C ‰
1	6-8	107.6± 1.7 PMC	-28.9
2	16-18	<410	-27.4
3	26-28	<455	-28
4	40-42	<485	-27.8

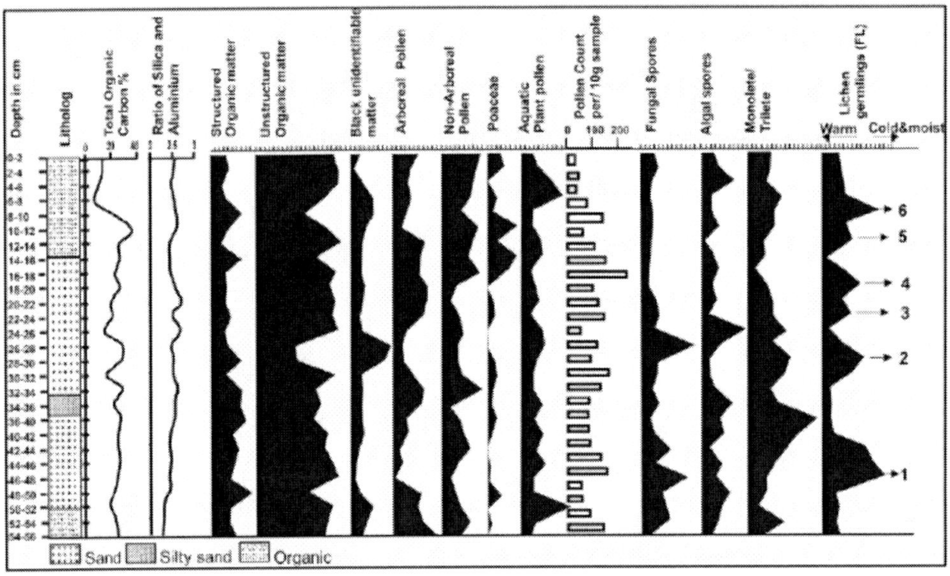

Figure 3. Organic matter and palynological spectrum of SW1-core (each division= 5%) 1 to 6= cold and moist climate.

Radiocarbon Chronology

Four sediment samples from core SW were prepared for conventional dating at the Department of Geosciences, Radiocarbon Laboratory, University of Arizona, Tucson, US, (Table 1). The upper 8 cm, of the sediment core is dated as 1,950 AD. The dates for remaining three specimens indicate a depositional interval not exceeding 500 years.

TOC, Silica and Alumina Contents

The TOC content is high along the entire depth of the core SW (Figure 3). It is higher in the upper level at 16 cm from the lake floor surface in the core N2 (Figure 4) and at 24-38 cm in the core SE3 (Figure 5). In the SW1 core, the Si/Al ratio shows two prominent peaks at 20 cm and 24 cm depth (Figure 3). Down the core, the Si/Al value decreases. In the N2 core, two prominent peaks are seen at about 6 cm and 16 cm (Figure 4) likewise higher Si/Al value is observed from about 26-38 cm. Here, again the Si/Al value decreases down the core. In core SE3, there is no prominent fluctuation in the Si/Al value (Figure 5).

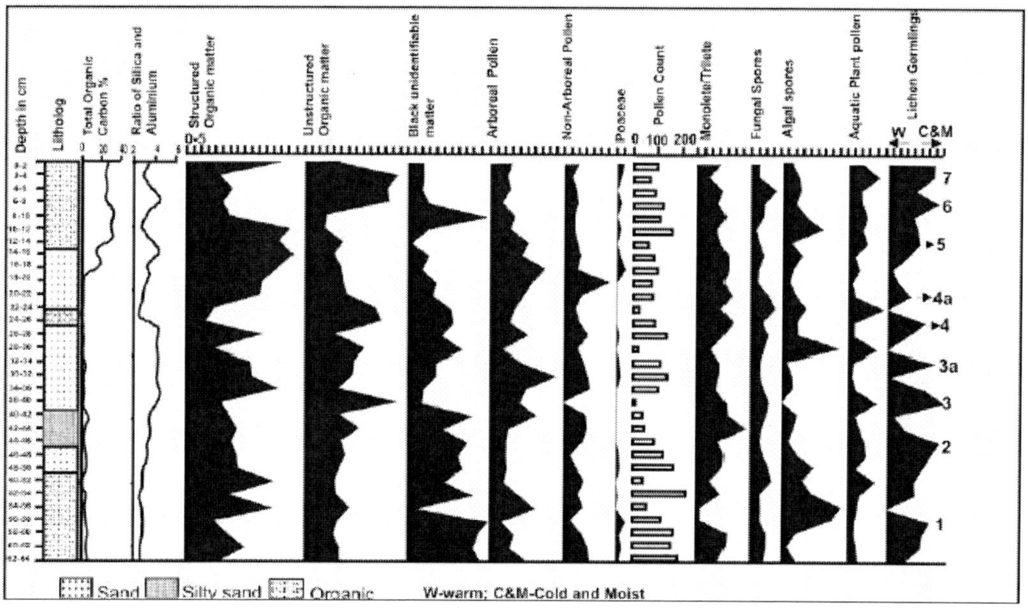

Figure 4. Organic matter and palynological spectrum of core N2 (each division=5%) 1 to 7= cold and moist periods.

Palynology

As Ganga Lake falls in the lowland, transitional, tropical-temperate rainforest vegetation area, it is ideal for examining the alternation of cold-humid and warm-humid climate regimes using high resolution study of pollen and spores. The sediments in all the three cores show well preserved organic matter. The fruiting bodies (apothecia) of Epiphyllous Lichen (EL) are of particular use because these are conspicuous on vegetation in cool-temperate, well illuminated habitats [Galloway, 1992; Aptroot and Sipman, 1997]. Nevertheless, recent inventories demonstrate that the lichen diversity in tropical rainforests surpasses that of temperate areas [Gradstein, 1992].

Core SW1 (0-56 cm Depth)

The organic matter (OM) in this sediment core includes 70-80% of dark brown, unidentifiable, irregular and unstructured organic matter (UOM) (Figure 3). The rest of the organic matter is represented as *in situ*, light brown to pale yellow coloured herb, and contains identifiable plant tissue as structured organic matter (SOM). Three horizons of carbonised parts, unidentifiable, lathe-shaped and black unidentified debris (BUD) occur at the depth of 50-52, 26-28 and 8-12 cm that correspond to the times/phases of clastic influx OM from the hinterland to the lake. The Poaceae pollen content is low indicating a dense forest cover that does not encourage their proliferation.

Pollen grains of non-arboreals (Figure 3) such as *Ageratum, Canscora, Cleome*, *Justicia,* Papaveraceae, *Physalis, Ranunculus,* Solanaceae, *Strobilanthes,* etc. exhibit periods of intense monsoon that support the growth of herbaceous annuals. The occurrence of high content of arboreal pollen, represented by *Adenia*, Araliaceae, *Artocarpus, Cassia, Campanula, Canarium, Casuarina, Corylus, Euonymous, Ilex, Lannea, Mallotus, Mangifera, Phyllanthus, Sterculia* and *Syzigium* indicate wet-evergreen to semi-deciduous flora adapted to subtropical-temperate transitional zones.

Phase-I is characterized by 4 high peaks of EL accompanied by the corresponding low percentage of arboreal pollen. The intermittent four warmer periods correspond to low EL content (Figure 3). High content of Monolete/Trilete fern spores at 36-40 cm depth indicate humid conditions. The fruiting bodies (Apothecia) of EL recorded are shown in Plate 1.

The proliferation of lichens recorded at various depths in the sediment core point towards their adaptation in cold and moist climate. The increase of aquatic pollen and algal spores with high percentage of fungal spores that belong to basidiomycetes and ascomycetes suggest very humid conditions. Comparatively enhanced Cereal pollen in the Phase-II exhibit intense anthropogenic activity during the last century. Pollen grains of *Acacia*, *Anogeissus*, *Ardisia,* Asteraceae, *Betula, Cassia, Commelina, Dendrocalamus, Diospyros, Elaeodendron, Lagerstroemia, Madhuca, Mallotus, Mangifera, Meliosma, Mitragyna, Oldenlandia, Pinus, Polygonum, Randia, Semicarpus, Symplocos, Terminalia* and *Vernonia* reveal diversification of hinterland flora including addition of deciduous taxa such as *Acacia*, *Cassia*, *Mitragyna, Lagerstroemia,* etc. perhaps are also influenced by anthropogenic activity.

The EL shows three intermittent high and low values indicating at least three events of cold and moist conditions during the last three centuries. The sediments from 8-12, 18-20, 26-30 and 48-50 cm reveal high peaks of EL, which show an inverse relationship with the arboreal pollen taxa. This indicates cold and moist climate favoured by well illuminated conditions which supports the proliferation of the lichens. The fungal spores of basidiomycetes and ascomycetes that are generally saprophyteson dead terrestrial plant tissues under very humid conditions are most abundant at 50-60 cm depth.

Core N2 (0-64 Cm Depth)

Unlike SW1, the entire sediment core is rich in identifiable SOM. It generally constitutes light brown to pale yellow colour, structured *in situ* identifiable slender, fibrous and plant cuticular tissue, attributed to either nearby area or *in situ* aquatic vegetation. Four peaks of BUD were recorded that imply clastic influx through streams and rivers during the Phase-I (Figure 4). These depths also correspond to high percentage of aquatic/swampy plant pollen like *Eicchornia, Typha, Ranunculus, Myriophyllum, Nymphaea,* etc. indicating nutrient influx in the lake. Most of the pollen belongs to tropical wet evergreen to semi-evergreen rainforest vegetation, confined to low and high altitude areas in the northern flank of the lake. These include arboreal pollen taxa, such as, *Adenia, Altingia,* Araliaceae, *Artocarpus, Campanula, Canarium, Cassia, Casuarina,*Celastraceae, *Corylus, Elaeocarpus, Euonymous, Eurya, Hypericum, Ilex, Lannea, Leucas, Mallotus, Mangifera, Phyllanthus, Pinus, Syzigium, Sterculia, Symplocos, Semecarpus, Tabernaemontana, Urticaceae,* etc. The non-arboreals include pollen taxa, e.g., *Argemone*, *Camelia, Cannabis, Canscora, Chlorophyton, Croton, Portulaca,* and Solanaceae. Algal spores largely comprise *Spirogyra* and *zygnematous* spores. The occurrence of low percentage of Poaceae pollen (*Cerealea*) throughout the sediment core signifies less anthropogenic activity around this flank of the lake. At 50-52 cm depth, occurrence of *Crucigenia* algae indicates Oligotrophic ecosystem, suggesting low clastic input during this period. Monolete/Trilete ferns show moderate percentage throughout the core except a very low percentage at 54-56 cm depth. This depth corresponds to high arboreal pollen taxa and terrestrial saprophytic fungal spores along with very low BUD indicating less fluvial clastics. The presence of high content of pollen/spore assemblage at this depth points towards the aeolian transportation. Intermittent four peaks of EL represented by diverse cleistothecia and apothecia indicate dominance of cold and moist conditions in the hinterland. However, events labelled as 3a and 4a (Figure 6) indicate shorter periods of climatic fluctuations. During Phase-II (during the last 2-3 centuries, the occurrence of three EL peaks suggests its transportation from the adjacent northern and northeastern Himalayan belt responding to a comparatively enhanced cold and wet climatic condition. Pollen grains comprise of hinterland low and high altitude taxa such as, *Acacia*, Acanthaceae, *Ardisia, Artemisia,* Asteraceae, *Azadirachta, Betula*, *Bombax, Careya, Cassia*, Campanulaceae, *Chorisia, Diospyros, Elaeocarpus, Eurya,* Ericaceae, *Justicia, Lagerstroemia, Madhuca, Mallotus, Mangifera, Meliosma, Mitragyna, Oldenlandia, Pinus, Polygonum, Randia, Sapotaceae, Semicarpus, Symplocos, Terminalia, Quercus and* Umbelliferae.

CORE SE3 (0-68 CM DEPTH)

The entire core is very rich in SOM that consists of *in situ* or nearby herbaceous plant cuticular fragments. The OM is light coloured, fibrous and of slender cellular character. The amorphous organic matter (AOM) is high as shown in Figure 5. The BUD is very abundant at 62-64 cm and 14 -16 cm. Arboreal, non-arboreal and aquatic pollen taxa (Figure 6) are *Arabis, Casearia, Adenia, Mangifera,* Sapotaceae, *Fabaceae, Moraceae, Meliaceae, Emilia, Lannea, Syzigium, Cassia, Cocculus,* Malvaceae, *Dalbergia, Primulaceae, Pinus, Bauhinia,* Rutaceae, Arecaceae, Arundariaceae (Cane*), Altingia,* Sapindaceae, *Strobilanthes,* Chenopodiaceae, Caryophyllaceae, *Diospyros, Myriophyllum, Eicchhornia, Typha*, etc. Other aquatic and algal spores are similar as in other two cores. In this core too, arboreal and non arboreal pollen taxa correlate with fungal spores. Four high peaks of EL indicate cold and moist conditions interrupted by short warmer events similar to as recorded in the other two cores. The Phase-II shows similar fluctuations as observed in Core N2. The arboreal pollen show a decline in Phase-II with increase in EL and Pteridophytic spores. The BUD shows a drastic decline suggesting restricted inflow of clastics and water in the lake resulting into low clastic organic matter from hinterland.

DISCUSSION

The geomorphic study around the Ganga lake and lithological observations indicate that neotectonic movements have created a lake-filled depression in an area with intense weathering (indicated by the Si/Al data) and high productivity of plant matter. The anoxic lake sediments provide a repository of plant material, largely from the forested flanks of the lake that can be related to climatic cycle over the past few hundred years (~500 years).

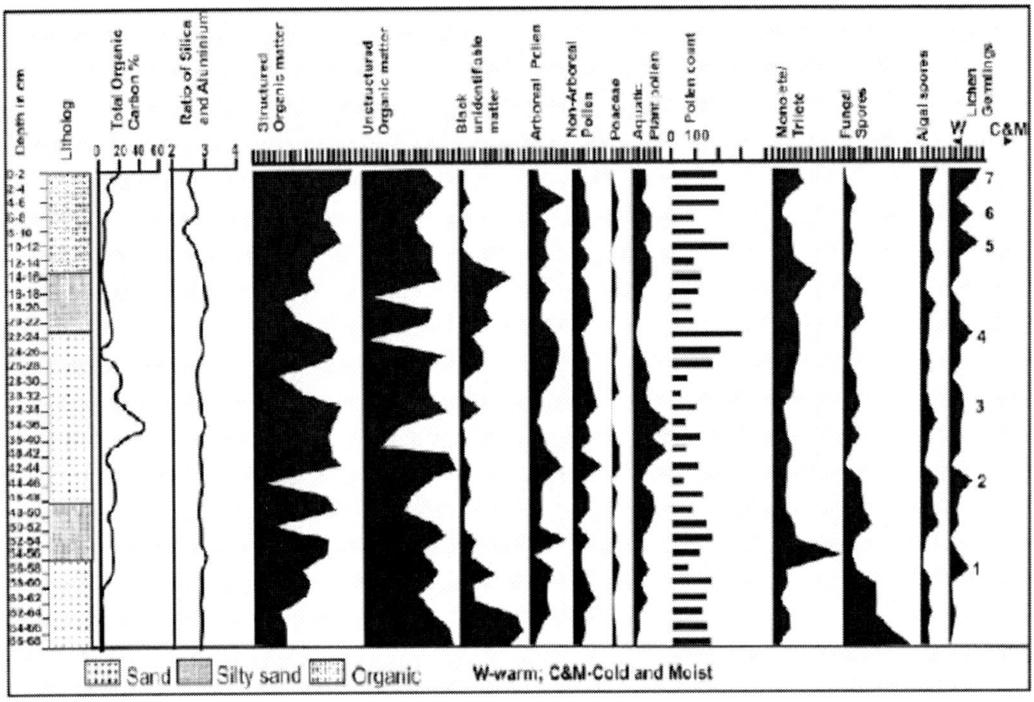

Figure 5. Organic matter and palynological spectrum of core SE (each division=5%) 1 to 5= cold and moist climate.

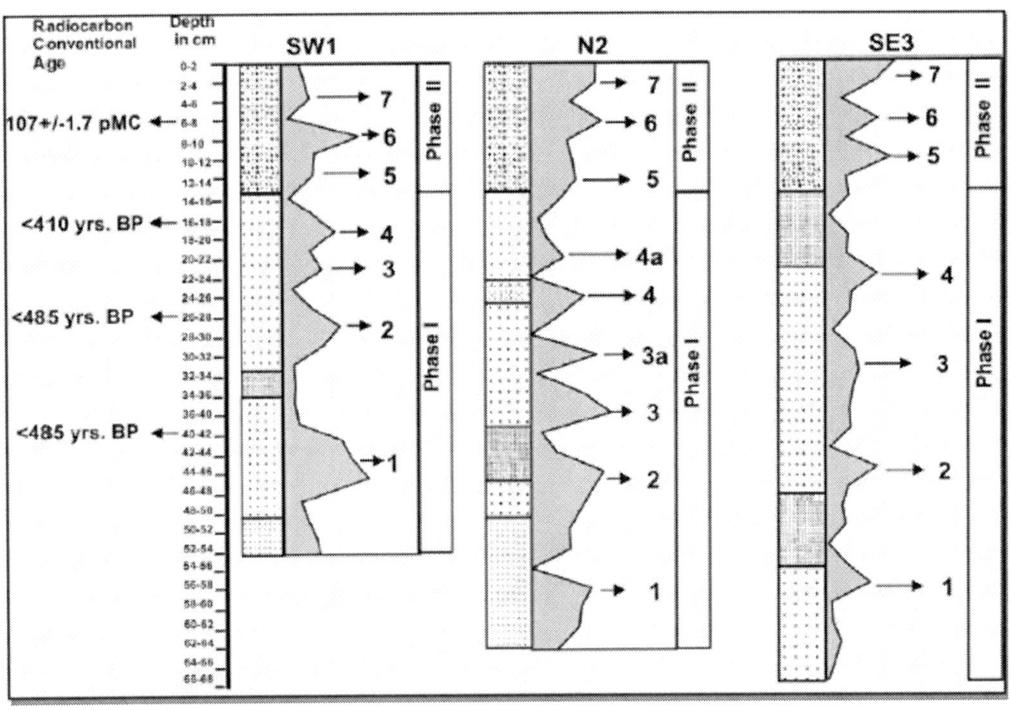

Figure 6. A comparative account of lichen germlings (ELP) showing cold and moist climate (1-7) conditions in three corers of Ganga lake.

All three cores show strong cycles in the abundance of EL, which provide a framework for interpretation of the whole palynological data set. The Epiphyllous lichen (EL) are acclimatized to cold and moist climate and are a diverse group of tiny crustose lichens that colonize the surfaces of leaves in humid tropical-temperate forests [Lucking, 2001]. They develop and colonize quickly, because their substrate leaf is short lived (6-12 months) in fast growing plants, and 12-36 months in other tropical rainforest plants.

The germlings and fruiting bodies of lichens have an advantage of long-distance (over 1,500 km) dispersal [Lucking and Lucking, 1995; Lucking and Kalb, 2001]. The fruiting bodies of EL recorded in the Ganga Lake sediments are environmental indicator and markers that have responded rapidly to changes in climatic conditions. In the present study, the down core variations in the percentage of EL in succession during the past ~500 yrs exhibit cool and humid climatic conditions that favoured the EL growth in lowland tropical rainforest and adjacent high altitude areas. It is likely that the abundant EL directly or indirectly influences the growth of vegetation. This is because a continuous layer of epiphylls intercepts 20-85% of the light penetrating the forest canopy and reduces photosynthetic performance by 20-30%, which is particularly significant considering the low light levels in the understory [Coley and Kursar, 1996]. However, an epiphyll cover of 50% or more is only found in very old leaves, which assumes functions other than photosynthesis; younger leaves are sparsely covered [Chabot and Hicks, 1982; Monge-Najera, 1989; Hartmann, 1993; Rogers et al., 1994; Monge-Najera and Blanco, 1995]. For example, considering all available phorophytes within an area of submontane rainforest in Costa Rica, average epiphyll cover (follicolous lichens and bryophytes combined) was estimated to range between 1.8 and 13.1%, while it amounted to 4.4-36.4% on the most densely covered leaves [Monge-Nájera, 1989]. The species richness of Folicollous (epiphyllous) lichen in undisturbed localities was twice that of disturbed localities in the montane forests of Rwanda [Serusiaux, 1989]. However, recent records show that species richness in tropical regions may surpass that of temperate areas [Aptroot, 1997].

In the Ganga Lake sediment cores, a low abundance of arboreal and non-arboreal pollen taxa corresponds to an increased abundance of fruiting bodies of EL in the sediment. Therefore, it is apparent from the present study that during these periods, EL abundance may have mechanically affected the forest vegetation. It is also likely that the dispersal and diffusion of air was slow during cold and humid climate which restricted the transportation of pollen/spores by wind. Due to the abundance of EL in such climatic conditions, its percentage probably outnumbered the pollen/spore count in the sediments. A similar situation is observed in cold seasons when the smoke from anthropogenic activity especially in urban areas appears to diffuse very slowly and smog (smoke+fog) like atmosphere is developed which may arrest the dispersal of various forms. The high arboreal and non-arboreal pollen assemblage in all the sediment cores corresponds to high percentage of saprophytic terrestrial fungal spores indicating humid conditions and strengthened monsoon. This increase does not show similarity in phases with high clastic influx through fluvial system. Thus, it is inferred that most of the pollen/spores were perhaps transported to the lake via wind and also through streams/rivers. The high and low percentage occurrence of pollen/ spores in the lake sediment suggests that its dispersal, transportation and deposition were controlled by the strengthened and weak monsoon. This factor is very well identified by a high and low percentage of pollen/spores corresponding to reduction and abundance of EL, respectively. In other words, percentage of pollen was inversely proportional to percentage of EL in the sediment core.

A high percentage of *Cereal* pollen was recorded in the SW flank showing intensity of anthropogenic activity. Percentage of green filamentous algal spores in the studied cores correspond to high BUD indicating rich mineral influx like Phosphorous, Nitrogen etc. from the land supporting the algal growth. All the ecological groups studied show a similarity in all the three cores. However, succession variation during the time period of sediment deposition is distinct and reveals total seven phases of cold and humid conditions represented by high EL percentage and six phases of warm and humid conditions (Figure 6). Comparatively, the periods of cold and humid phases are prolonged punctuated by short periods of warm and humid conditions. It is likely that because of such conducive environment, i.e., in cold humid conditions, the growth of EL is conspicuous and hence in such a study area and climatic conditions, EL has an advantage and best suited proxy-record for understanding high resolution moist climatic conditions at century to decadal scale. Among all these, EL recorded for the first time in the lake sedimentary section, reveals its high sensitivity to changing cold/humid and warm/humid climates.

LITTLE ICE AGE (LIA) AND CLIMATE CHANGE

The LIA (16th to mid 19th century) is recognized as a period of cooling [Mann 2003; Yadav and Singh, 2002] occurring after a warmer era known as the Medieval Climate Optimum [Lamb, 1965; Crowley and Lowery, 2000]. Inter Governmental Panel for Climate Change (IPCC) describes LIA as a period during which there was a modest cooling of the northern hemisphere less than $1^{o}C$. Scientists have identified two causes of LIA both internal and external, i.e., decreased solar activity and increased volcanic activity. During this period, there was also decreased agricultural output and reforestation taking up more carbon from the atmosphere that may have prolonged the LIA. However, the LIA was highly variable from region to region and was not a globally synchronous cold period [Bradley and Jones, 1993; Mann et al., 1999]. The δ13C values since the last 485 years as observed in different sediment layers ranging from 28 to 29‰ (Table 1), indicate cool and less moist conditions. This range of δ13C values indicates that the climate did not drastically change to support C4 vegetation-a typical signature of drier and arid environment. The δ13C values also indicate that the glacial conditions during the LIA were not extensive and that it did not extend to great heights in the eastern Himalaya. This is supported by the observations made by Yadav et al. [1999] and Bhattacharyya and Chaudhary [2003] who suggested that the cooling during the LIA was a discrete phenomenon with intermittent short warmer episodes. There has been an overall increase in steppe vegetational cover in the alpine region of the Himalaya and this may be due to the impact of low temperatures spanning longer time during the LIA. This phenomenon is also observed from the sediment cores studied from the eastern and western Himalaya [Yadav et al. 1999; Bhattacharyya and Chaudhary, 2003;].

CONCLUSION

The occurrence and setting of the Ganga Lake in Arunachal Pradesh with its unusual triangular shape and steep-sided banks is a unique geomorphic feature amidst the rugged and hilly Siwalik terrain. The Ganga Lake is sensitive to landscape impact and its origin is due to ponding of sediments caused by neootectonism. The high rate of sedimentation and physicochemical properties recorded in the three cores reflects the chronology, ecology and vegetation influenced by climate since ~500 yrs. BP. Abundant fruiting bodies of EL in the sediments is inversely proportional to pollen/spores indicating cool and moist climatic conditions. Thus, EL serves as a dependable proxy for understanding the cold and moist conditions on short-term scale. This study reveals 5 warmer phases, intercepted by cold moist phases which occurred throughout the LIA. The warmer events are of shorter duration during LIA. Thus, the overall result highlights that the LIA was highly variable from region to region and was not a globally synchronous cold period.

ACKNOWLEDGMENTS

The authors thank the reviewers for their constructive comments that helped in the presentation of the manuscript.

REFERENCES

Acharya, S. K., 1980. Stratigraphy and tectonics of Arunachal Himalaya. In: Valdiya, K. S., Bhatia, S. B., (Eds.*), Stratigraphy and correlation of Lesser Himalaya Formations, Hindustan Pub. Corp.*, Delhi, pp. 231-241.

Aptroot, A., 1997. Lichen biodiversity in Papua New Guinea, with the report of 173 species on one tree. *Bibliotheca Lichenologica* 68, 203-213.

Aptroot, A., Sipman, H. J. M., 1997. Diversity of lichenized fungi in the tropics. In: Hyde, K. D., (Ed.), Biodiversity of tropical microfungi, *Hong Kong University Press*, Hong Kong, pp. 93-106.

Bhattacharyya, A., Chaudhary, V., 2003. Late-summer temperature reconstruction of the Eastern Himalayan region based on tree-ring data of Abies densa. *Arctic Antarctic Alpines* 35, 196-202.

Bhattacharyya, A., Sharma, J., Shah, S. K., Chaudhary, V., 2007. Climatic changes during the last 1800 years B. P. from Paradise Lake, Sela Pass, Arunachal Pradesh, Northeast Himalaya. *Current Science* 93, 983-987.

Bradley, R. S., Jones, P. D., 1993. Little Ice Age summer temperature variations: their nature and relevance to recent global warming trends. *Holocene* 3, 367–376.

Burden, E. T., McAndrews, J. H., Norris, G., 1986. Geochemical indicators in lake sediments of upland erosion caused by Indian and European farming, Awenda Provincial Park, Ontario. *Canadian Journal of Earth Sciences* 23, 55-65.

Carver, R. E., 1971. Procedures in Sedimentary Petrography, *John Wiley and Sons*, New York, pp. 47–86.

Chabot, B. F., Hicks, D. J., 1982. The ecology of leaf life spans. *Annual Reviews Ecology and Systems* 13, 229-259.

Chowdhery, H. J., 1999. Floristic diversity and conservation strategies in India, Botany Survey of India, Calcutta, India.

Chowdhery, H. J., Giri, G. S., Pal, G. D., Pramanik, A., Das, S. K., 2008. Materials for the Flora of Arunachal Pradesh. Vol. II. *Botanical Survey of India*, Calcutta. pp. 693.

Coley, P. D., Kursar, T. A., 1996. Causes and consequences of epiphyll colonization. In: Mulkey, S. S., Chazdon, R. L., Smith, A. P., (Eds.), *Tropical E'orcst Plant Ecophvsiology*, Chapman and Hall, London, pp. 337-362.

Crowley, T. J., Lowery, T. S., 2000. How warm was the Medieval Warm Period? *Ambio.* 29 (1), 51–54.

Devi., M., Singh. T., 2006. Morphotectonic setting of the Ganga Lake, Itanagar capital complex, Arunachal Himalaya. *Geomorphology* 76, 1-11.

Edwards, K. J., Whittington, G. W., 1993. Aspects of the environmental and depositional history of a rock basin lake in eastern Scotland, U. K. In: McManus, J., Duck, R. W., (Eds.), *Geomorphology and Sedimentology of Lakes and Reservoirs.* Wiley, Chichester, pp. 155-180.

Engstrom, D. R., Wright, H. E. Jr., 1984. *Chemical stratigraphy of lake sediments as a record of environmental history.* Leicester University Press, Leicester, pp. 11-67.

Erdtman, G., 1943. An Introduction to Pollen Analysis. *Waltham Mass.*, pp. 239.

Galloway, D. J., 1992. Biodiversity: a lichenological perspective. *Biodiversity and Conservation* 1, 312-323.

Gradstein, S. R., 1992. The vanishing tropical rainforest as an environment for bryophytes and lichens. In: Bates, J. W., Farmer, A. M., (Eds.), *Bryophytes and Lichens in a Changing Environment*, Clarendon Press, Oxford, pp. 243–258.

Hartmann, C., 1993. dkologische Untersuchungen an foliikolen Flechten in Costa Rica, Zentralamerika. Diplom Thesis, Faculty of Natural Science, University of Ulm, Germany.

Jackson, M. L., 1958. *Soil chemical analysis*. Prentice Hall, New York, pp 485.

Kumar, S., Singh, T., 1980. Tectono-stratigraphy set-up of the Subansiri District, Arunachal Pradesh. In: Valdiya, K. S., Bhatia, S. B., (Eds.), *Correlations of Lesser Himalayan Formations*, Hindustan Publication Corporation of India, Delhi, pp. 267-279.

Lamb, H., 1965. The early medieval warm epoch and its sequel. *Palaeogeography Palaeoclimatology Palaeoecology* 1, 13–37.

Lücking, A., Lücking, R., 1995. Anpassugen und konvergenzen inder phyllosphäre am Beispiel von epiphyllen Flechten, Moosen und Isekten. 8. *Jahrestagung der gatö February*, 1995, Hamburg, Posterabst, pp. 31.

Lucking, R., 2001. Lichens on leaves in tropical rain forests: life in a permanently ephemerous environment. *Dissertatioties Boranicae* 346, 41-77.

Lucking, R., Kalb, K., 2001. New Caledonia, foliicolous lichens and island biogeography. *Bibliotheca Lichenologica* 78, 247-273.

Mackereth, F. J. H., 1965. Chemical investigation of lake sediments and their interpretation. *Proceedings Royal Society B Series* 161, 295-309.

Mackereth, F. J. H., 1966. Some chemical observations on post-glacial lake sediments. *Philosophical Transaction Royal Society B. Series* 250, 165-213.

Mann, M. E., Bradley, R. S., Hughes, M. K., 1999. Northern Hemisphere temperatures during the past millennium: inferences, uncertainties and limitations. *Geophysical Research Letters* 26, 759–762.

Mann, M. E., Jones, P. D., 2003. Global surface temperatures over the past two millennia. *Geophysical Research Letters* 30(15), 1820, doi: 10.1029/2003GL017814.

Monge-Najera, J., Blanco, M., 1995. The influence of leaf characteristics on epiphyllic cover: a test of hypothesis with artificial leaves. *Tropical Bryozoa* 11, 5-9.

Monge-Nájera, J., 1989. The relationship of epiphyllous liverworts with leaf characteristics and light in Monteverde, Costa Rica. *Cryptogamie, Bryologie-Lichénologie* 10(4), 345-352.

Oldfield, F., 1977. Lakes and their drainage basins as units of sediment-based ecological study. *Proceedings Physical Geography* 1, 460-504.

Pearsall, W. H., Gay, J., Newbould , J., 1960. Post-glacial sediments as a record of regional soil drifts. *Journal Soil Science* 11, 68-76.

Rogers, R. W., Barnes, A., Conran, J. G., 1994. Lichen succession on Wilkea macrophylla leaves. *Lichen* 26, 135–147.

Sérusiaux, E., 1989. Foliicolous lichens: ecological and chorological data. *Botanical Journal Linnean Society* 100, 87–96.

Shapiro, L., Brannock, W., 1952. Rapid analysis of silicate rocks. *US Geol. Survey Circular*. 65, pp.17.

Singh, A. K., 2007. Chemical Characteristics of Alkaline Basalt from the Abor Volcanics of Arunachal Pradesh. *Journal of Geological Society of India* 69 (6), 1189-1194.

Thompson, R., Batterbee, R. W., O'Sullivan, P. E., Oldfield, F. 1975. Magnetic susceptibility of lake sediments. *Limnology and Oceanography* 20, 687-698.

Walker, A., 1947. A Critical examination of a rapid method for determining organic carbon in soil. *Soil Science* 63, 251-263.

Yadav, R. R., Park, W. K., Bhattacharyya, A., 1999. Spring-temperature variations in western Himalaya, India, as reconstructed from tree rings: A. D. 1390-1987. *Holocene* 9, 85-90.

Yadav, R. R., Singh, J., 2002. Tree-ring-based spring temperature patterns over the past four centuries in western Himalaya. *Quaternary Research* 57, 299-305.

In: Holocene
Editor: Bahadur Singh Kotlia
ISBN: 978-1-62257-722-4

Chapter 12

RECONSTRUCTION OF AUGUST-SEPTEMBER TEMPERATURE IN NORTH-WESTERN HIMALAYA SINCE AD 1773, BASED ON TREE-RING DATA OF *PINUS WALLICHIANA* AND *ABIES PINDROW*

Vandana Chaudhary[1*], Amalava Bhattacharyya[2#], Joel Guiot[3†], Santosh K. Shah[2], Sangeet K. Srivastava[4‡], Jean-Louis Edouard[5] and André Thomas[5¶]

[1]Ministry of Earth Sciences, Prithvi Bhawan, New Delhi, India
[2] Birbal Sahni Institute of Palaeobotany, University Road, Lucknow, India
[3]C. E. R. E. G. E., C. N. R. S./ Université Paul Cézanne U. M. R. 6635, Cedex, France
[4] Department of Mathematics, I. T. M. Gurgaon, India
[5] I. M. E. P., C. N. R. S./Université Paul Cézanne U. M. R. 6336, Cedex, France

ABSTRACT

A study of tree-ring data of Silver fir (*Abies pindrow* [Royale] Spach.) and Himalayan pine (*Pinus wallichiana* A.B. Jackson) from the sub-alpine forest of Northwestern Himalayan region signifies the importance of August and September temperature in controlling the growth of these trees. In this chapter, the mean temperature for these two months has been reconstructed from AD 1773 to AD 1995 based on ring-width and density data. The calibration model explains 44% variance in the instrumental data (1902-1987). The reconstructed temperature series shows annual to multiyear fluctuations suggesting that AD 1830-1852 and AD 1961-1972 were the coldest and

[*]Ministry of Earth Sciences, Prithvi Bhawan, Lodhi Road, New Delhi, 110003, India.
[#] Birbal Sahni Institute of Palaeobotany, 53, University Road, Lucknow, 226 007, India.
[†] CEREGE, CNRS/ Université Paul Cézanne UMR 6635, 13545 Aix-en-Provence, Cedex 4, France.
[‡] Department of Mathematics, ITM Gurgaon, 122 017, India
[¶]IMEP, CNRS/Université Paul Cézanne UMR 6336, 13545, Aix-en-Provence Cedex 4, France. E mail: *amalava@yahoo.com*.

warmest periods respectively. The Little Ice Age cooling and the 20th century warming are recorded as discrete phenomena, with intermittent smaller phases of warming and cooling. Our reconstruction show a cooling phase since AD 1973 contrary to the recent global warming. The sequential change-point analysis was used for the first time in the dendroclimate studies from this region and the results indicate the major regime shifts at AD 1783, 1794, 1805, 1827, 1862, 1873, 1898 and 1971.

Keywords: Western Himalaya, tree-rings, *Abies pindrow*, *Pinus wallichiana*, Sequential change

INTRODUCTION

The Himalayan region plays an important role in both local and extra-regional atmospheric circulation system [Khandekar, 1991; Prell and Kutzbach, 1992]. Yet, the instrumental records from this province are very few and far between and barely cover the last 100 years, except for a few stations. Therefore, the long-term high-resolution proxy climate records seem essential to supplement the meteorological records. In this context, the forests of the Himalaya offer a possibility to generate long records of high-resolution tree-ring data which could be useful in reconstructing synoptic scale climate variations. Several dendroclimatological reconstructions [Hughes, 1992; Borgaonkar et al., 1994, 1996; Yadav et al., 1997, 1999; Yadav and Singh, 2002; Bhattacharyya and Chaudhary, 2003; Cook et al., 2003; Singh and Yadav, 2006, Yadav, 2011] from the Himalaya have contributed in extending proxy seasonal climate records of this region. However, in the northwestern part of the Himalaya, most of the reconstructions have been done only for pre-monsoon temperature and precipitation, except for Kashmir [Hughes, 1992, 2001] where besides the pre-monsoon, late summer temperature has also been reconstructed using ring-width and density data. In another study, the annual precipitation was reconstructed for over the last 1,000 years from high mountains of northern Pakistan, using oxygen isotope record from tree-rings [Tredyte et al., 2006]. Inclusion of new sites and additional tree-ring parameters (ex. density, isotope) may increase the possibility of reconstructing other climate variables. The reconstruction of these variables related to the monsoon would be significant in understanding the dynamics of circulation patterns. Earlier densitometric studies [Pant, 2000; Borgaonkar et al., 2001; Hughes, 2001] from Northwestern Himalaya have shown encouraging results. This chapter deals with the reconstruction of August and September temperature (AS) using two different conifer species (*P. wallichiana* and *A. pindrow*) growing close to the upper tree line of the northwestern Himalaya. Besides the ring width data, the density data of *P. wallichiana* from one site have also been used in the present study.

MATERIAL AND METHODS

Tree-Ring Data

Tree-ring samples from two conifer taxa viz. P. wallichiana (PIWA) and Abies pindrow (ABPI), constituent of sub-alpine forest of the western Himalaya were analyzed (Table 1).

Tree-ring cores of both the species were collected from nine sites where trees are currently growing under ecologically stressed conditions and are close to the glaciers (Figure 1).

Table1. Description of tree ring sampling sites along with general statistics of the chronologies analyzed in this chapter. The species are Pinus wallichiana (PIWA) and Abies pindrow (ABPI). Lat.: Site Latitude; Long.: Site Longitude; Elev: Site elevation in meters; NC: Number of cores; MS: Mean sensitivity; COR: Series Inter-Correlation; MSL: Mean segment length in years

No.	Site Code	Species	Lat.	Long.	Elev.	Chronology length	NC	MS	COR	MSL
1	GAN	PIWA	78°56'E	31°04'N	3050	304 (1685-1988)	55	0.26	0.53	109
2	CHI	PIWA	79°01'E	30°01'N	3600	305 (1695-1999)	68	0.21	0.58	153
3	DBG1	ABPI	78° 47'E	30° 52'N	3700	165 (1831-1995)	16	0.2	0.45	119
4	DBG2	ABPI	78° 47'E	30° 51'N	3650	170 (1826-1995)	12	0.19	0.35	114
5	DBG3	ABPI	78° 46'E	30° 51'N	3680	269 (1727-1995)	28	0.23	0.52	143
6	DBG4	ABPI	78° 46'E	30° 50'N	3600	400 (1596-1995)	59	0.18	0.39	209
7	DBG5	ABPI	78° 45'E	30° 50'N	3650	370 (1626-1995)	31	0.18	0.39	196
8	DBG6	ABPI	78° 45'E	30° 51'N	3670	223 (1773-1995)	15	0.24	0.38	147
9	DBG7	ABPI	78° 44'E	30° 51'N	3620	381 (1615-1995)	48	0.21	0.35	207

Figure 1. Map showing location of tree-ring sampling sites and meteorological stations (see text for site description).

Table 2. Correlation among all standard ring width chronologies for the period 1831-1988 AD (n=158)

	DBG1-2	DBG3	DBG4	DBG5	DBG6	DBG7	GAN	CHI
DBG1-2	1							
DBG3	0.476^{**}	1						
DBG4	0.405^{**}	0.707^{**}	1					
DBG5	0.336^{**}	0.595^{**}	0.825^{**}	1				
DBG6	0.386^{**}	0.572^{**}	0.592^{**}	0.636^{**}	1			
DBG7	0.389^{**}	0.645^{**}	0.777^{**}	0.786^{**}	0.705^{**}	1		
GAN	0.23^{**}	0.203^{*}	0.298^{**}	0.006	0.197^{*}	0.221^{**}	1	
CHI	0.175^{*}	0.214^{**}	0.285^{**}	0.24^{**}	0.337^{**}	0.4^{**}	0.538^{**}	1

** Correlation is significant at the 0.01 level (2-tailed). Correlation is significant at the 0.05 level (2-tailed).

PIWA samples were collected from two sites, near Gangotri temple (3,048 m asl) and Chirbasa (3,600 m asl) of Gangotri glacier valley. Near Gangotri temple (GAN), 55 cores from 35 trees were collected. In Chirbasa (CHI), 68 cores from 38 trees were collected from pure, mixed age stand, growing in a mesic tree-line site with even terrain, occasionally with gentle slope.

ABPI samples were collected from seven different sites in Din Gad Valley growing on several glaciated features, several hundred meters below the present position of the snout of Dokriani Bamak glacier (mentioned in map and elsewhere as DBG1-DBG7). The details of these sites are described in Bhattacharyya et al. [2001]. Most of the glaciated area in Din Gad valley is largely covered by A. pindrow intermixed with Betula utilis and Rhododendron near the timberline and broad- leaved mixed conifer forest in lower levels. For the present study, 209 cores from 133 trees were collected from these sites.

Generally two cores per tree, one each from the opposite directions at chest height, were collected with an increment borer. For the tree-ring width data, the samples were mounted and processed using standard procedure of tree ring analysis following Fritts, [1976] and Cook and Kairiukstis [1990] and evaluated by "Cross-dating" technique [Stokes and Smiley, 1968]. The rings in each sample were dated to the calendar year of their formation and the ring widths of each dated core were measured using an increment-measuring stage with 0.01 mm precision, coupled with a microcomputer. These measurements and dates were then checked using the computer program COFECHA [Holmes, 1992]. The cores for which errors appeared were re-examined to evaluate the sources of the errors and corrections were made accordingly.

For the acquisition of density data, the cores of PIWA from CHI were first processed to extract water-soluble substances and resin. Thereafter for each sample a lath of constant thickness of 1 mm was cut perpendicular to the direction of fibers with the help of a twin-blade circular saw [Schweingruber, 1988]. The samples were irradiated for 40 minutes with a soft radiation (8 kV and 18 mA) and a distance tube to film of 2 m; the atmosphere of the X-ray room was controlled to 20°C and 50% RH. The films were developed in an automatic processor to control homogeneity as much as possible. The densitometric measurement of the X-ray radiographs was carried out on a DENDRO-2003 tree-ring workstation. All the densitometry analyses were carried out in IMEP laboratory. Six parameters were measured on each tree-ring sample from transmitted image of x-ray film to produce time series of early wood width, latewood width, minimum density, maximum density, early wood density and late wood density. The boundary between early wood and late wood was identified as the mid point between the maximum and minimum density of each ring.

CHRONOLOGY

The ring width data of each site and species were detrended and standardized to form tree-ring indices using ARSTAN program [Holmes, 1992]. It removes growth trends related to age and stand-dynamics while retaining the maximum common signal. For each ring-width series, we applied double detrending method for standardization. In this way, the ring width series were detrended with a negative exponential growth curve using least squares. The detrended tree-ring series were then filtered with a cubic smoothing spline [Cook and Peters, 1981] using 50% frequency response function with cut-off equal to two-third of the length of the series. For density data, we used only a negative exponential growth curve for detrending. The standardized tree-ring indices were then averaged together to form tree-ring chronology of a site. Since the two sites (DBG-1 and DBG-2) were comparatively very close to each other and had only a few samples, so we clubbed them together to one chronology. Thus, we developed eight ring-width chronologies of ABPI and PIWA. The length of the chronologies ranges from 165 to 400 years (Table 1). Six 300-year long chronologies of PIWA from CHI were also developed, one for each parameter: early wood width, latewood width, minimum density, maximum density, early wood density and late wood density. These chronologies were based on only 10 cores as the other cores were not found suitable for densitometric analysis. The correlation among the different ring-width standard chronologies for the common period (1831-1988) is presented (Table 2). All the chronologies have positive correlation (at 0.05 significance level) amongst each other, except for GAN with DBG5 site-chronology.

CLIMATE DATA

Long records of climate around Dokriani Bamak and Gangotri are not available. Regular meteorological monitoring was started only a few years ago and is inadequate for detailed analysis of tree-growth and climate relationship of the region. Even in the Himalayan region, in general, the meteorological stations are still very few and are situated mostly at lower elevations. Moreover, the stations available in close by areas do not have long records and have lot of missing data thereby constraining their use for dendroclimate analyses. Most of the earlier studies from this region [e.g., Yadav et al., 1997, 1999; Yadav and Singh, 2002, 2006] have used the data from Shimla (77°17'E, 31°10'N; 2,205 m.a.s.l.) and Mukteshwar (79°39'E, 29°28'N; 2,311 m.a.s.l.) or the mean series by averaging the data of these two stations. Since climate in Himalayan region varies widely within short distances because of the complex topography, we have taken the interpolated temperature and precipitation data (1901 to 2000) of the nearest grid point (78°75'E, 30°75'N, 3,337 m.a.s.l.) of CRU-TS 2.0

dataset [Mitchell et al., 2004] to analyze the response function for the trees at different sites. This grid point is more or less at the same elevation as the tree-ring sampling sites. The correlation between the temperature of Shimla, Mukteshwar and grid point for the common period 1901-1988 is 0.75 and 0.84 (p=0.01) respectively whereas for precipitation, it is 0.24 (p=0.05) and 0.62 (p= 0.01) respectively. The difference in correlation is because of geographical complexity as well as the grid interpolation involved in the analysis.

In general, the Gangotri area receives a good amount of precipitation during winter but summer monsoon is low in comparison to the sites close to Dokriani Bamak glacier. In those sites, a good amount of precipitation occurs both during summer and winter. In both the areas, the coldest and hottest months of the year are January and August respectively.

Calibration, Verification and Reconstruction

We used bootstrap regression using principal components [Guiot, 1991] to estimate the tree-growth and climate relationship and for the reconstruction of climate parameters. This standard method involves random generation of hundreds of data matrix out of the original one. First, a year is randomly selected and is stacked to form a new data matrix. Afterwards, another year is randomly picked up and the corresponding data vector is stacked next in this matrix. In this randomized combination, called "pseudo-dataset" matrix, same year may be drawn several times. A regression is then performed on this pseudo-dataset. This methodology is repeated 200 times, giving 200 regressions and 200 predictions. The summary statistics on the regression coefficient and multiple correlations were obtained by calculating mean and standard deviations on all the 200 samples.

A large number of regression functions (called response functions) were tested between each standardized tree-ring series and different climate variables for the common period of 1902-1987. Out of these, August-September (AS) temperature was found to be significantly correlated and thus was selected for reconstruction. It has been noted that the increase in temperature during post summer is favourable for net photosynthesis at the high elevation sites. The optimum temperature for photosynthesis at tree-line is around 13-18^{0} C [Tranquillini, 1964]. As a result, the increased late-summer temperature may reach to optimal point for higher rate of photosynthesis and may enhance latewood formation. Moreover, a surplus food production has a carry-over effect of stored food on the growth of the subsequent years [Fritts, 1976].

A total of 42 regressors, eight ring-width chronologies and six density series for the previous, current and next year growth (hereafter called as T-1, T and T+1 respectively) were put together to form the predictor matrix for the estimation of AS temperature (transfer function). A subset consisting of 14 tree-ring predictors, i.e., ring width chronologies (T and T+1) of sites DBG 1-2,3,4,6 and late wood width, late wood density and maximum density of PIWA from CHI were found to be significant. The reliability of the reconstruction model was tested by cross-validation. Firstly, the calibration was made on 70% of the dataset and the independent verification was done on the other 30%, and secondly, it is calibrated on the last 70% and verified on the first 30%. Lastly, for the final reconstruction of August-September temperature (Figure 2a), the regression coefficients were calculated for the entire period of 1902-1987 (Table 3).

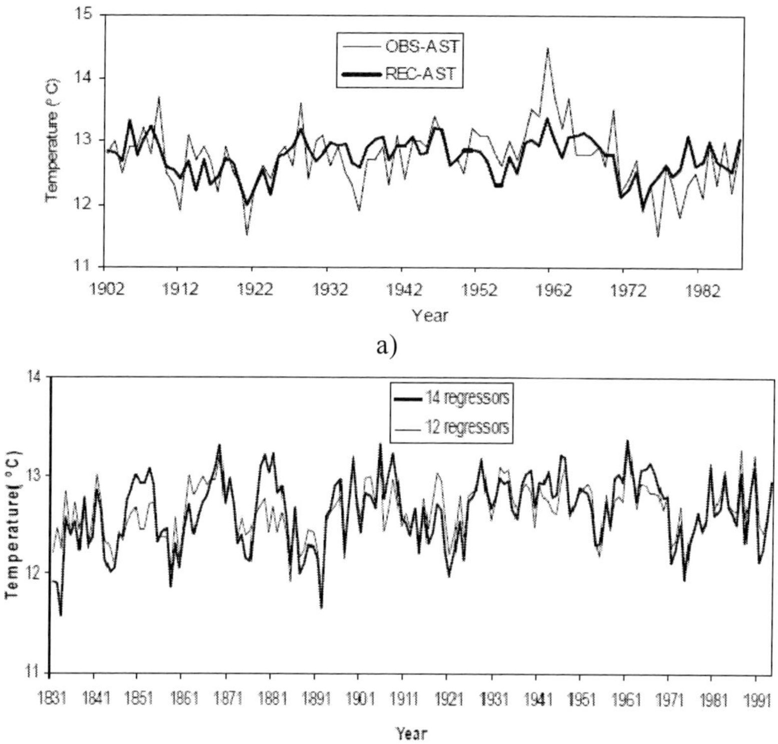

a)

b)

Figure 2. a) Observed and reconstructed mean August-September temperature for the period 1902-1987 AD. b) Reconstructed mean August-September temperature for the common period (AD 1831-1994), using 14 regressors and 12 regressors.

Table 3. Summary of the calibration and verification statistics. R= Pearson Correlation Coefficient, R^2 = variance accounted for by the calibration model, RMSE= Root mean Square Error

Calibration			
Period	1902-1961	1928-1987	1902-1987
R	0.702	0.667	0.661
R^2	49%	44%	44%
Verification			
Period	1962-1987	1902-1927	1902-1987
Mean_obs/est	12.62/12.64	12.65/12.56	12.74/12.74
Std. Dev._obs/est	0.54/0.30	0.43/0.43	0.49/0.30
Corr._obs/est	0.468	0.551	0.646
RMSE	0.48	0.42	0.38

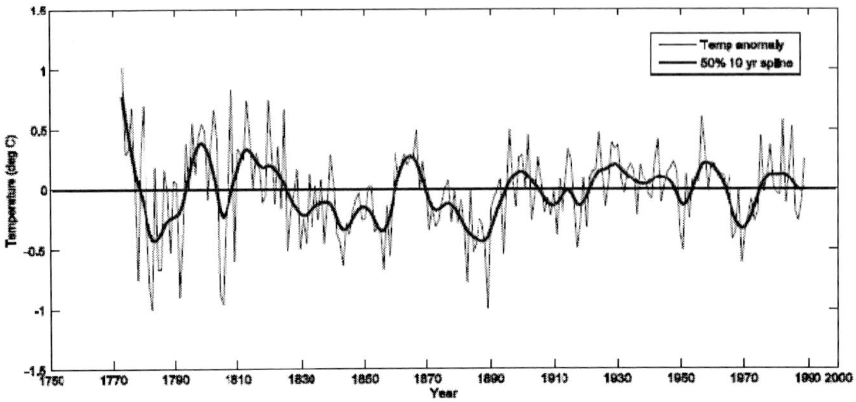

Figure 3. Mean late summer (AS) temperature anomalies for AD 1773-1994 relative to the 1902-1987 mean. Bold line represents a smoothing line with 50% frequency cut off of 10 years.

The final transfer function is based on the most significant 14 regressors to provide a better reconstruction of August-September temperature up to 1831 AD. The regression equation accounts for 44% of the variance in calibration for the period 1902-1987. Since out of these 14 significant regressors, two were only up to AD 1831, the calibration was, therefore, done again on the remaining 12 significant regressors. There is a good agreement between the reconstructions with 12 and 14 regressors for the common period of AD 1831-1994 (Figure 2b). Therefore, based on these 12 regressors, which explained 32% of the variance, the temperature was reconstructed for the period AD 1773-1830. Although a few other regressors were available before AD 1773, the reconstructed variance was not significant enough to provide good estimates of temperature before this year. Finally, various verification statistics, specifically, mean and standard deviation were calculated depending on the calibration and verification periods (Table 3). The final reconstruction was calculated as the median of the 200 bootstrapped reconstructions with a 90% confidence level explained by the 5th and 95th percentiles.

Analysis of the Reconstructed August-September Temperature Data

The anomaly for the reconstructed mean AS temperature of the 223 years (AD 1773-1995) were calculated relative to the mean for the period AD 1902-1987 (see Figure 3) with 50% 10-yr filtered smoothening, superimposed on the anomaly. The anomaly indicates the following significant cold AS temperatures, AD 1783-1797, 1830-1852, 1859-1868, 1888-1900, 1914-1929, and 1973-1983. On the other hand, relatively warm AS temperatures were observed during the following periods of AD 1773-1782, 1798-1807, 1814-1829, 1901-1913, 1930-1955, and 1961-1972 (Figure 3).

To identify the major climate shifts in the studied sites, a sequential change-point analysis [Rodionov, 2004] was also performed on the reconstructed AS temperature anomaly. A change point occurred at some point in the given time series if all the values up to and including it shared a common statistical distribution. There are many methods in use, starting from single change-point detection to sequential detection of multiple change-points. The analysis adopted in this study with respect to the mean, is a test of discontinuity in the site data due to various forcing, namely change in elevation or microclimatology of the site. In brief, the methodology consists of choosing a window-size, a significance level and performing a moving student t-test based on these parameters. First we estimated the variance for the whole series and then mean for the first window. Thereafter, we tested the next point if it belongs to the statistical distribution of the first window, and if it does not, we kept testing other points. Once we encountered a point which might belong to the previously defined distribution, we marked it as a probable change point and calculated standardized t-value with respect to the estimated variance and mean of the previous window. Subsequently, we started from this point as a new regime and tested for other possible change points. Here, we used 10-year window with p-value 0.05 (for two-tail student t-test) to calculate the Regime Shift Index (RSI) values. The RSI values (not shown here) were measures of departure from the previous climate regime's mean. This analysis showed major regime shifts at AD 1783, 1794, 1805, 1827, 1862, 1873, 1898 and 1971 (Figure 4).

The most salient finding in the reconstruction was the cooling during the major part of the nineteenth century compared to the warming in the 20th century, particularly from AD 1925 to 1972. These findings are well corroborative with the study of D'Arrigo et al. [2001] that also suggests a coherent 20th century warming and 19th century cooling over the boreal tree-ring sites. The present reconstruction of AS temperature based on tree-ring data of Abies is common with Kashmir and Eastern Himalayan regions. However, the correlation between the chronologies from these two regions and the one used in the present study was found to be poor. This heterogeneity is due to the fact that these sites are located in different climate regime. The Kashmir region is strongly influenced by extra tropical disturbances and is outside the influence of monsoon currents unlike the sites near Dokriani and Gangotri that receive both monsoon and western disturbances though with varying intensities.

Interestingly, many of the cool and hot summer years recorded here have also been found in other temperature reconstructions from Tibet [e.g., Braeuning and Mantwill, 2004] and Himalayan region [e.g., Cook et al., 2003; Yadav et al., 2004]. The coolest episode (AD 1783-1893), noticed in the present reconstruction is in agreement with cooling in Central Asia and Karakorum [Briffa et al., 2001; Esper et al., 2002].

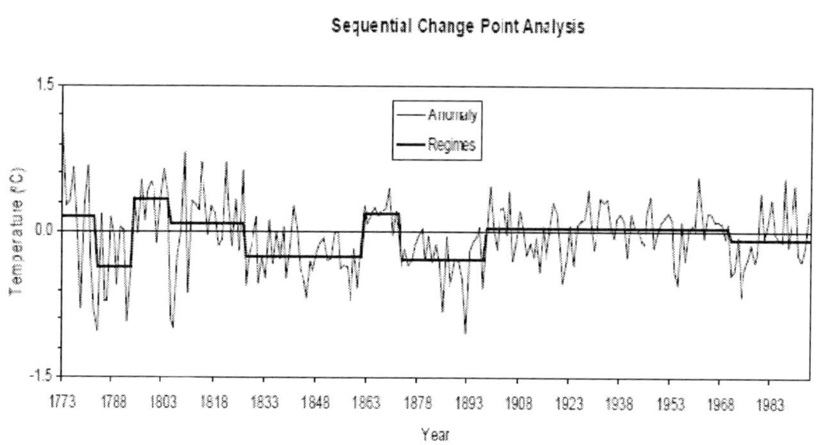

Figure 4. Sequential change point analysis (with p=0.05, window = 10 yr) on the mean August-September temperature anomalies from AD 1773 to 1994.

Another interesting feature recorded in the present reconstruction (Figure 3) is warming during beginning of the 19th century, followed by a colder episode. This cooling might be linked with a drop in temperature in the northern Hemisphere with effect of Tambora eruption, which is well documented in climate reconstruction from Nepal [Cook et al., 2003]. However, it is not a major event here. Even cooling of AD 1830 to 1900 is in less agreement with the Nepal data. The cooling during 1820s-1850s in the western Himalaya coincides with that in central Asia and Karakorum [Briffa et. al., 2001; Esper et al., 2002]. However, Nepal and Tibet experienced colder episodes during first two decades of the 19th century [Briffa et al., 2001; Cook et al., 2003]. The tree-ring proxies from Mangolia and high-latitude of northern hemisphere indicate the coldest phase of "Little Ice Age" (LIA) during 19th century [Mann et al., 1999; D'Arrigo et al., 2001] which is also reflected in our study. The coldest episode during AD 1830-1852 in the last 223 years in the Western Himalaya is well in agreement with the spatially coherent Little Ice Age period in the Northern Hemisphere [Grove, 1988].

In our reconstruction cooling has been recorded during the latter part of the 20thcentury (AD 1973 onwards) on the contrary to the other high-resolution proxies from high-latitudes of northern and southern hemispheres that indicate unprecedented warming in the 20th century. The cooling during the last decades of the 20th century has also been documented in other tree-ring based reconstructions from Kashmir [Hughes, 2001], Nepal [Cook et al., 2003; Sano et al., 2005], Tibet and Central Asia [Briffa et al., 2001].

Conclusion

The tree-ring data of Pinus wallichiana and Abies pindrow of the sub-alpine belt of the Western Himalaya exhibit good coherence with the climate changes both on annual and multiyear scales. August and September temperature has been found most significant climate variable limiting the growth of trees at the upper tree line. There is a good correspondence with the evidence of temperature variability linked with the Little Ice Age cooling of 19th century and warming in the 20th century.

However, both these episodes were not long lasting and were rather interrupted by smaller phases of warming and cooling in between. In contrary to the recent global warming, August-September temperature has been recorded as cold phases over the region since AD 1973. The sequential change-point analysis which has been used for the first time in the dendroclimate studies from this region also substantiates the present findings.

Acknowledgments

This study was supported by DST, Government of India, vide grant No. ESS/91/34/95 and ESS/91/018/97. We are grateful to the Director, Birbal Sahni Institute of Palaeobotany for providing necessary facilities to carry out the work. A part of the work was carried out at CEREGE and IMEP, Aix-en-Provence, France under the long-term fellowship awarded to the first author by the Ministry of Scientific Research, France.

REFERENCES

Bhattacharyya, A., Chaudhary, V., 2003. Late summer temperature reconstruction of the Eastern Himalayan region based on tree-ring data of Abies densa. *Arctic, Antarctic and Alpine Research* 35(2), 196-202.

Bhattacharyya, A., Chaudhary, V., Gergan, J. T., 2001. Tree-ring analysis of Abies pindrow around Dokriani Bamak (Glacier), Western Himalaya, in relation to climate and glacier behaviour: Preliminary results. *Palaeobotanist* 50, 71-75.

Borgaonkar, H. P., Pant, G. B., Rupa Kumar, K., 1994. Dendroclimate reconstruction of summer precipitation of Srinagar, Kashmir, India since the late 18th century. *Holocene* 4, 299-306.

Borgaonkar, H. P., Pant, G. B., Rupa Kumar, K., 1996. Ring width variations in Cedrus deodara and its climate response over the Western Himalaya. *International Journal of Climatology* 16, 1409-1422.

Borgaonkar, H. P., Rupa Kumar, K., Pant, G. B., Okada, N., Fuziwara ,T., Yamashita, K., 2001. Climate implications of tree-ring density variations in Himalayan conifers. *Palaeobotanist* 50, 27-34.

Braeuning, A., Mantwill, B., 2004. Summer temperature and summer monsoon history on the Tibetan plateau during the last 400 years recorded by tree-rings. *Geophysical Research Letters* 31, L24205, doi 10.1029/2003GL020793.

Briffa, K. R., Osborn, T. J., Schweingruber, F. H., Harris, I. C., Jones, P. D., Shiyatov, S. G., Vaganov, V. G., 2001. Low frequency temperature variations from northern tree-ring density network. *Journal of Geophysical Research* 106, 2929-2941.

Cook, E. R., Kairiukstis, L. A., 1990. Methods of Dendrochronology: *Applications in the Environmental Sciences*. Kluwer Academic, Dordrecht, pp. 394.

Cook, E. R., Krusic, P. J., Jones , P. D., 2003. Dendroclimate signals in long tree-ring chronologies from the Himalaya of Nepal. *International Journal of Climatology* 23, 63-79.

Cook, E. R., Peters, K., 1981. The smoothing spline: a new approach to standardizing forest interior tree-ring width series for dendroclimate studies. *Tree-Ring Bulletin* 41, 45-53.

D'Arrigo, R., Jacoby, G., Frank , D., Pederson, N., Cook, E. R., Buckley, B., Nachin, B., Mijidorj, R., Dugarjav, C., 2001. 1738 years of Mangolian temperature variability inferred from a tree-ring width chronology of Siberian pine. *Geophysical Research Letters* 28, 543-546.

Esper, J., Schweingruber, F., Winiger, M., 2002. 1300 years of climate history for western central Asia derived from Tree-rings. *Holocene* 12, 267-277.

Fritts, H. C., 1976. Tree rings and Climate. Cambridge University Press, Cambridge. London, pp. 567.

Grove, J. M., 1988. The Little Ice Age. Meuthen, London, pp. 498.

Guiot, J., 1991. The boot-strapped Response Function. *Tree-Ring Bulletin* 15, 39-41.

Holmes, R. L., 1992. Program ARSTAN (Version B-1992). Laboratory of Tree-Ring Research, University of Arizona, Tucson, US, pp. 81.

Hughes, M. K., 1992. Dendroclimate evidence from the Western Himalaya. In: Bradley, R. S., Jones P. D., (Eds.), Climate since A. D. 1500. Routledge, London.

Hughes, M. K., 2001. An improved reconstruction of summer temperature at Srinagar, Kashmir since 1660 AD, based on tree-ring width and maximum latewood density of Abies pindrow {Royle} Spach. *Palaeobotanist* 50, 13-19.

Khandekar, M. L., 1991. Eurasian snow cover, Indian monsoon and El-Ninõ/ Southern Oscillation: A synthesis. *Atmosphere Ocean* 29, 636–647.

Mann, M. E, Bradley, R. S.,Hughes, M. K., 1999. Northern Hemisphere temperatures during the past millennium: Inferences, uncertainties and limitations. *Geophysical Research Letters* 26, 759-762.

Mann, M. E., Jones, P. D., 2003. Global surface temperatures over the past two millennia. *Geophysical Research Letters* 30(15), 1820, doi: 10.1029/2003GL017814.

Mitchell, T. D., Carter, T. R., Jones, P. D., Hulme, M., New, M. 2004. A comprehensive set of high-resolution grids of monthly climate for Europe and the globe: the observed record (1901-2000) and 16 scenarios (2001-2100). *Tyndall Centre Working Paper* No. 55.

Pant, G. B., Rupa Kumar, K., Borgaonkar ,H. P., Okada, N., Fuziwara ,T., Yamashita, K. 2000. Climate response of Cedrus deodara tree-ring parameters from two sites in the Western Himalaya. *Canadian Journal of Forest Research* 30, 1127-1135.

Prell, W. L., Kutzbach, J. E. 1992. Sensitivity of the Indian monsoon to forcing parameters and implications for its evolution. *Nature* 360, 647-652.

Rodionov, S. N., 2004. A sequential algorithm for testing climate regime shifts. *Geophysical Research Letters* 31, L09204, doi: 10.1029/2004GL019448.

Sano, M., Furuta, F., Kobayashi, O., Sweda, T. 2005. Temperature variations since the mid-18th century for western Nepal, as reconstructed from tree-ring width and density of Abies spectabilis. *Dendrochronologia* 23(2), 83-92.

Schweingruber, F. H., 1988. Tree Rings: basics and applications of Dendrochronology. Kluwer Academic Publishers, Dordrecht, Netherlands, pp. 276.

Singh, J., Yadav, R. R., 2006. Tree-ring based hydrological records for Western Himalaya, India, since A. D. 1560. *Climate Dynamics* 26, 295-303.

Stokes, M. A., Smiley, T. L., 1968. *An introduction to tree ring dating.* University of Chicago Press, Chicago, US, pp. 73.

Tranquillini, W., 1964. The physiology of plants at high altitudes. *Annual Review of Plant Physiology* 15, 345-362.

Treydte, K. S., Schleser, G. H., Helle, G., Frank, D. C., Winiger, M., Haug, G. H., Esper, J., 2006. The twentieth century was the wettest period in northern Pakistan over the past millennium. *Nature* 440, 1179-1182.

Yadav, R. R., Bräuning, A., Singh, J., 2011.Tree-ring inferred summer temperature variations over the last millennium in western Himalaya, India. *Climate Dynamics*, doi:10.1007/s00382-009-0719-0.

Yadav, R. R., Park, Won-Kyu, Bhattacharyya, A., 1997. Dendroclimate reconstruction of April- May temperature fluctuations in the Western Himalaya of India since A. D. 1698. *Quaternary Research* 48, 187-191.

Yadav, R. R., Park, Won-Kyu, Bhattacharyya, A., 1999. Spring variations in Western Himalaya, India as reconstructed from tree rings: 1390-1987. *Holocene* 9(1), 85-90.

Yadav, R. R., Park, Won-Kyu, Singh, J. and Dubey, B., 2004. Do the Western Himalaya defy global warming? *Geophysical Research Letters* 31, L17201, doi: 10.1029/2004GL02021.

Yadav, R. R., Singh, J. 2002. Tree-ring based Spring Temperature Patterns over the past Four Centuries in Western Himalaya, *Quaternary Research* 57, 299-305.

In: Holocene
Editor: Bahadur Singh Kotlia

ISBN: 978-1-62257-722-4

Chapter 13

SNOW MELT EPHEMERAL STREAMS IN THE GANGOTRI GLACIER AREA, GARHWAL HIMALAYA, INDIA

Dhruv Sen Singh*
University of Lucknow, India

ABSTRACT

The Gangotri Glacier area is covered with snow during winter (accumulation period) and the ephemeral streams are initiated as the snow melts during summer (ablation period). The ephemeral streams erode the fine-grained sediments from the pre-existing landforms, facilitating also a facies change from matrix-supported boulders to clast-supported boulders. Therefore, the matrix- and clast-supported facies indicate the accumulation/cold stage and ablation/warm stage, respectively and can be used as a proxy for the interpretation of similar events in the geological record. The preserved geomorphic features and the associated sedimentary record provide information about the palaeoclimate, which is continuously being modified by the snow-melt streams in a glaciated terrain. Therefore, it is important to understand the snow-melt ephemeral streams and their role in the modification and evolution of landform. This chapter presents the influence of snow-melt streams on the landscape evolution of the Gangotri glacier area.

Keywords: Snow-melt streams, Gangotri glacier, Himalaya, Landscape

INTRODUCTION

Glaciers are sensitive indicators of climate change as they expand and shrink in direct response to the changes in the temperature and amount of snowfall. This dynamic behavior of

* Centre of Advanced Study in Geology, University of Lucknow, 226007, India. E mail: *dhruvsensingh@rediffmail.com*.

glacier contributes to the development of the landforms, which provides clues for the reconstruction of palaeoclimate. The reconstruction and inferences are ideal if the surface processes and their associated landforms have a one to one correspondence. However, the coexistence of various geomorphic features and their history of modification suggest that the region is subjected to complex interplay of more than one processes. In this case, the glacier is the main surface modification agent but the snow-melt stream that comes into existence later accelerates the mass movement and thus contributes significantly in shaping the landscape of the area.

The distribution and style of glaciations throughout central Asia is strongly influenced by SW Indian Monsoon, mid latitude Westerlies and the El-Niño southern oscillation (ENSO) [Owen et al., 1998]. Rapid differential uplift of the Himalaya during late Tertiary and Quaternary times may also have been important in controlling the style of glaciations [Fort, 1995]. In addition to uplift of the Himalaya, the Tibetan Plateau may have had profound effects on the global climate throughout the late Cenozoic [Ruddiman and Kutzbach, 1991]. Sufficient snowfall, shady northern slopes, low temperature, and gentle slope gradient of the valley are the favorable conditions for glaciations over the Himalaya.

The Himalaya, an abode of snow, ice, and glaciers, is one of the most important regions outside the polar region to have glaciers. It contains about 11,000 glaciers spreading over 43000 km^2,which varies in size from less than a km to as long as Siachen glacier, which is 72 km long [GSI, 1999]. The Gangotri glacier is one of the longest valley glaciers of the Kumaun and Garhwal Himalaya. Glaciers in the Himalaya are formed due to successive accumulation of snow over long period of time and have evolved in direct response to the climate and Himalayan orogeny. Snow layer is formed when snow is sufficient to persist on the ground till the next spell of snowfall. The melting of snow during ablation period, initiates the ephemeral streams, which erode and deposit the sediments and thus modify the pre-existing landforms. However, the glaciated regions have not been analyzed in the light of ephemeral streams. It has been studied for its Quaternary glacial history [Sharma and Owen, 1996], morpho-sedimentary processes [Singh, 2004], retreat [Singh and Mishra, 2001; Tangri, 2002], general characteristics using GIS technique [Singh and Mishra, 2002a], morphometry [Prasad and Naithani, 2003], vegetation and climate change [Ranhotra et al., 2002] and role of tributary glaciers in the modification of landforms [Singh and Mishra, 2002b]. Its snout was monitored by Geological Survey of India since 1935 with the description of its tongue fluctuations [Puri and Shukla, 1996]. The control of snow-melt stream is of great significance for understanding the evolutionary history and palaeoclimate reconstruction of this region.

STUDY AREA

The Gangotri glacier is situated between $30^0 44'$-$30^0 56'$N latitude and $79^0 04'$-$79^0 15'$E longitude in the Uttarkashi District of Uttaranchal state. It originates from the Chaukhambha group of peaks at an elevation of 7100 m and is about 30 km long and 0.5-2.5 km wide [GSI, 1999] (Figure 1).

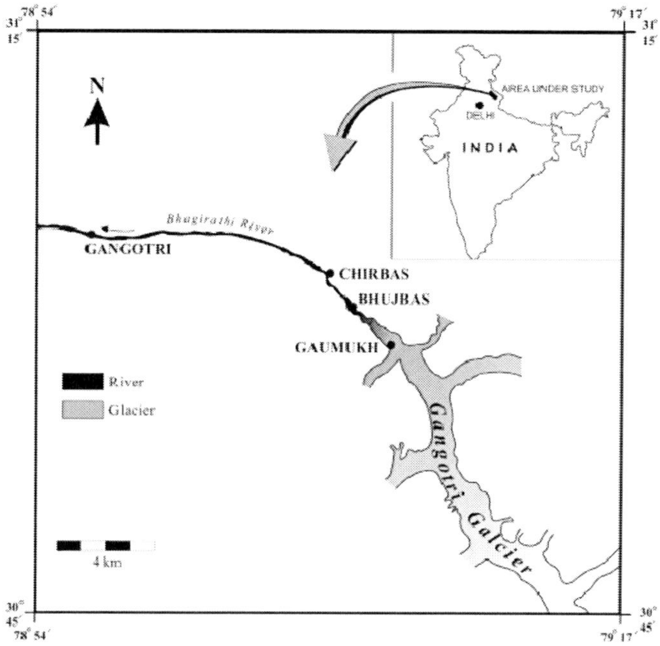

Figure 1. Location map of the study area.

Figure 2. Snout of the Gangotri glacier with its left bank lateral moraine and recessional moraine.

The Gangotri glacier is characterized by tributary glaciers, supraglacial moraines, supraglacial lakes, crevasses, moraines, and heavy snow fall during winter [Singh, 2004]. The ice melts continuously in the ablation zone so that the debris in this area increases and may eventually cover the glacier surface.

The Gangotri glacier is like a river of ice where many tributaries make their own contribution to the trunk stream. Numerous small-sized glaciers, designated as tributary glaciers feed and contribute to the main glacier to form the Gangotri group of glaciers. The tributary glaciers are smaller in size (length 3-20 km) but larger in number (30), so their cumulative effect is very high and cannot be ignored. Depending upon their present day attachment and detachment with respect to the main glacier, the tributary glaciers are named as active and inactive tributary glaciers respectively [Singh and Mishra, 2002 b]. The end point of the glacier, i.e., snout (Figure 2) (known as Gaumukh for this glacier) is at a distance of about 18 km from the Gangotri township. Gaumukh is about 10-20 m in height and 15-30 m in width. The Bhagirathi River originates from Gaumukh and is situated more towards left valley wall due to tilt of the valley.

FACIES AND LANDFORMS ASSOCIATED WITH SNOW-MELT STREAMS

The climatic changes have resulted in the retreat of the glacier body and its snout, which is evidenced by the presence of lateral and recessional moraines occupying different elevations and positions. The positions of moraines show several major and minor fluctuations in the spatial extent of the Gangotri glacier (Figure 2).

The melting of snow generates the ephemeral streams and contributes to the discharge of the Bhagirathi River. The Gangotri glacier area is completely covered with snow during winter/accumulation period. Remotely sensed satellite data indicate that the snowfall during winter (accumulation period) is about 10 times higher than the permanent snow and ice in the Himalaya [Tangri, 2002]. It has been observed that building up of snow cover starts from October to December, develops during January to March and starts ablating from April onwards, reaching a minimum in September. The snow accumulates and melts first below the snout and afterwards above the snout.

The snow cover decreases from 57% of the watershed area during winter to as low as about 27% of the watershed area during summer [Tangri, 2002]. The decrease in snow covers due to increase in temperature and change of season generates the ephemeral streams. The snow-melt streams on the top surface of a glacier are called supraglacial streams. These streams erode the finer sediments from the pre-existing landforms, mainly the diamictons. As a result, the facies changes from the matrix-supported boulders to the clast-supported one (Figure 3).

At some places, the clasts of boulder size cap the fine-grained sediments lying below it. Large boulders that are too heavy to be transported by ephemeral streams and precipitation are bypassed during erosion and protect the fine-grained sediments from removal, while the surrounding slope surface is lowered. The sediments in between the boulders are removed and the poorly-sorted substrate remains as pillar structures with a boulder cap (Figure 4). These structures are the standing portion of poorly-sorted material consisting of sediments from clay to boulder. These 2-10 feet long and 1-4 feet wide features are designated as pillar like structures because they look like a pillar. These features are present only on the right bank of the valley and are absent on the left bank due to steep slope gradient of the valley wall. The facies analysis and location of the pillar like structures explain that these landforms are the remnant morainic features, formed by differential weathering. The fine material of the pillar is removed with time and the cap boulder falls. Later, the fine-grained materials are also removed and the pillar structures are washed out completely. These structures are very common in the Gangotri glacier region but their evolution is explained for the first time.

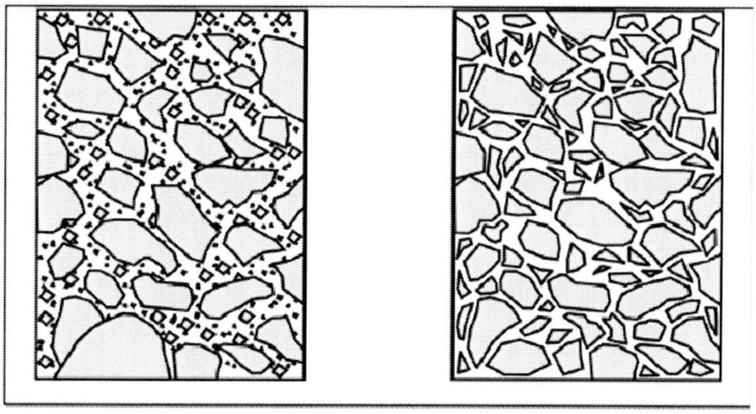

Figure 3. Schematic model for the matrix-supported moraine (with dots) and clast-supported moraine (without dots) on the basis of field observations.

Figure 4. Showing the evolution of pillar like structures.

Figure 5. Left and right bank debris cones near Bhujbas.

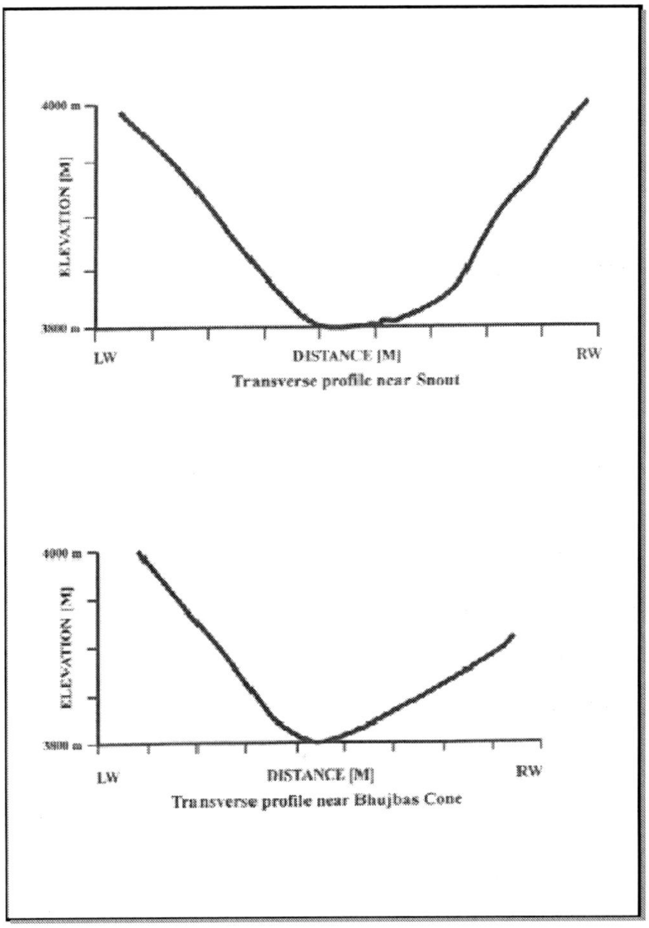

Figure 6. Transverse profile indicating the shape of the valley near the snout and Bhujbas.

Recession of the Gangotri glacier vacates the Bhagirathi valley and provides the accommodation space. Ephemeral streams accelerate the mass movement process and forms the debris cone on both sides of the valley. The abundance of debris cone in this region is due to the snow-melt streams. The debris cone is a tongue like feature which may be triangular, fan or delta in shape. The debris cones of the left bank are smaller in size, whereas that on the right bank are larger in size (Figure 5). The debris cones formed on the right side of the valley are large because of the gentle slope gradient, large accommodation space and high supply of sediments. It is 50-100 m in length and 20-50 m in width. Larger debris cones have forced the river to shift towards the left bank at many places and have controlled the sinuosity of the river up to some extent. Left side debris cones are 10-30 feet in length and 5-20 feet in width.

The sediments removed from the pre-existing landforms are deposited in the vacated valley. The deposition of sediments by snow-melt streams and mass movement keeps on modifying the shape of the valley. Profiles have been constructed near snout (Gaumukh) and near Bhujbas, 4 km downstream to the Gaumukh to observe the shape of the valley. The transverse profile is U shaped near snout and V shaped near Bhujbas (Figure 6). The profile indicates that the stream processes are active near Bhujbas which alters the glacial U shaped valley into the fluvial V shaped valley.

CONCLUSION

The landscape of the study area have been carved by the last period of glacial activity and modified by the snow-melt streams that evolved during the interglacial period. The snow-melt stream initiates the fluvial activity, erodes the finer sediments and accelerates the mass movement processes and evolves the new landforms through modification of the pre-existing ones. The clast-supported facies of the moraines indicate the interglacial stage, whereas the matrix-supported one indicates the glacial stage. This facies model can be used as an analogue to identify the glacial and interglacial stages in the rock record.

The snow-melt ephemeral streams are affecting this region in various ways (1) removes the fine-grained sediments from pre-existing glacial landforms and changes its facies from matrix-supported boulder to clast-supported boulder (2) accelerates the mass movement process to form the debris cones (3) evolves the pillar-like structures due to differential weathering, and (4) converts the transverse profile of the valley from U shaped glacial to V shaped fluvial.

ACKNOWLEDGMENTS

Financial assistance from Government of the Uttar Pradesh, Lucknow and also from the Department of Science and Technology, New Delhi are highly acknowledged. The Head, Department of Geology, University of Lucknow, Lucknow, India, is thanked for providing the working facilities. Prof. B.S. Kotlia is highly acknowledged for the invitation to write this paper.

REFERENCES

Fort, M., 1995. The Himalayan Glaciation: Myth and Reality. *Journal Nepal Geological Society Special Publication issue* 11, 257-272.

G. S. I. 1999. Inventory of Himalayan Glaciers. *Geological Survey of India Special Publication Number* 34. 1-164.

Owen, L. A., Derbyshire, E., Fort, M., 1998. The Quaternary glacial history of the Himalaya. *Quaternary Proceedings* 6, 91-120.

Prasad, C., Naithani, A. K., 2003. Glacier Morphometry: A case study of Gangotri Group of Glaciers, Garhwal Himalaya, India. *Journal of Geological Society of India* 61, 325-334.

Puri, V. M. K., Shukla, S. P., 1996. Tongue fluctuation studies on Gangotri glacier, Uttarkashi Distt., Uttar Pradesh. *Geological Survey of India Special Publication Number* 21 (2), 289-291.

Ranhotra, P. S., Bhattacharya, A., Kar, R., Sekar, B., 2002. *Vegetational and Climatic changes around Gangotri Glacier during Holocene Geological Survey of India Special Publication Number* 65 (III), 2001: 67-71.

Ruddiman, W. F., Kutzbach., J. E., 1991. Plateau uplift and climate change. *Scientific American* 264(3), 42-50.

Sharma, M. C., Owen, L. A., 1996. Quaternary glacial history of the Garhwal Himalaya, India. *Quartenary Science Review* 15, 335-365.

Singh, D. S., 2004. Late Quaternary Morpho-Sedimentary Processes in the Gangotri Glacier area Garhwal Himalaya, India. *Geological Survey of India Special Publication Number* 80, 97-103.

Singh, D. S., Mishra, A., 2001. Gangotri glacier characteristics, retreat and processes of sedimentation in the Bhagirathi valley. *Geological Survey of India Special Publication Number* 65 (III), 17-20.

Singh, D. S., Mishra, A., 2002 a. Gangotri glacier System: An analysis using G. I. S. technique. In: Pant, C. C., Sharma, A. K. (Ed.), Aspects of Geology and Environment of the Himalaya. *Gyanoday Prakashan, Nainital*, pp. 349-358.

Singh, D. S., Mishra, A., 2002 b. Role of Tributary glaciers on landscape modification in the Gangotri glacier area, Garhwal Himalaya, India. *Current Science*, 82 (5), 101-105.

Tangri, A. K., 2002. Shrinkage of the Himalayan Glaciers: Hope for the future. Aspects of Geology and Environment of the Himalaya. In: Pant, C. C., Sharma, A. K. (Ed.), Aspects of Geology and Environment of the Himalaya. *Gyanoday Prakashan, Nainital*, pp. 359-367.

In: Holocene
Editor: Bahadur Singh Kotlia

ISBN: 978-1-62257-722-4

Chapter 14

SOFT SEDIMENT DEFORMATION STRUCTURES IN THE EARLY HOLOCENE LAKE DEPOSITS, SPITI VALLEY, HIMACHAL PRADESH, INDIA

B. S. Kotlia*, L. M. Joshi and Naval Kishore[1]

Centre of Advanced Study in Geology, Kumaun University, Nainital, India
[1]Centre of Advanced Studies in Geology, Panjab University, Chandigarh, India

ABSTRACT

The Atargoo palaeolake in the Spiti basin, Himachal Tethys Himalaya, NW India seems to have formed as a result of neotectonic activity in the zone of the Spiti fault during the early Holocene period. In this chapter, an attempt is made to reveal the importance of the soft sediment deformational structures (seismites) and their imprints on a lowermost 3.48 m part of a 37m thick exposed profile which comprises the lacustrine and fluvial deposits and belongs to the Early Holocene period.

The studied part of the profile has preserved the flame structures, folds and faults including micro-faulting, load and pseudo-nodule structures in the mud, silt and sands. Such deformed levels, having a considerable lateral extent, are bounded by completely undeformed strata. We suggest that the seismites horizons may have formed due to thixotropy and fluidization causing visco-plastic failure of the mud, and the density inversion between mud and sand due to liquefaction, triggered perhaps by past shaking events. The study of deformed layers may be a key to understand the pre-historic shaking events in the Spiti basin.

Keywords: Holocene lake sediments, Seismites, Shaking events, Spiti fault, Himachal Pradesh, Himalaya

* E-mail: bahadur.kotlia@gmail.com.

INTRODUCTION

This chapter describes the soft sediment deformational structures (seismites), formed during the Early Holocene in the Spiti Himalaya. The geodynamically active Himalaya orogene provides a classical example of the subduction and collision-related mountain building process (Molnar and Tapponnier, 1975), associated with the ongoing crustal deformation and present-day seismicity of different magnitudes (Hodges, 2000; Jain et al., 2002; Singh and Jain, 2007).

The Himalayan frontal arc is one of the seismically active regions of the world (Jade, 2004; Bilham, 2005; Arora, 2006). The Lahaul-Spiti-Kinnaur basin in the Himachal Pradesh is bound by the Vaikrita group of the Central Crystalline Zone on the SW and by the Indus Tectonic zone on the NE (Srikantia and Bhargava, 1998). The neotectonic movements on active faults/thrusts have triggered the large landslides and glacial debris avalanches on an extensive scale.

This phenomenon has resulted in the damming up of rivers' courses and formation of lakes throughout the Higher Himalayan belt, e.g., Ladakh (Burgisser et al., 1982; Fort et al., 1989; Kotlia et al., 1997a; Upadhyay, 2003; Phartiyal, and Sharma 2009), Kashmir (Burbank and Johnson, 1982; Agrawal et al., 1989), Spiti (Banerjee et al., 1997) and Kumaun (Kotlia et al., 1997b, 1998, 2000; Kotlia and Rawat, 2004; Basavaiah et al., 2004; Juyal et al., 2004; Rawat and Kotlia, 2006). Although the palaeoseismic investigations in the Higher Himalayan regime are carried out by a number of researchers, this study aims at using primary and secondary palaeoseismic evidences for establishing the past shaking records and behavior of reactivation of the major/minor faults. Such evidences and their correlation firstly on local scale and subsequently on regional basis may help to establish and understand the local and the major seismic behavior of the past earthquakes.

The Quaternary deposits (especially the lake sediments) form an important source of information on the frequency and distribution of the past earthquakes. The largest of India's earthquakes, however, occur on the northern boundary of the Indian plate where it descends beneath the southern Tibet. At least four major Himalayan earthquakes (1505, 1803, 1934 and 1950) have been recorded so far (Bilham, 2004) with magnitude exceeding 8 and several more with magnitude 7.

STUDY AREA

The Spiti valley in Himachal Pradesh lies between the Higher Himalaya in the south and the Tibetan plateau in the north and has cold arid to semi-arid climate with annual temperature fluctuating between +30 ^{0}C and -30 ^{0}C. The present investigated area is located in the lower Spiti valley (Figures 1a, b) at Atargoo village where about 37 m thick fluvio-lacustrine sediments are exposed (Figure 2).

Prasad et al. (2008), suggested that these sediments were deposited in a lake formed by the landslide damming of the Spiti river during the Early Holocene intensified monsoon phases.

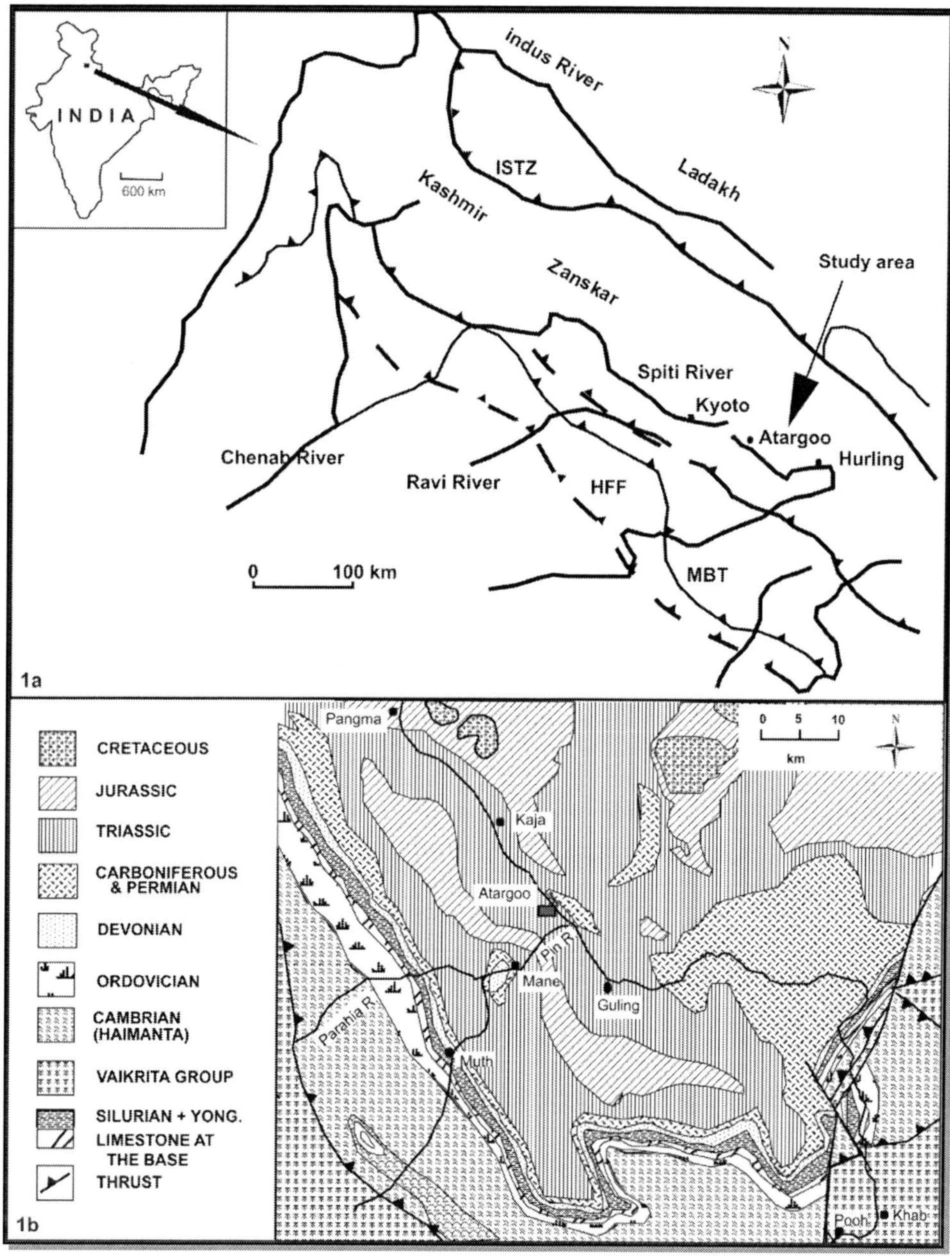

Figure 1. a, Location and tectonic map of the Spiti valley (b) Rocks exposed around Atargoo (after Bagati, 1991).

Regional Geological Setting and Tectonics

The Spiti-Zanskar Phanerozoic belt represents a major parautochthon and comprises two major structural units, the Kuling- Lilang schuppen zone and the Mesozoic- Cenozoic parautochthon. The Mesozoic-Cenozoic parautochthon comprises mainly Lilang (Triassic-Jurassic), Spiti formation (Jurassic) and the Kanji group (Eocene).

Figure 2. Panoramic view of a 37 m thick fluvio-lacustrine profile at Atargoo.

The Spiti rocks belong to the Lilang group (Stoliczka, 1865). At Atargoo, the Pin river meets the Spiti river and the massive limestone horizons of Kioto-Jurassic formation crosses towards NE over the Dankar Gompa (Sinha, 1998). There are major faults running along and across the Spiti river (Spiti fault and Takche fault).

The N–S-trending Kaurik–Chango (K–C) fault in the lower Spiti valley and NNE-trending west-dipping Leo-Pargil Horst fault are mostly responsible for the neotectonic activity of the area (Singh and Jain, 2007) and blocking the river to form a number of lakes in and around the area. Near Atargoo, the valley is narrow and the Triassic carbonates dominate on either side of the river (Bagati and Suresh, 1991). Bhargava (1990) has been of the opinion that the Spiti fault (Figure 3) generated the landslides and formed a lake between Atargoo and Po due to the Holocene tectonics. The study area is characterized by significant geomorphic features such as well preserved river terraces (Figure 4), alluvial fans, palaeolake sediments (see Figure 2) as well as braded river channels (Figure 5), hinting towards the neotectonic activity.

Lithology of Atargoo Lake Sediments and Chronology

The palaeolake sequence at the Atargoo village (our study area) is dominated by sands, muds, clays and silts showing well detectable laminations and a number of mm-cm scale horizons of the rock fragments with sandy matrix. The laminated sediments exhibit alternating dark and light colored laminae and may represent seasonal deposition as has been suggested by Ashley (1975).

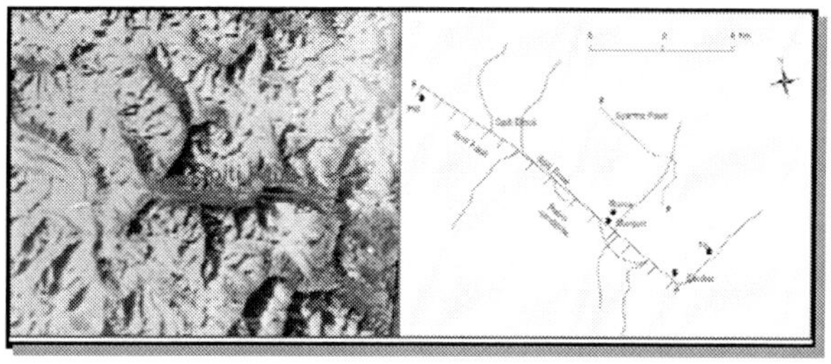

Figure 3. Location of the Spiti Fault along the Spiti River.

Figure 4. The river terrace along the lower Spiti valley.

Figure 5. Braided river at confluence of Spiti and Pin rivers.

The profile has dark colored quartz dominated and light colored calcite dominated sub-laminae (Prasad et al., 2008). Generally, in such a situation, the darker laminae consist of clay representing winter deposition whereas silt/silty clays constituting the lighter laminae are marked by summer deposition. The organic sample at the base is dated to 8,781 cal yr BP (Bookhagen et al., 2005) and a wood fragment at 6.55-6.57 m height (from the base) of the profile is assigned an age of 7,397±75 cal yr BP (Anoop et al., 2012). See Table 1.

Table 1. AMS radiocarbon dates from the Atargoo lake sediments

Height from base (m)	Material	Lab no	^{14}C age (yr)	^{14}C age cal yr BP (1σ)	Ref.
6.55-6.57	Wood fragment	Poz-33176	6,490±80	7,397±75	Anoop et al., 2012
0-0.1	Wood fragment	-	7,926±38	8,781±104	Bookhagen et al., 2005

Figure 6. Lithology of the 3.48 m profile of Atargoo deposits, with 8 seismite layers.

A number of studies (e.g., Mohindra and Bagati, 1996; Singh and Jain, 2007; Phartiyal et al., 2009) have pointed towards the palaeoseismic activity within the Spiti basin. The soft sediment deformation structures (seismites) discussed here were observed between the basalmost horizon to 3.48 m height of the profile (Figure 6). Most likely, the syn-sedimentary deformation structures were developed due to liquefaction and fluidization, triggered by the past seismic activity as they are overlain and underlain by the horizontally laminated undeformed layers. In general, the deformed horizons show fining upward sequence with penecontemporaneous deformation, caused by triggering of liquefaction due to seismic activity as has been suggested by Kuenen (1958), Lowe (1975, 1976) and Mills (1983).

Soft Sediment Deformation Structures

The observations of the earthquakes occurred in the recent times and associated increasing pore fluid pressure resulting in the sediment deformation allow us to study more precisely the past seismic activity, i.e., palaeoseismicity. The significance of the soft sediment deformation to understand the palaeoseismicity has been discussed in detail by Seilacher (1969), Sims (1973, 1975), Mills (1983), Hempton and Dewey (1983), Mohindra and Bagati,

(1996), Rossetti (1999), and Sukhija et al. (1999) etc. The seismites are formed due to the disturbance of sediments during or shortly after the deposition and are common in sand and sandstone deposits (Allen, 1982a; Maltman, 1999). The seismite layers observed in the Spiti basin are a result of local, vigorous fluidization and hydroplastic deformation during liquefaction (Singh and Jain, 2007). The silty clay are hydroplatically injected into the overlying coarser unit, forming complex flame structures with coarsening upward cycle and the sediments are injected upwards due to the effect of fluidization (Singh and Jan, 2007). A total of 8 deformed layers, which are bounded by the undeformed layers are found in the lowermost 3.48 m part of the 37 m sequence of the Atargoo profile. A variety of deformational structures range from load casts to convolute structures and micro-faulting. These structures are also developed in the sandy overburden and clayey substratum. The convolute flame structures are observed in the different heights of the profile. Genetically, the flame structures are developed due to the mechanical instability within the sediment layers (Kelling and Walton, 1957; Kuenen, 1958; Anketell et al., 1970; Allen, 1982b; Owen, 1987; Collinson, 1994). The flame structures are associated with the psuedonodules. The ball and pillow, psuedonodules and flame structures (Figures 7-8) are the most common structures. The ball and pillow structures and psuedonodules are interpreted to be a result of water flowing over freshly deposited material in the hydroplastic state (Collinson and Thompson, 1989). These occur at the interface between coarse grained (sands) and fine grained sediment unit. The sediment units exhibiting flame development are confined to the planer and undisturbed sediments layers. The Flame load cast structures are associated with mud and clayey sand/sand where the coarse grained units overlie fine grained sediments.

The convolute flame structures are observed at various depths of laminated sediments. They have been developed in the mud unit which is overlain by clayey sand.

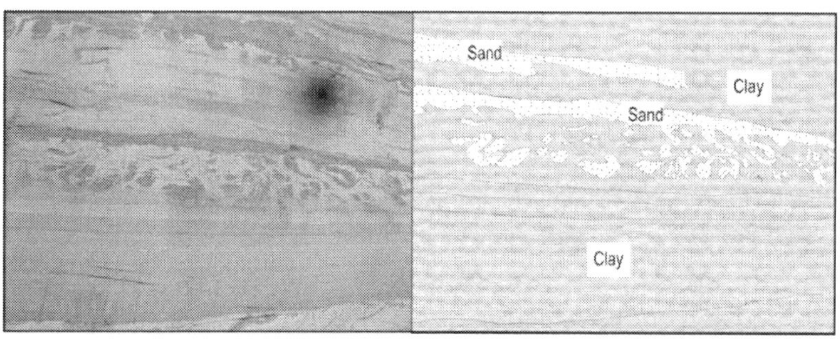

Figure 7. Psuedonodule structures in the multilayered sediments of alternate varve and sands.

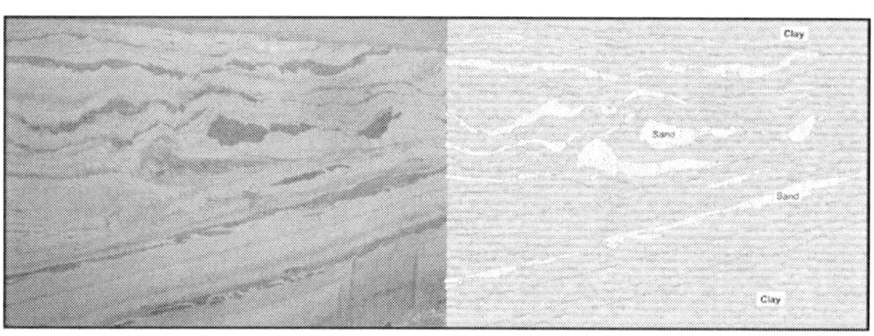

Figure 8. Ball and Pillow structures and psuedonodules in alternate clay and sand layers.

Flames are a few cm in size and occur irregularly (Figures 9-11). The laterally persisting flames exhibit broader synclinal and sharp anticlinal parts. The convolute laminations have the structural styles, generally occurring within a single bed, usually of silt or sand grade material (Collinson and Thompson, 1989; Maltman, 1999). The brittle structures such as faults are observed in our study and most micro-faults are parallel to the Spiti fault.

These structures are generally of mm to cm scale and restricted to the deformed units only. These brittle structures take place due to differential compaction of the sediments during the folding and plastic flowage (Mayall, 1983; Anand and Jain, 1987). The faults affecting several sediments units may develop during compaction of the inter-stratified sediments such as mud-silt under the influence of the overburden, liquefaction or synsedimentary shocks (Reineck and Singh, 1973; Sukhija et al., 1999).

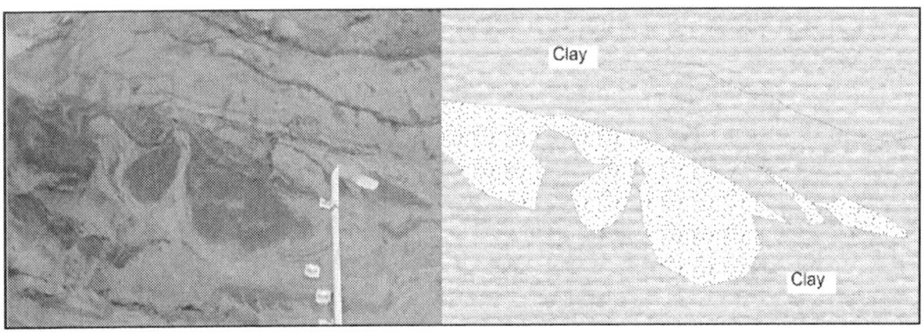

Figure 9. Convolute flame structure developed in carbonaceous muds confined within clayey sand.

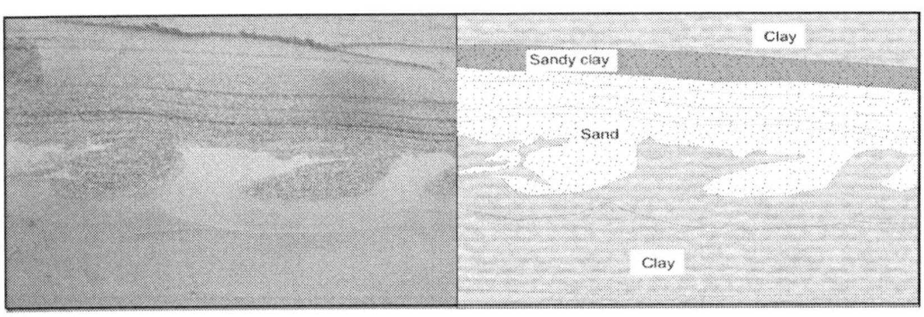

Figure 10. Convolute flame structures.

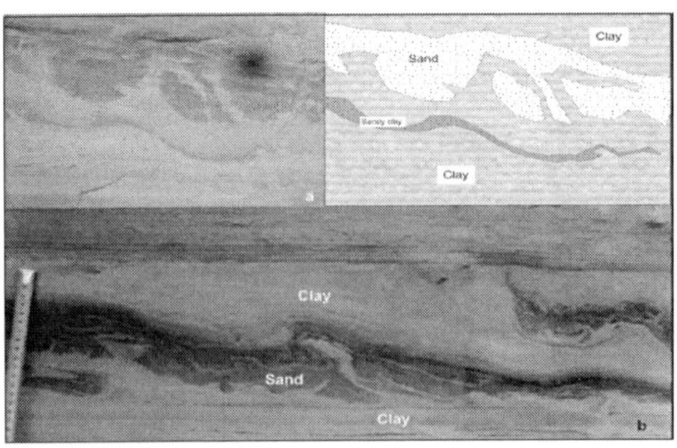

Figure 11. a, Convolute flame structures with sand sagging, (b) Deformed layer underlain and overlain by the undeformed layer.

The liquefaction is induced by a triggering agent such as pore pressure build up during the seismic shaking or groundwater movements (Allen, 1882a, Owen, 1987).

DISCUSSION

The soft sediment deformation structures from the Himalayan arc, caused by episodic earthquake recurrences during the Holocene, are produced by the ongoing tectonic processes. According to Kuribayashi and Tatsuoka (1975), a relationship exists between the Richter earthquake magnitude and range of liquefaction area, viz., an earthquake of magnitude 5 would cause negligible or little liquefaction while a 6 magnitude earthquake may result in the liquefaction of the sediments within 4 km radius of the epicentral distance. The liquefaction phenomenon in the present area may be linked to the active Kaurik Chango fault. We opine that several earthquakes of very high magnitude too place by the fault reactivation. The soft sedimentary features may have been formed due to past seismic activity and are linked to the reactivation of the Spiti fault. The synsedimentary soft sedimentary structures are the evidence of paleoseismicity and are mainly brittle and visco-plastic structures which were formed as a result of liquefaction and fluidization. All the deformed layers are bounded by the undeformed horizons. The identified micro faults are parallel to the active NW–SE-trending Spiti fault. The nature of the deformation structures in the study area is very similar to those described from other areas, (Anand and Jain, 1987; Kotlia and Rawat, 2004; Rawat and Kotlia, 2006; Singh and Jain, 2007; Phartiyal and Sharma, 2009), especially in the Higher Himalayan belt, implying that the whole belt has been influenced by the shaking events during the Holocene.

ACKNOWLEDGMENTS

The authors are grateful to the ISRO-GBP (Ahmedabad, India), UGC and CSIR (New Delhi), and CAS in Geology and Uttarakhand Centre on Climate Change (Kumaun University, Nainital) for providing financial assistance, infrastructure and other necessary facilities.

REFERENCES

Agrawal, D. P., Dodia, R., Kotlia, B. S., Razdan, H., Sahni, A., 1989. The Plio-Pleistocene geologic and climatic record of the Kashmir valley: A review and new data. *Palaeogeography Palaeoclimatology Palaeoecology*, 73, 267–286.

Allen, J.R.L., 1982a. Developments in Sedimentology. In: Allen, J.R.L. (Eds.), sedimentary structures their character and physical basis. Vol II. Elsevier, Amsterdam,180-198.

Allen, J.R.L., 1982b. Sedimentary Structures: their character and physical basis. Elsevier, Amsterdam, 213 p.

Anand, A., Jain, A. K., 1987. Earthquakes and deformational structures (seismites) in Holocene sediments from the Himalayan-Andaman Arc, Indian. *Tectonophysics*, 133, 105-120.

Anketell, J. M., Cegla, J., Dzulynski, S., 1970. On the deformation structures in systems with reversed density gradient. *Am. Geol. Soc. Pol.,* 40, 3-30.

Anoop, A., Prasad, S., Basavaiah, N., Brauer, A., Shahzad, F. Deenadayalan, K., 2012. Tectonic versus climate influence on landscape evolution: A case study from the upper Spiti valley, NW Himalaya. *Geomorphology,* 145–146, 32–44.

Arora, B. R., 2006. Himalaya-the Nagadhiraj. ISG, Newsletter 12 (3 and 4), 414 p.

Bagati, T.N., 1991. Evolution of the Tethyan sedimentary basin in the western Himalaya (in) Sedimentary basin of India: Tectomics context. In: Tandon, S.K., Pant C.C and Casshyap, S.M. (Eds.), Gyanodaya Prakashan, Nainital, 218-235.

Bagati, T.N., Suresh, N., 1991. Sedimentology of the Lingti Lacustrine sequence, Spiti valley, Himachal Pradesh. *Journal of Himalayan Geology*, 2(2), 125-132.

Banerjee, D., Singhvi, A. K., Bagati, T. N., Mohindra, R., 1997. Luminescence chronology of seismites at Sumdo (Spiti valley) near Kaurik-Chango fault, northwestern Himalaya. *Current Science,* 73(3), 276-281.

Basavaiah, N., Juyal, N., Pant, R.K., Yadava, M. G., Singhvi, A. K., Appel, E., 2004. Late Quaternary climate changes reconstructed from mineral magnetic studies from proglacial lake deposits of higher central Himalaya. *Journal of Indian Geophysical union,* 8(1), 27-37.

Bhargava, O.N., 1990. Holocene tectonics south of the Indus suture, Lahaul- Ladakh Himalaya. India: A consequence of Indian plate motion, *Tectonophysics,* 174, 315-320.

Bilham, R., 2004. Earthquakes in India and the Himalaya: tectonics, geodesy and history, *Annals of Geophysics,* 74 (2), 839-858.

Bilham, R., 2005. A flying start followed by slow slip. *Science,* 308 (5725), 1126.

Bookhagen, B., Thiede, R.C., Strecker, M.R., 2005. Late Quaternary intensified monsoon phases control landscape evolution in the northwest Himalaya. *Geology*, 33, 149–152.

Burbank, D. W., Johnson, G. D., 1982. Intermontane-basin development in the past 4 Myr in the north-west Himalaya. *Nature,* 298, 432–436.

Burgisser, H. M., Gansser, A., Pika, J., 1982. Late glacial lake sediments of the Indus valley area, northwestern Himalaya. *Eclogae. Geol. Helv.,* 75, 51–63.

Collinson, J.D., 1994. Sedimentary deformation structures. In: Maltman, A. (Eds.), The Geological deformation of sediments. Chapman and Hall, London, 95-125.

Collinson, J.D., Thompson, D.B., 1989. Sedimentary structures, Unwin Hyman, 102-106.

Fort, M., Burbank, D. W., Freytet, P., 1989. Lacustrine sedimentation in a semiarid Alpine setting: An example from Ladakh, northwestern Himalaya. *Quaternary Research,* 31, 332–350.

Hempton, M. R., Dewey, J.F., 1983. earthquake induced deformation structure in the young lacustrine sediments, East Anatolian Faults, Southeast Turkey. *Tectonophysics*, 78, 7-14.

Hodge, K. V., 2000. Tectonics of the Himalaya and southern Tibet from two decades perspectives. *Geological Society of American Bulletin*, 112, 324-350.

Jade, S., 2004. Estimates of plate velocity and crustal deformation in the Indian subcontinent using GPS geodesy, *Current Science*, 86, 1343-1348.

Jain, A.K., Singh, S., Manickavasagam, R.M., 2002. Himalayan collision tectonics, Bhagirathi valley Garhwal. Gondwana Research Group Memoir 7, 114p.

Juyal, N., Pant, R.K., Basavaiah, N., Bhushan, R., Jain, M., Saini, N.K., Yadava, M.G., Singhvi, A.K. 2009. Reconstruction of Last Glacial to early Holocene monsoon variability from relict lake sediments of the Higher Central Himalaya, Uttarakhand, India. *Journal of Asian Earth Sciences,* 34, 437–449.

Juyal, N., Pant, R.K., Basavaiah, N., Yadava, M. G., Saini, N.K., Singhvi, A. K., 2004. Climate and seismicity in the higher Central Himalaya during 20–10 ka: evidence from the Garbyang basin, Uttaranchal, India. *Palaeogeography Palaeoclimatology Palaeoecology*, 213, 315– 330.

Kelling, G., Walton, E.K 1957. Load cast structures, their relationship to upper surface structures and their mode of formation. *Geological Magazine*, 94, 481-490.

Kotlia, B. S., Bhalla, M. S., Shah, N., Sharma, C., Rajagopalan, G., 1998. Palaeomagnetic results from the Pleistocene-Holocene lake deposits of Bhimtal and Bhowali (Kumaun Himalaya) and Lamayuru (Ladakh Himalaya) with reference to reversal events. *Journal of Geological Society of India,* 51, 7-20.

Kotlia, B. S., Bhalla, M. S., Sharma, C., Rajgopalan, G., Ramesh, R., Chauhan, M. S., Mathur, P. D., Bhandari, S., Chacko, S. T., 1997a. Quaternary fluvio-lacustrine deposits of the Lamayuru basin, Ladakh Himalaya, preliminary multidisciplinary investigations. *Geological Magazine,* 134(6), 807-812.

Kotlia, B. S., Bhalla, M. S., Sharma, C., Ramesh, R., Rajagopalan, G., Mathur, P. D., Bhandari, S., Chacko, S. T., 1997b. Palaeoclimatic conditions in the Upper Pleistocene and Holocene Bhimtal-Naukuchiatal lake basin in south-central Kumaun, North India. *Palaeogeography Palaeoclimatology Palaeoecology,* 130, 307-322.

Kotlia, B.S., Rawat, K.S., 2004. Soft sediment deformation structures in the Garbyang palaeolake: evidence for the past shaking events in the Kumaun Tethys Himalaya. *Current Science*, 87(3), 377-379.

Kotlia, B. S., Sharma, C., Bhalla, M. S., Rajagopalan, G., Subrahmanyam, K., Bhattacharyya, A. and Valdiya, K. S., 2000. Palaeoclimatic conditions in the Late Pleistocene Wadda lake, eastern Kumaun Himalaya (India). *Palaeogeography Palaeoclimatology Palaeoecology*, 162: 105-118.

Kuenen, P. H., 1958. Experiments in Geology. *Transactions of the Geological Society of Glasgow*, 23, 1- 28.

Kuribayashi, E., Tatsuoka, F. (1975). Brief review of liquefaction during earthquake in Japan. *Soils Found.,* 15 (4), 18-25.

Lowe, D. R., 1975. Water escape structures in coarse-grained sediments. *Sedimentology*, 22, 157-204.

Lowe, D. R., 1976. Subaqueous liquefied and fluidized sediments flow and their deposits. *Sedimentology,* 23, 285-308.

Maltman, A., 1999. Introduction and overview. The Geological Deformation of Sediments. In: Maltman, A. (Eds.), Chapman and Hall.

Mayall, M.J., 1983. An earthquake origin for synsedimentary deformation in late Triassic (Rhaetain) lagoonal sequence, Southwest Britain. *Geological Magazine,* 120, 613-622.

Mills, P. C., 1983. Genesis and diagnostic value of soft sedimentation structures: a review. *Sedimentary Geology*, 35, 83-104.

Mohindra, R., Bagati, T. N., 1996. Seismically induced soft-sediment deformation structures (seismites) around Sumdo in the lower Spiti valley (Tethys Himalaya). *Sedimentary Geology,* 101, 69-83.

Molnar, P., Tapponnier, P., 1975. Cenozoic tectonics of Asia: effects of a continental collision. *Science,* 189, 419-426.

Owen, G., 1987. Deformation processes in unconsolidated sands. In: Jones, M.E., Preston, R.M.F. (Eds.), Deformation of sediments and sedimentary rocks. *Special Publication, Geological Society of London*, 29, 11-24.

Patriat, P., Achache, J., 1984. India-Eurasia collision chronology has implications for crustal shortening and driving mechanism of plates. *Nature,* 311, 615-621.

Phartiyal, B., Sharma, A., 2009. Soft-sediment deformation structures in the Late Quaternary sediments of Ladakh: Evidence for multiple phases of seismic tremors in the North western Himalaya region. *Journal of Asian Earth Sciences,* 34, 761-770.

Prasad, S., Brauer, A., Garber, S., Basavaiah, N., Kotlia, B.S., Anoop, A., 2008. Holocene climate variability in the Himalayan region: an overview and preliminary results from lake sediments in the Spiti valley. *Himalayan Geology,* 29(3), 73-75.

Rawat, K.S., Kotlia, B.S., 2006. Some examples of flame structures from the Garbyang palaeolake, Kumaun Tethys Himalaya, *Journal of Geological Society of India,* 68, 639-647.

Reineck, H.E., Singh, I.B., 1973. Depositional Sedimentary Environments. Springer, Berlin, 439 p.

Rosseti, D. F., 1999. Soft sediment deformation structures in late Albian to Cenomanian deposits, Sao Luis Basin, Northern Brazil: evidence for palaeoseismicity. *Sedimentology,* 46, 1065-1081.

Searle, M. P., Cooper, D. W. J., Rex, A. J., 1988. Collision tectonics of the Ladakh-Zanskar Himalaya, Philosophical Transactions of Royal Society, London, Series A 326, 117–150.

Seilacher, A., 1969. Fault graded beds interpreted as seismites. *Sedimentlogy,* 13, 155-159.

Sims, J. D., 1973. Earthquake induced structures in sediments of Van Normal Lake, San Fernando, California. *Science,* 182, 161-163.

Sims, J. D., 1975. Determining earthquake recurrence intervals from deformational structures in young lacustrine sediments. *Tectonophysics,* 29, 141-152.

Singh, S., Jain, A. K., 2007. Liquefaction and fluidization of lacustrine deposits from Lahaul-Spiti and Ladakh Himalaya: Geological evidences of palaeoseismicity along active fault zone. *Sedimentary Geology*, 196, 47-57.

Sinha, A.K., 1998. Geo-traverses through Higher and Tethys zones of Lahaul and Spiti sector of NW Himachal and Adjoining part of Ladakh Himalaya. *Palaeobotanist*, 47, 1-15.

Srikantia S.V., Bhargava O.N., 1998. Geology of Himachal Pradesh, Geological Survey of India, 1-406.

Stoliczka, F., 1865. Geological section across the Himalaya mountain range from Wangtu Bridge on the river Sutlej to Sungdeo with an account of formations in Spiti accompanied by a revision of all known fossils from that district. Memoirs of Geological Survey of India 5, 153p.

Sukhija, B. S., Rao, M. N., Reddy, D. V., Nagabhushanam, P., Hussain, S., Chadha, R. K., Gupta, H. K., 1999. Timing and return period of major palaeoseismic events in the Shillong plateau, India. *Tectonophysics,* 308, pp. 53-65.

Upadhyay, R., 2003. Earthquake-induced soft-sediment deformation in the lower Shyok river valley, northern Ladakh, India. *Journal of Asian Earth Sciences,* 21, 413–421.

In: Holocene
Editor: Bahadur Singh Kotlia

ISBN: 978-1-62257-722-4

Chapter 15

Coseismic Subsidence due to Yoro Fault Activity Detected in the Holocene Delta Sequence of the Western Nobi Plain, Central Japan: An Example of Off-Fault Paleoseismological Study in the Coastal Plain

***Yuichi Niwa*[1,2•] *and Toshihiko Sugai*[1]**

[1]Department of Natural Environmental Studies,
Graduate School of Frontier Sciences,
The University of Tokyo, Kashiwanoha, Kashiwa, Chiba, Japan
[2]Research Fellow of the Japan Society for the Promotion of Science, Japan

Abstract

Coastal plains have been formed by the influence of sea-level changes, tectonic vertical displacement and fluvial process during the Holocene. Thus, the Holocene coastal sediments can record the information about tectonic vertical displacement (e.g., coseismic subsidence). In addition, these records can be reconstructed with high temporal resolution because of application of radiocarbon dating. Based on the timing of tectonic vertical displacement recorded in the Holocene coastal sediments, paleoseismicity can be also reconstructed.

In this chapter, we review the results of paleoseismic study based on the detection and timing of coseismic subsidence recorded in the Holocene delta sequence in the Nobi plain, central Japan. Firstly, we will introduce whole image of the Holocene delta sequence on the Nobi plain influenced by sea-level changes. Secondly, we will present the results of detection of coseismic subsidence in the western part of the plain during the past 6,000 years. Additionally, we review the results of reconstruction of paleoseismicity on the Yoro fault fringing the Nobi plain, based on the results of detection of coseismic subsidence.

• E-mail: niwayu@nenv.k.u-tokyo.ac.jp.

Keywords: Coseismic subsidence, Yoro fault, Delta, Nobi plain, Central Japan

INTRODUCTION

Coastal plains have been formed by the influence of sea-level changes, tectonic vertical displacement, and fluvial process during the Holocene. Information from sedimentary sequences can contribute to understanding of vertical tectonic displacements on the coastal plains. The sedimentary records in the Holocene coastal sediments including the influence of vertical tectonic displacements can be reconstructed with high temporal resolution because of application of radiocarbon dating. Based on the timing of tectonic vertical displacement recorded in the Holocene coastal sediments, paleoseismicity can also be reconstructed.

Regarding fault paleoseismological studies, such as trench surveys, have provided detailed records of the timing and magnitude of ancient fault displacements (e.g., Machette et al., 1992; McCalpin and Nishenko, 1996; Sebrier et al., 1997; Rubin et al., 1998; Liu et al., 2004, 2006; Ran et al., 2010; Langridge et al., 2011).

However, these studies have lacuna, for example, in order to find the record of multiple ruptures in the past, the trenches on dip-slip faults should be deep, and secondly, the reconstruction of past displacements can become difficult because of subaerial erosion of specific stratal layers on the upthrown side that could define the offset precisely (e.g., Burbank and Anderson, 2001). Identification of paleoseismic events on the basis of on-fault paleoseismology is often difficult where active fault is located on the sea floor or in and around large alluvial plains (Yamazaki et al., 2002).

On the other hand, the paleoseismic evidence can be recorded in a location away from the earthquake source fault (McCalpin and Nelson, 1996). Some examples of paleoseismic evidence include sand boiling caused by liquefaction (e.g., Tuttle et al., 1999, 2002, 2005; Tuttle, 2001; Takada and Atwater, 2004; Obermeier, 2009), sandy deposits transported by strong stream of tsunami (e.g., Williams and Hutchinson, 2000; Nelson et al., 2004, 2008; Kelsey et al., 2005; Jankaew et al., 2008), marine terraces which reflect coseismic uplift (e.g., Ota et al., 1991; Palyvos et al., 2008; Shishikura et al., 2009; Hsieh and Rau, 2009), and buried soil which reflects coseismic subsidence (e.g., Atwater, 1992; Nelson et al., 1996, 2009; Sherrod, 2001; Knudsen et al., 2002; Nichol et al., 2007). These paleoseismic evidences can be recorded in the coastal geomorphology and sedimentary sequences, formed during the Holocene.

Thus, to clarify and precisely reconstruct paleoseismic activity, both on-fault and off-fault evidence is desired. Especially, focusing attention on off-fault paleoseismic evidence can contribute to understanding the paleoseismicity more precisely.

In this chapter, we review the results of off-fault paleoseismic study in the coastal plain. In particular, we introduce studies on off-fault paleoseismology based on the detection and timing of coseismic subsidence recorded in the Holocene delta sequence in the Nobi plain, central Japan (Niwa et al., 2009, 2010, 2012).

First, we present as to why the Nobi plain is significant in the off-fault paleoseismological study. Second, we introduce whole image of the Holocene delta sequence on the Nobi plain, influenced by sea-level changes. Third, we describe the results of detection of coseismic subsidence in the western part of the plain during the past 6,000 years.

Additionally, we review the results of reconstruction of paleoseismicity on the Yoro fault fringing the Nobi plain, based on the results of detection of coseismic subsidence.

REGIONAL SETTING AND SIGNIFICANCE OF THE NOBI PLAIN AS OFF-FAULT PALEOSEISMIC STUDY SITE

In this sub-section, we elucidate the significance of the Nobi Plain (Figure 1) as an example of off-fault paleoseismic study site based on geomorphological and geological setting.

Geomorphological and Geological Features and Paleoseismic Events on the Yoro Fault System

The Yoro fault system fringes the western margin of the Nobi Plain, and consists of the Yoro, Kuwana, and Yokkaichi faults from north to south in ascending order, with a total length of about 55 km (Research Group for Active Faults of Japan, 1991; Sugai et al., 1999; Ikeda et al., 2002; Figure 1).

The Yoro fault runs along the boundary between the Yoro Mountains and the Nobi Plain. The Yoro Mountains consist of Mesozoic and Paleozoic sedimentary rocks (Takada et al., 1979). The Neogene and Quaternary strata with a thickness of over 1500 m are reported to be beneath the Nobi Plain (Sugai and Sugiyama, 1998).

Thus, the Yoro fault forms the geomorphic and geologic western boundary of the Nobi Plain (Kuwahara, 1968; Ishiyama et al., 2007). Sugai et al. (1999) and Sugai (2011) suggested that slip on the Yoro fault generated both the AD 745 Tempyo (Usami, 1996) and AD 1,586 Tensho (Iida, 1987) earthquakes on the basis of vertical offsets of depositional surfaces in shallow marine sediments. The Kuwana and Yokkaichi faults displace middle to late Pleistocene terrace surface (Kaizuka, 1950; Matsuda et al., 1977; Okada, 1979; Research Group for Active Faults of Japan, 1981, 1991; Ota and Sangawa, 1984) and are indicated to displace the Holocene terrace (Awata and Yoshida, 1991).

It is also pointed out that the Kuwana fault generated both the Tempyo earthquake in AD 745 (Usami, 1996) and Tensho earthquake in AD 1586 (Iida, 1987) on the basis of the accumulation of vertical offset shown by the depositional surfaces of shallow marine sediments (Sugai et al., 1998). Detailed analyses of changes in the accumulation rate of inner bay deposits at the Kuwana fault (Naruhashi et al., 2004, 2008) indicate that the fault has been active five times between 7,000 and 2,000 yr BP (ca. 6,600, 5,700, 4,000, 3,600, and 2,100 yr BP).

The accumulation pattern of vertical displacement during the past 7,000 years and timing of the last three faulting events of the Yokkaichi fault are in agreement with those of the Kuwana fault (Ogami and Sugai, 2006), even though the Yokkaichi fault has a right-stepping relationship to the Kuwana fault (Figure 1).

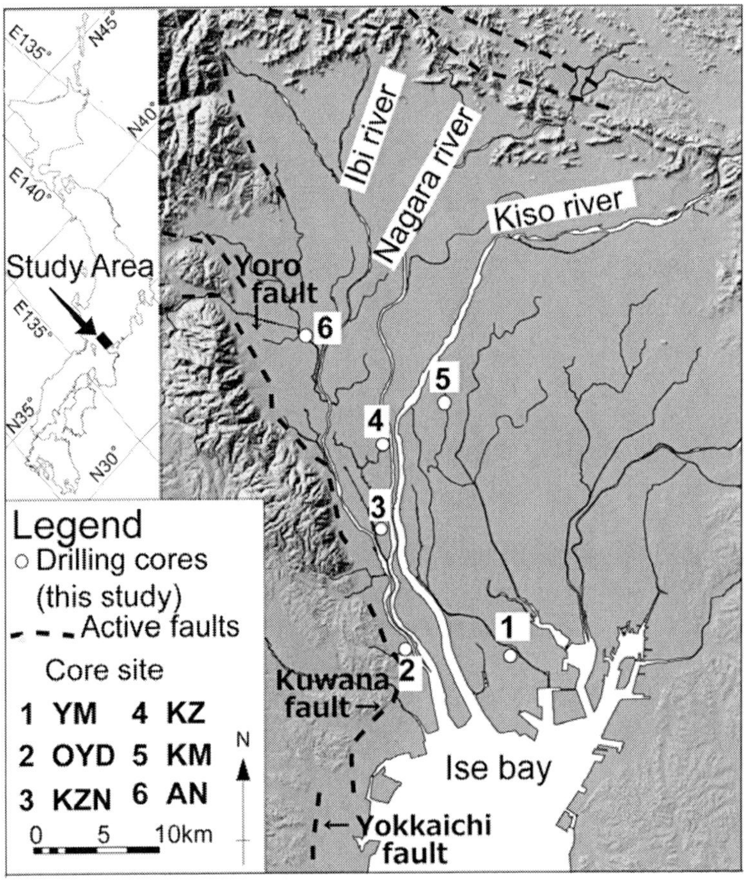

Figure 1. Map showing the location of Yoro, Kuwana and Yokkaichi faults.

Stratigraphy and Sedimentary Facies of the Latest Pleistocene–Holocene Deposits of the Nobi Plain

The Nobi Plain is located on the footwall side of the Yoro fault system (Figure 1). The Nobi Plain tilts down to the west as a result of activity on the Yoro fault system (Kuwahara, 1968), tilting and subsiding at the mean rates of 0.86×10^{-4} /ka and 1 m/ka, respectively in the southwestern part of the plain (Sugai and Sugiyama, 1999).

The Holocene deposits are thick on the plain (maximum thickness of the latest Pleistocene–Holocene deposits on the plain is more than 50 m; Yamaguchi et al., 2006a) because of the postglacial marine transgression and huge amount of sediments supplied by the Kiso River from the central mountain range of Japan, that has the world's highest denudation rate (Ohmori, 1983). The stratigraphy and sedimentary facies of the latest Pleistocene-Holocene deposits in this plain have been studied, and depositional environment and process of delta formation have been reconstructed (Iseki, 1962; Umitsu, 1992; Masuda and Iwabuchi, 2003; Yamaguchi et al., 2003; Ogami et al., 2009). Yamaguchi et al. (2003) and Ogami et al. (2009) divided the latest Pleistocene–Holocene sequence into five sedimentary units; 1) braided river deposits, 2) fluvial to intertidal deposits, 3) inner bay to prodelta deposits, 4) delta front deposits, and 5) delta plain and fluvial deposits in ascending order. These studies also clarified process of delta progradation and estimated progradation rate of the delta at 5 to 10 m/yr during the last 6,500 years. Yamaguchi et al. (2006b) have discussed the process of formation of the uppermost part of the sequence (delta plain and fluvial deposits).

Significance of the Nobi Plain as Off-Fault Paleoseismic Study Site

As described above, the Nobi plain is a coastal plain where environmental changes from terrestrial environment to marine environment can take place in response to the relative sea-level rise. The formation of the Holocene delta sequence is reconstructed with high temporal resolution based on analyses of many sediment cores and over 100 ^{14}C ages (Yamaguchi et al., 2003; Ogami et al., 2009). Since the normal formation process of the Holocene sediments is explained as mentioned above, evidence for sudden events caused by coseismic subsidence (e.g., subsidence of terrestrial sediment under the sea-level, and temporal deepening in the shallow marine) can be recorded in the Holocene delta sequence. Additionally, the timing of the faulting on the Yoro fault system are estimated based on the on-fault paleoseismological study (e.g., Sugai et al., 1999; Naruhashi et al., 2008) which can enable us to understand probability of the results obtained from off-fault paleoseismological study. In this regard, the Nobi Plain is suitable as an example of the off-fault paleoseismological study area.

Materials and Methods

We used lithofacies and grain-size data, electrical conductivity (EC), C/N ratio and radiocarbon ages for sediments from the uppermost 20 m of each of the six sediment cores drilled in the Nobi Plain.

We have described grain size, sedimentary structure and existence or nonexistence of bioturbation, shells and plant fragments. We measured grain size distribution at depth intervals of 5-10 cm in sediments of the four cores (YM, KZN, KM, and AN) by using a laser diffraction particle-size analyzer (SALD-3000S; SHIMADZU Corporation, Kyoto, Japan). The sediment samples used in the grain size analysis were obtained from the parts of core sediments which show typical lithofacies without bioturbation. For OYD and KZ cores, we used grain size data measured under the same condition, following Yamaguchi et al. (2005).

We measured EC at depth intervals of 5-200 cm in sediments of four cores (YM, KZN, KM, and AN) based on the method described in Yokoyama and Sato (1987). The samples were first parched at 110 °C for 48 hr and then milled in a vibrating sample mill (TI-100; CMT Co. Ltd, Tokyo, Japan). Afterward, 5 g of each sample was stirred with 60 ml of distilled water in a sample vial. The samples were stored at room temperature for 7 days in the laboratory. The EC of samples was measured with a conductivity meter (Cond Meter ES-51; Horiba, Kyoto, Japan) 7 days after the samples had been mixed with the distilled water. For the OYD and KZ cores, we used EC data measured under the same condition following Yamaguchi et al. (2001). For the KZN core, C/N ratio was calculated as ratio of total organic carbon (TOC) to total nitrogen (TN). TOC and TN were measured by elemental analyzer (FLASH EA1112; Thermo Finnigan), after perching at 110 °C for 48 hr and milling.

A total of 46 radiocarbon ages on fossil molluscan shells and wood fragments were obtained from the six cores by accelerator mass spectrometry (AMS) method (including ages obtained by Yamaguchi et al., 2003) and Ogami et al., 2009). The calendar years were calibrated with Calib version 6.0 (Stuiver and Reimer, 1993; software MARINE09; Reimer et al., 2009) for fossil molluscan shells and INTCAL09 (Reimer et al., 2009) for wood fragments respectively. The local reservoir effect for the marine samples was not calculated. Because of this, ΔR is defined as 0, following the recent study in this area (e.g., Hori et al., 2008; Naruhashi et al., 2008).

We identified sedimentary units and interpreted depositional environments based on the lithofacies and EC. Thereafter, for detection of coseismic subsidence, we discussed the reason of temporal facies changes which cannot be explained by normal depositional process. Further, we discussed about paleoseismicity based on the correlation of detected coseismic subsidence events with previously known faulting events. In the following sections, we refer to the results of Niwa et al. (2009, 2010, 2012).

Sedimentary Unit Segmentation and Depositional Environments

Based on the lithofacies and EC, the sedimentary unit was divided and depositional environment was estimated. On the basis of the interpretations by Niwa et al. (2011), we made a comparative review of EC with the facies results and diatom analysis. It is interpreted that the EC lower than 0.4 mS/cm represent the fresh water environment, EC higher than 0.4 mS/cm and lower than 0.9 mS/cm the brackish water environment, and EC higher than 0.9 mS/cm the marine water environment respectively. For our interpretation of the origin of organic material, we followed Kumon et al. (2003). That is, for sediments with C/N ratios of about 6, we considered the origin of organic material to be plankton; for those with C/N ratios higher than 15, we considered the origin of the organic material to be terrestrial higher plants. Figures 2 - 7 show core logs, sedimentary units, median grain size, sorting, EC, C/N ratio and calendar ages.

Prodelta Mud Deposits

The prodelta mud deposits (PD in Figures 2-4, 6-7) is apparent in the core depth of 20 – 16.64 m in the YM core, 20 – 15.25 m in the OYD core, 20 – 14.43 m in the KZN core, 20 – 18 m in the KM core and 20 – 17.80 m in the AN core respectively. This unit comprises structureless mud. The median grain size and sorting is 5.6 to 7.0 ϕ and 1.6 to 2.1 respectively. The EC is higher than 1.1 mS/cm. For the core KZN, the C/N ratio is 9 to 13. This unit has been interpreted as prodelta mud by Yamaguchi et al. (2003) and Ogami et al. (2009) on the basis of its fine and homogeneous grains and marine shells. EC greater than 1.1 mS/cm, which suggests a marine environment supports this interpretation. The calendar ages of this unit were 3,314–3,542 cal BP (core OYD), 2,682–2,845 cal BP (core KZN), and 6,736–7,086 cal BP (core AN), respectively.

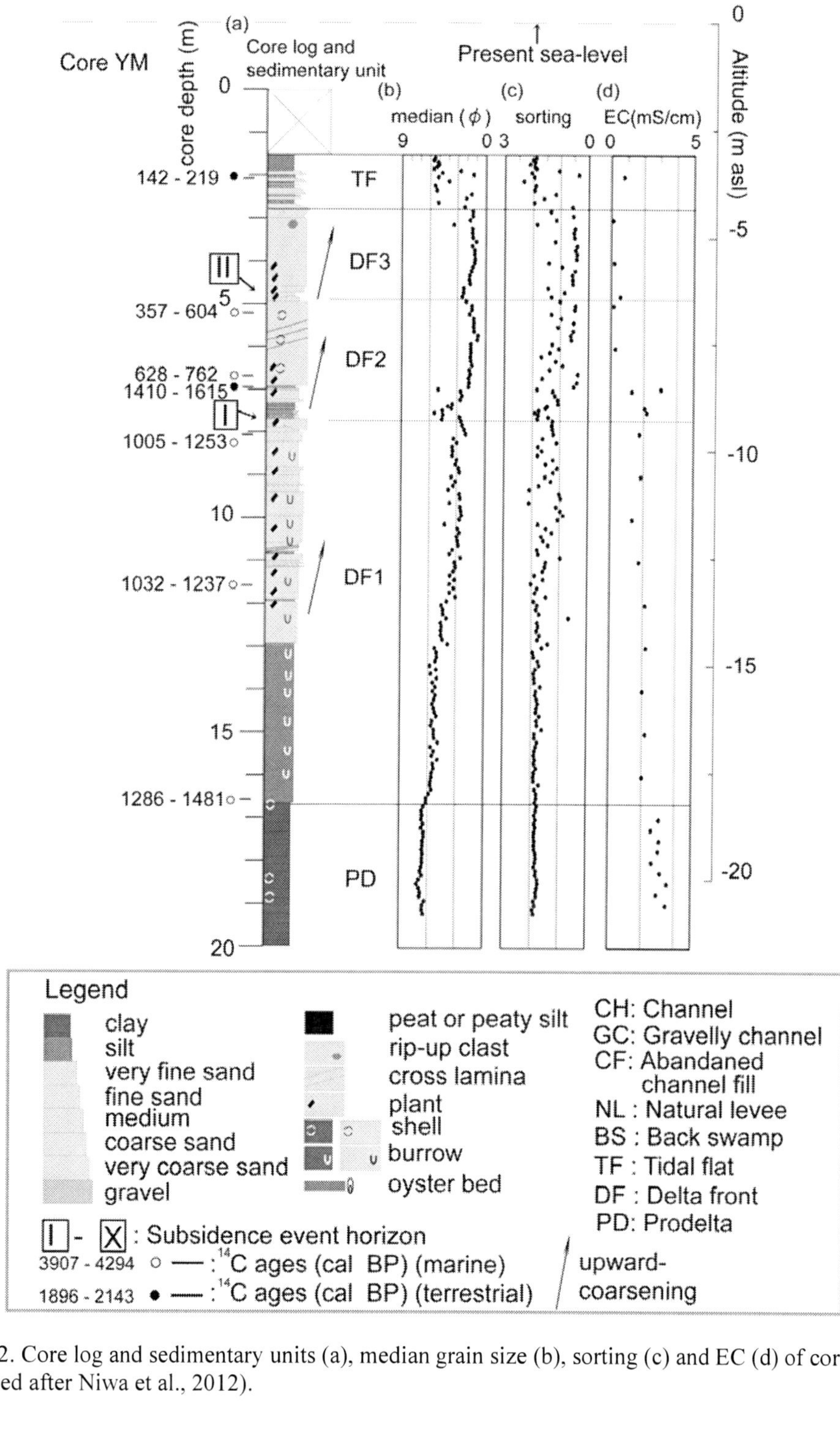

Figure 2. Core log and sedimentary units (a), median grain size (b), sorting (c) and EC (d) of core YM (modified after Niwa et al., 2012).

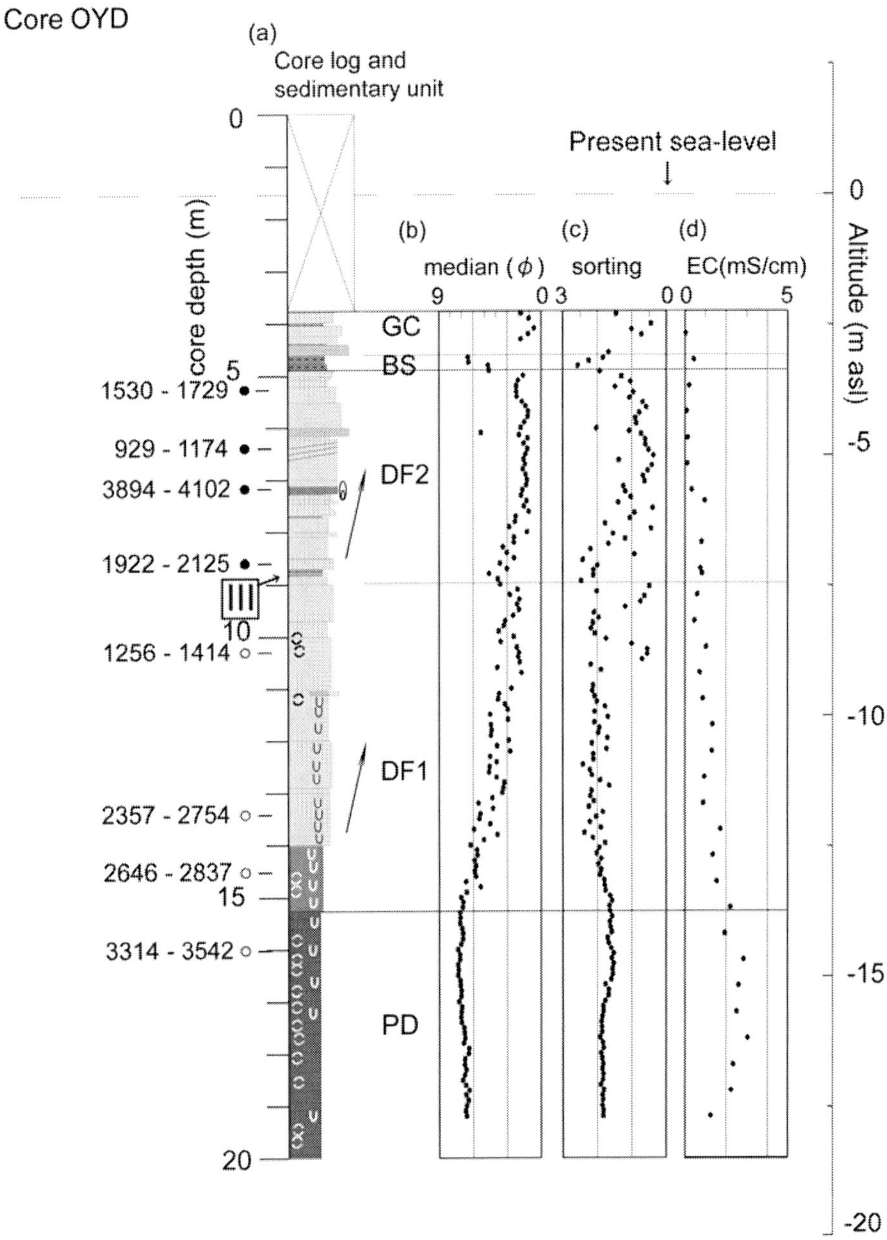

Figure 3. Core log and sedimentary units (a), median grain size (b), sorting (c) and EC (d) of core OYD (modified after Niwa et al., 2012). Legend is shown in Figure 2.

Delta Front Sandy Silt and Sand Deposits

The delta front sandy silt and sand deposits (DF in Figures 2 – 7) is apparent in the core depth of 16.64 – 2.80 m in the YM core, 15.25 – 4.90 m in the OYD core, 14.43 – 4.47 m in the KZN core, 20 – 7.90 m in the KZ core, 18.00 – 7.89 m in the KM core and 17.80 – 9.19 m in the AN core respectively.

This unit coarsens upward from sandy silt to sand. The boundary with the underlying prodelta mud deposits is sharp but not erosional. The median grain size and sorting change from 6 ϕ to 1 ϕ, from 1.8 – 2.2 to 0.5 respectively with coarsening upward. The bioturbation is apparent in the lower to middle part of this facies, whereas the plants fragments and cross laminations are apparent in the middle to upper part.

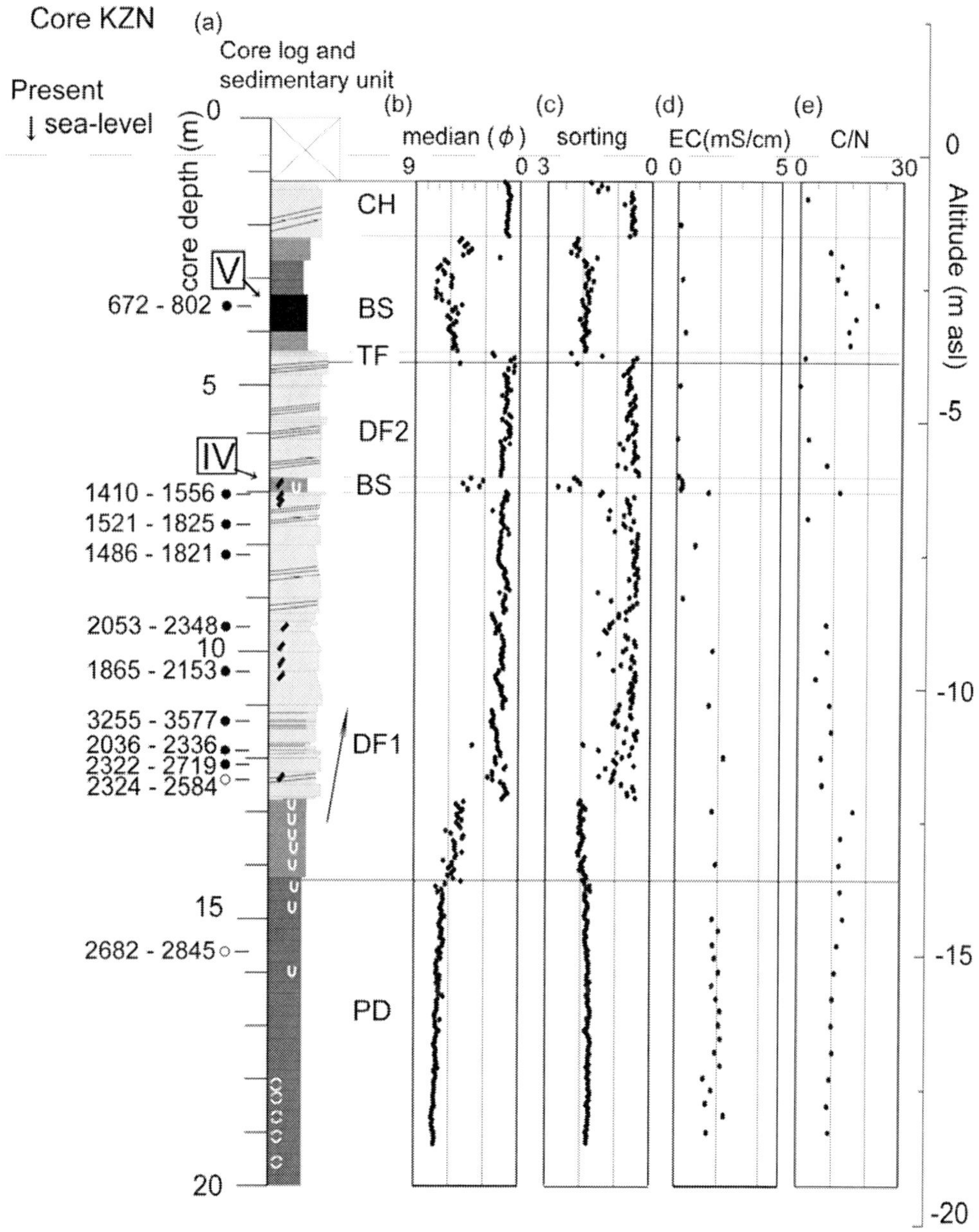

Figure 4. Core log and sedimentary units (a), median grain size (b), sorting (c), EC (d) and C/N ratio (e) of core KZN (modified after Niwa et al., 2012). Legend is shown in Figure 2.

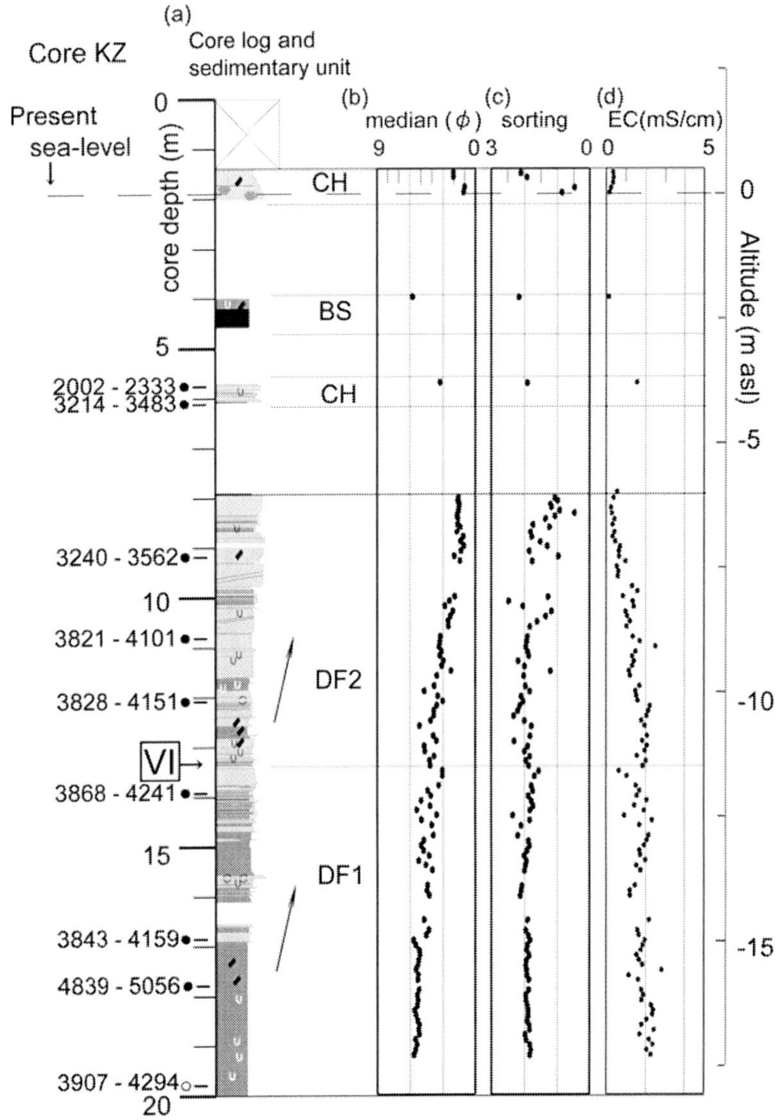

Figure 5. Core log and sedimentary units (a), median grain size (b), sorting (c) and EC (d) of core KZ (modified after Niwa et al., 2012). Legend is shown in Figure 2.

EC is 0.3–3.0 mS/cm in core YM, 0.1–2.2 mS/cm in core OYD, 0.05–2.2 in core KZN, 0.3–2.4 mS/cm in core KZ, 0.5–1.9 mS/cm in core KM and 0.2–1.6 mS/cm in core AN. Broadly, EC decreases upwards. For core KZN, C/N is 0.2–16.0. The temporal fining and increase in EC is identified (I, II, Figure 2; III, Figure 3; VI, Figure 5; VII, Figure 6; IX, X, Figure 7). In the core depth of 7.00 to 6.74 m in KZN core, very fine sand to silt layer with median grain size and sorting of 3.2 ϕ to 4.9 ϕ and 2.1 to 2.7 overlays the medium sand layer with median grain size and sorting of 1.0 to 1.5 ϕ and 0.5 to 1.5. Above the boundary at 6.74 m depth (Figure 4 – IV), medium sand layer with median grain size and sorting of 1.5 to 1.6 ϕ and 0.4 to 1.0 was again deposited. In the fining upward strata at 7.00–6.74 m depth in core KZN, EC is 0.2–0.3 mS/cm.

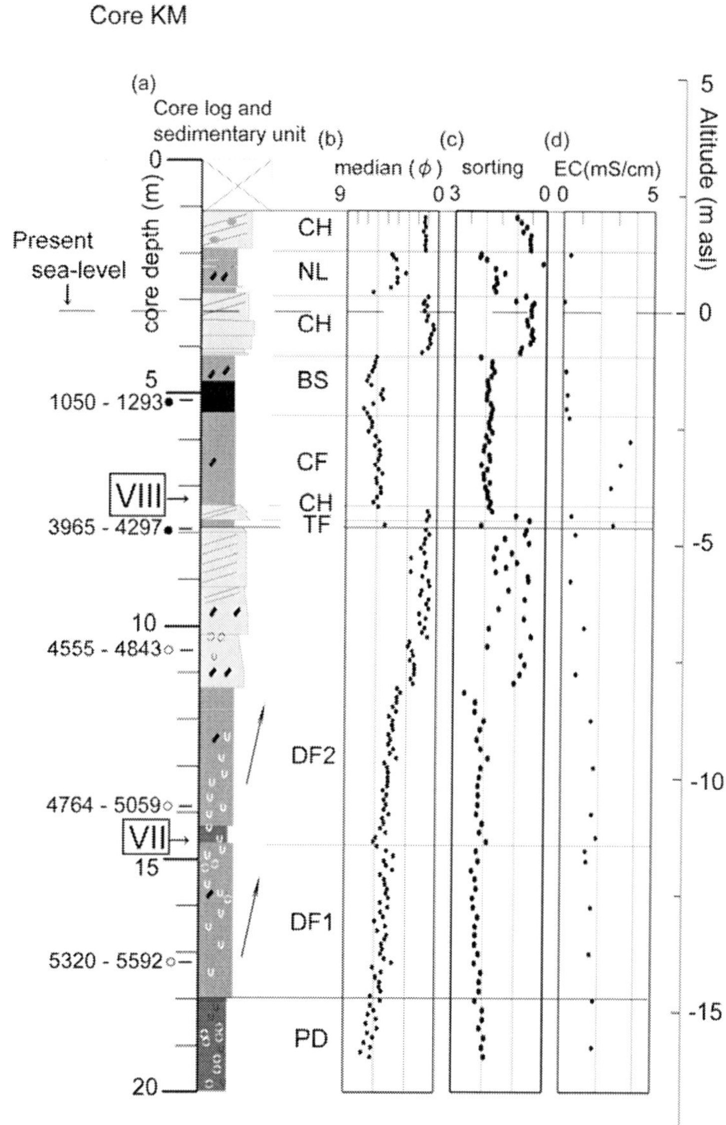

Figure 6. Core log and sedimentary units (a), median grain size (b), sorting (c) and EC (d) of core KM (modified after Niwa et al., 2012). Legend is shown in Figure 2.

This unit was previously interpreted as delta front sandy silt and sand (Yamaguchi et al., 2003; Ogami et al., 2009) because it conformably overlies prodelta mud deposits and coarsens upwards. The upward decreasing trend of EC and the appearance of plant fragments in the middle to upper part of this unit support an upward shallowing environment. We divided this facies into three upward coarsening sub-units (DF1, DF2, and DF3; Figures 2–7), separated by the fining upward strata described above (sub-facies boundaries I, II, III, VI, VII, IX, and X in Figures 2 – 7). On the basis of the dominance of fresh water diatom species (Saegusa et al., 2011) and low EC, we interpreted the fining upward strata at 7.00–6.74 m depth in core KZN as back swamp deposits (BS in Figure 4). We ascribed the sub-units below and above the fining upward strata as DF1 and DF2, respectively.

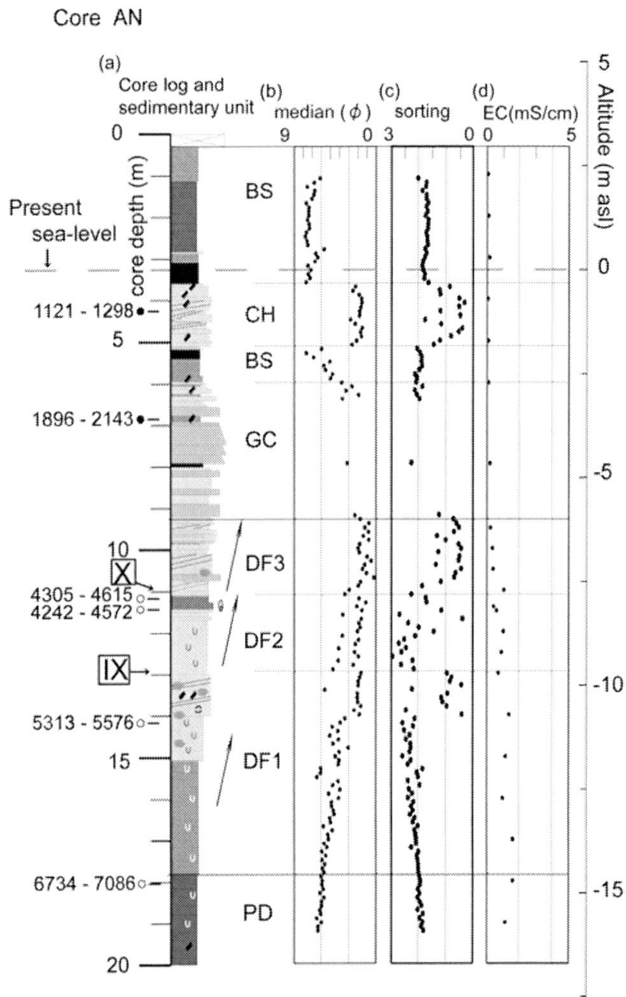

Figure 7. Core log and sedimentary units (a), median grain size (b), sorting (c) and EC (d) of core AN (modified after Niwa et al., 2012). Legend is shown in Figure 2.

Calendar ages of this unit were 1,286–,481 to 357–604 cal BP (YM), 2,646–2,837 to 929–1,174 cal BP (OYD), 2,324–2,584 to 1,410–1,556 cal BP (KZN), 3,907–4,294 to 3,240–3,562 cal BP (KZ), 5,320–5,592 to 3,965–4,297 cal BP (KM) and 5,313–5,576 to 4,242–4,572 cal BP (AN) respectively. Calibrated ages obtained from the core depth of 6.96 m in YM, from the core depth of 8.61, 7.18, and 5.28 m in OYD, from the core depth of 11.32 and 9.55 m in KZN, and from the core depth of 17.80 m in KZ show 200 to 1200 years older than those obtained in the lower layers. Thus, we interpreted these samples (wood or plant fragment) as reworked.

Delta Plain and Fluvial Sand and Silt Deposits

The delta plain deposits are apparent in the core depth of 2.80 – 1.51 m in the YM core, 4.90 – 3.71 m in the OYD core, 4.47 – 1.18 m in the KZN core, 7.90 – 1.40 m in the KZ core,

7.89 – 1.13 m in the KM core, and 9.19 – 0.35 m in the AN core, respectively. This unit comprises silt layers with organic material and fine to medium cross-laminated sand layers. EC is lower than 0.8 mS/cm except for a few exceptions (1.6 mS/cm in the core depth of 5.70 m of KZ, and 2.6 to 3.7 mS/cm in the core depth of 7.80 to 6.00 m of KM). For the KZN core, C/N ratio is 2.0 to 22.5.

This unit was previously interpreted as delta plain and fluvial sand and silt facies (Yamaguchi et al., 2003; Ogami et al., 2009) because it overlies delta front sandy silt and sand deposits. This interpretation is supported by the generally lower EC which suggests a fresh to brackish water environment. We divided this unit into six sub-units on the basis of lithofacies and EC.

Tidal Flat Deposits

The tidal flat deposits (TF in Figures 2, 4, 6) are evident in the core depth of 2.80 – 1.51 m in the YM core, 4.47 – 4.36 m in the KZN core and 7.89 – 7.74 m in the KM core. This sub-unit overlies the delta front deposits and comprises coarse sand and silt. Median grain size is vertically- varying largely from 0.5 to 5.2 ϕ.

The sorting is 0.4 to 2.2. EC is 0.8 to 2.8 mS/cm. Yamaguchi et al. (2006b) reported that the TF overlies delta front deposits and its grain size varies greatly with depth. Thus, we interpreted this sub-unit as tidal flat deposits. This interpretation is supported by the EC (> 0.8 mS/cm), suggesting a brackish to marine environment.

Sandy Channel Deposits

The sandy channel deposits (CH in Figures 4-7) is clear in the core depth of 2.23 – 1.18 m in the KZN core, 6.08 – 5.70 m and 2.00 – 1.40 m in the KZ core, 7.74 – 7.41 m, 4.20 – 2.86 m, and 1.91 – 1.13 m in the KM core, and 5.12 – 3.65 m in the AN core. This sub-unit comprises an upward fining succession of fine to coarse sand layers. Median grain size and sorting is 0.4 to 2.8 ϕ and 0.3 to 1.8 respectively. Several sand layers contain cross laminations and rip-up clasts.

EC is lower than 0.6 mS/cm except for the core depth of 5.70 m of KZ. The upward-fining succession, cross laminations, and rip-up clasts are noticeable in sandy channel deposits (Yamaguchi et al., 2006b). Thus, this sub-unit is interpreted here as sandy channel deposits.

Gravelly Channel Deposits

The gravelly channel deposits (GC in Figures 3, 7) is apparent in the core depth of 4.62 - 3.71 m in the OYD core and 9.19 – 5.94 m in the AN core. This sub-unit is composed of sediment from very fine sand to pebble size. The sand to gravel layers show fining upward and gravel layers are clast-supported. Some sand layers with plant fragment are sandwiched between the gravel layers.

These features are clear in the gravelly channel deposits (Yamaguchi et al., 2006b). Thus, this sub-unit is interpreted as gravelly channel deposits.

Back Swamp Deposits

The back swamp deposits (BS in Figures 3-7) are clear in the core depth of 4.90 – 4.62 m in the OYD core, 4.36 – 2.23 m in the KZN core, 4.58 – 4.00 m in the KZ core, 5.43 – 4.20 m in the KM core, and 5.94 – 5.12 m and 3.65 – 0.35 m in the AN core.

This sub unit is composed of clay to silt layers containing peat and organic material. Median grain size and sorting is 4.8 to 8.0 ϕ and 1.6 to 2.6 respectively. EC is lower than 0.6 mS/cm. This sequence has been previously interpreted as back swamp deposits (Yamaguchi et al., 2006b). The boundary at 3.30 m depth in core KZN (V, Figure 4) marks a facies change from peat to inorganic silt with an abrupt decrease of C/N ratio from 23 to 13.

Abandoned Channel Fill Deposits

The abandoned channel deposits (CF in Figure 6) are noticeable in the core depth of 7.41 – 5.43 m in the KM core. This sub-unit comprises a poorly sorted silt layer overlying sandy channel deposits (separated by boundary VIII in Figure 6).

Median grain size and sorting is 5.6 to 6.7 ϕ and1.8 to 2.1 respectively. This sub-unit contains plant fragments. Sulfur deposition is also obserbed in this sub-unit. EC is 2.6–3.7 mS/cm. The sequence of sandy channel deposits overlain by a poorly sorted silt layer has been interpreted as abandoned channel fill deposits (Yamaguchi et al., 2006b).

Natural Levee Deposits

The natural levee deposits (NL in Figure 6) is evident in the core depth of 2.86 – 1.91 m in the KM core. This sub-unit is represented by a layer of silt to very fine sand. This facies is generally poorly sorted.

Median grain size and sorting is 3.3 to 6.1 ϕ and 1.4 to 2.2 respectively. At 2.20 m depth in core KM, there is a thin coarse sand layer (median ϕ, –0.8) with EC lower than 0.4 mS/cm. The poor sorting and variable grain size were interpreted by Yamaguchi et al. (2006b) as indicative of natural levee deposits. This interpretation is supported by the EC (< 0.4 mS/cm) which suggests a fresh water environment.

Detection of Coseismic Subsidence

Normal Depositional Process of Core Sediments

Schematic geologic cross-section of the western Nobi plain was described based on results of core analyses (Ogami, 2009; Ogami et al., 2009, see Figure 3). This shows that in the Kiso River delta, the delta front reached AN core site at ca. 7,000 cal BP, KM core site at ca. 5,500 cal BP, KZ core site at ca. 4,100 cal BP, KZN core site at ca. 2,800 cal BP, and YM site at ca. 1,300 cal BP. After delta front passed, the fluvial deposits overlay delta front deposits. Broadly, the core sediments show prograding delta sequence (i.e., environmental change from shallow marine to terrestrial). However, in detail, the temporal facies changes cannot be explained by normal delta progradation.

Cause of Temporal Facies Changes which Cannot Be Explained by Normal Depositional Process

During the Deposition of Delta Front Deposits

The upward-coarsening succession in the delta front deposits shows increase in coarse sediment supply because of river mouth approach of the core sites during delta progradation. During the delta progradation process, the temporal fining and EC enhancement is identified (I, II, Figure 2; III, Figure 3; VI, Figure 5; VII, Figure 6; IX-X, Figure 7). The reasons for temporal fining in the delta front deposits are either a temporary increase of the distance of the core site from the sediment source (river mouth), or by an increase in the supply of muddy sediments during flooding. If it is due to the flooding, there should be good amount of the fresh water flow in the core site. However, this cannot be explained by increasing trend in EC (Niwa et al., 2009). Thus, it is likely that the reason for temporal fining was responses to temporary increase in the distance of the core sites from the sediment source (river mouth).

Further, in the core depth of 7.00 to 6.74 m in the KZN core, the fine layer is sandwiched between the sand layers (Figure 4). This layer can be interpreted as BS because of dominance of fresh water diatom species and low EC. However, DF2 which suggests shallow marine environment overlies this layer (i.e., the facies changed from BS to DF2 across the boundary IV in Figure 4), conflicting upward-shallowing. This indicates that the temporal relative sea-level rise occurred after the deposition of the fine sediment. The process by which the characteristic changes of facies, described above may have occurred, is explained below.

In the delta front, the upward-coarsening sediments were deposited, because of the direct approach of the river mouth to the core sites and EC decreased as influence of fresh water became strong (core site 1, Figure 9a). At inland site, the delta front platform became terrestrial back swamp because of the delta progradation (core site 2, Figure 9a).

A temporary rise in relative sea level caused coastline retreat, so that the core site temporarily receded from the river mouth. At seaward site, decreasing of coarse sediment supply and increasing of influence of marine water made grain size fine and increase in EC (core site 1, Figure 9b). At inland site, the back swamp subsided into sea-water because of temporal relative sea-level rise (core site 2, Figure 9b).

After the temporary rise of relative sea level, the delta progradation occurred again. The upward-coarsening sediments (with upward-decreasing EC) were deposited (core site 1, Figure 9c), whereas the delta front platform sand layer was deposited again at the inland site (core site 2, Figure 9c). The episodes of temporal relative sea-level rise identified in core data (labeled I , II, III, IV, VI, VII, IX, and X in Figures 2 – 7) are shown as events I, II, III, IV, VI, VII, IX, and X respectively. Based on the calendar ages obtained from core YM (Figure 2), core OYD (Figure 3), core KZN (Figure 4), core KZ (Figure 5), core KM (Figure 6), core

AN (Figure 7), these events are interpreted to have occurred at 1,300 to 600 yr BP (event I), at 600 to 200 yr BP (event II), at 1,400 to 900 yr BP (event III), at 1,500 to 700 yr BP (event IV), at 4,200 to 3,800 yr BP (event VI), at 5,600 to 4,700 yr BP (event VII), at 5600 to 4200 yr BP (event IX) and at 4600 to 1900 yr BP (event X) respectively.

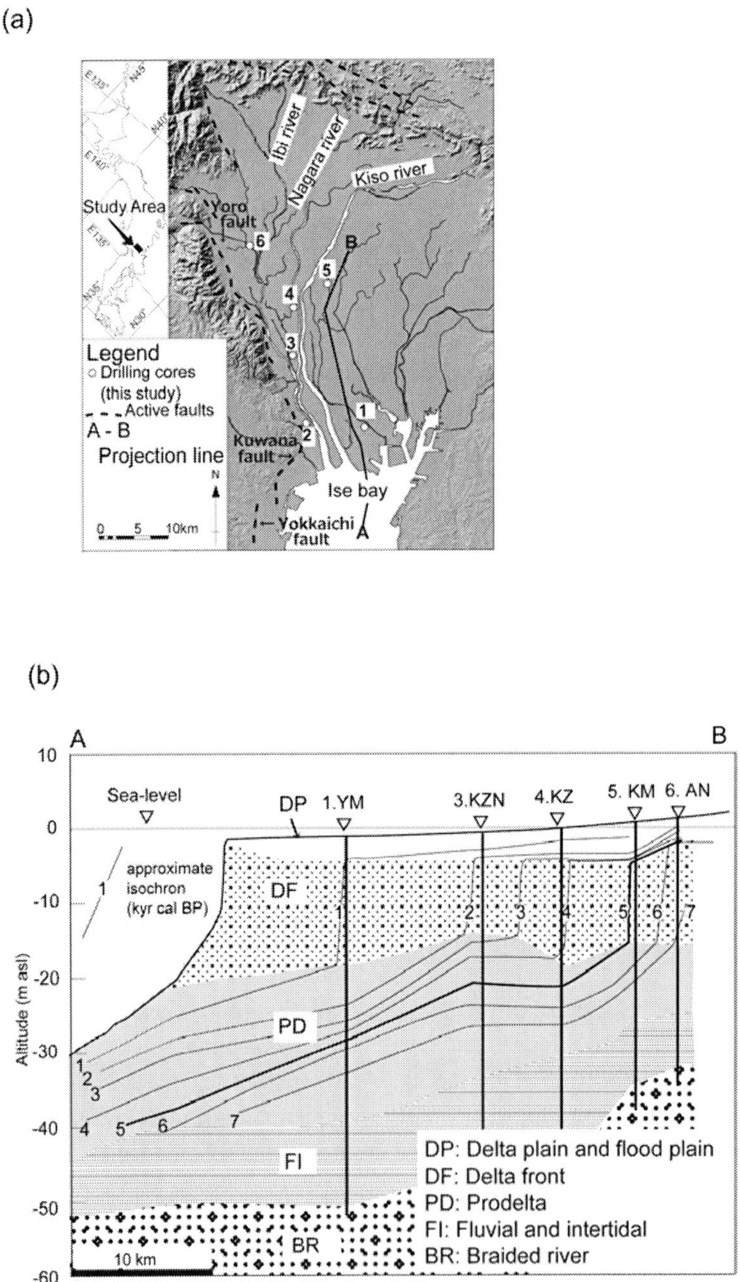

Figure 8. The location of cores and section line (A – B) (a), and schematic geologic cross-section across the western Nobi Plain (b), modified from Ogami et al. (2009).

During the Deposition of Delta Plain and Fluvial Deposits

Abandoned Channel Fill Deposits

The abandoned channel fill deposits (CF) identified at 7.41–5.43 m depth in core KM (VIII, Figure 6) were deposited in the terrestrial environment based on the feature of sedimentary facies, composed of poorly sorted silt layer overlying sandy channel deposits with plant fragments.

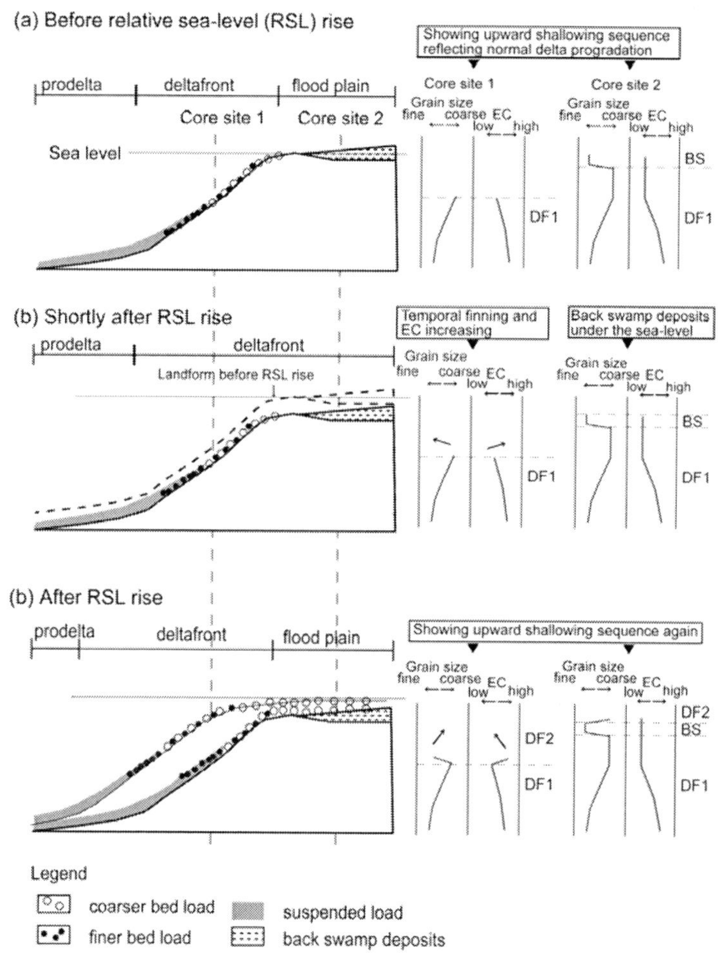

Figure 9. Sedimentary model showing the development of the Kiso River delta during an episode of rapid relative sea-level rise (modified after Niwa et al., 2012), (a) before relative sea-level (RSL) rise, (b) shortly after RSL rise, (c) after RSL rise.

On the basis of a comparison of EC with the facies characteristics and diatom analysis (Niwa et al., 2011), the fine-grained terrestrial sediment would show low EC value (i.e., less than 0.9 mS/cm), if a terrestrial environment continued without episodes of marine water invasion. However, EC is 2.6 to 3.7 mS/cm, suggesting a marine water environment. The process by which a temporary increase in EC may have occurred is described below. After the

KM core site became terrestrial due to delta progradation, the change in river channel took place and the channel located at the KM core site was abandoned (i.e., the facies changed from CH to CF; Figures 6, 10a). At that time, the paleo sea-level is expected to the altitude of TF (i.e., below facies CH). As the temporal relative sea-level rise occurred, the marine water may have invaded the abandoned channel. Presumably, this marine water invasion made EC of CF deposits high (Figures 6, 10b). Consequently, the sediments with high EC value were deposited until the KM core site became a terrestrial environment (Figure 10c).

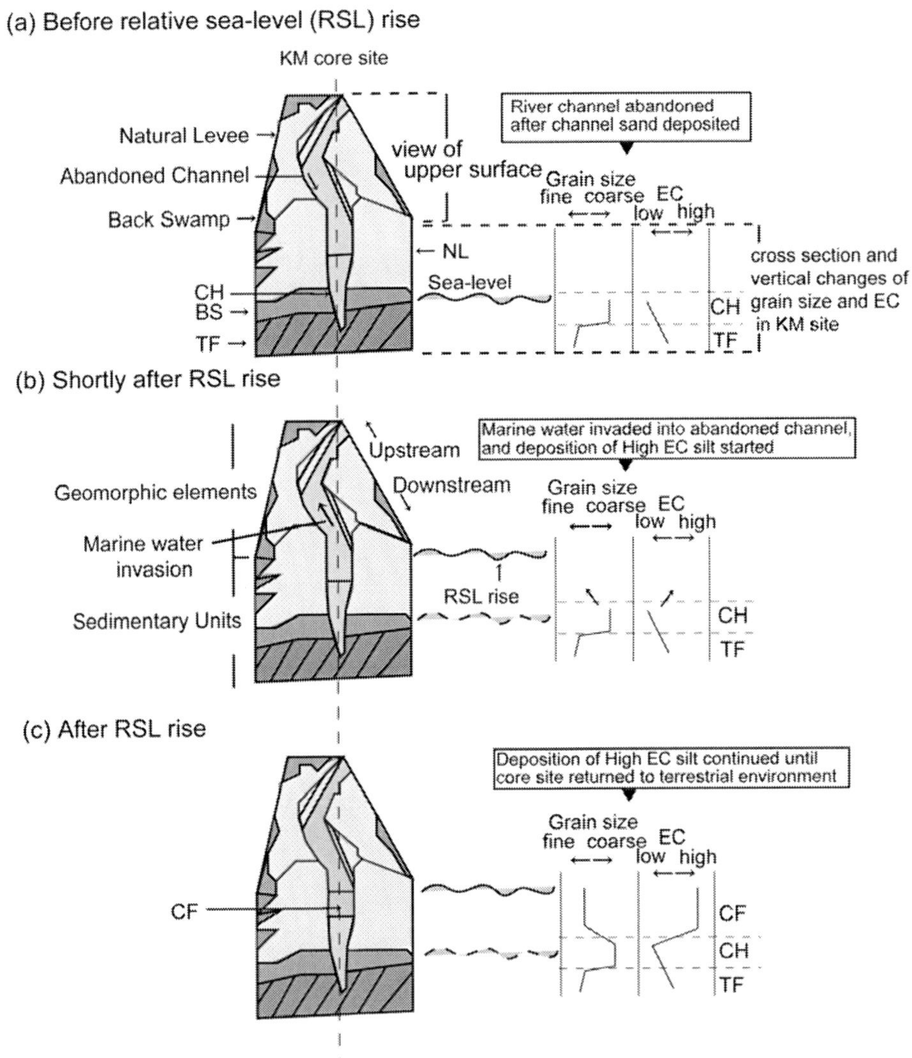

Figure 10. Sedimentary models explaining high EC value in the abandoned channel fill deposits during an episode of rapid relative sea-level rise (modified after Niwa et al., 2012), (a) before relative sea-level (RSL) rise, (b) shortly after RSL rise, (c) after RSL rise.

This episode of temporal relative sea-level rise identified in core data (labeled VIII; Figure 6) is shown as event VIII. Based on the calendar ages obtained from core KM (Figure 6), this event is interpreted to have occurred at 4,300 to 1,000 yr BP.

Back Swamp Deposits

A sudden change from peat layer to inorganic silt layer identified in BS in the core depth of 3.30 m of KZN (V, Figure 4) is accompanied by sudden lowering of C/N ratio, indicating that the origin of organic material was decreased. The amount of production of plant residue may have suddenly decreased in the back swamp. One basis for sudden decrease in production of plant residue is that water level increase in the back swamp, accompanied by land subsidence may have caused removal of vegetation, and the other reason is that back swamp was covered by mud due to flooding (Niwa et al., 2009). This episode of temporal water level rise in the wetland identified in core data (labeled V; Figure 4), is shown as event V. Based on the calendar ages obtained from core KZN (Figure 4), this event is interpreted to have occurred between 800 yr BP and present.

Coseismic Subsidence as a Cause of Detected Relative Sea Level Rise Events

The timings of relative sea-level rise (or water level rise) events I to X are shown in a schematic time–space diagram (Figure 11) across the delta plain in the region of the Yoro and Kuwana faults. The illustration indicates that the timing of event II corresponds to that of event V. The timing of event I corresponds to those of events III and IV. Further, the timing of event VII corresponds to the event IX. These data suggest that the episodes of relative sea-level rise on the western Nobi Plain occurred at 600-100 yr BP, 1,300-900 yr BP, 4,200 to 3,800 yr BP and 5,600-4,700 yr BP.

Considering that the relative sea-level rise events described above occurred on the footwall side of the active fault during the middle to late Holocene, there are four possible causes; (1) eustatic sea-level rise, (2) hydro-isostatic sea-level rise, (3) subsidence due to compaction, and (4) coseismic subsidence. During the middle to late Holocene, the eustatic changes of sea level were minimal or with a slight decline (e.g., Nakada and Lambeck, 1988). In the areas around Japan, hydro-isostasy caused a lowering trend of relative sea-level during the middle to late Holocene (Nakada et al., 1991, 1998; Sato, 2008). Thus, the temporal relative sea-level rise events we detected cannot be explained by eustasy or hydro-isostasy.

Although accurate estimation of influence of land subsidence by compaction on relative sea-level is difficult, the cumulative subsidence over most of the Nobi Plain has been estimated to be only a few centimeters since control of groundwater abstraction commenced in 1985 (Investigation Committees for land subsidence at the Tokai three prefectures, 2009). Thus, the temporal relative sea-level rise events recorded in core data cannot be explained by subsidence due to compaction. Furthermore, the land subsidence due to compaction occurs continuously, and does not explain temporal and rapid rise of relative sea level.

If the origin of relative sea-level rise events is coseismic subsidence, vertical changes in the sedimentary facies and EC, and synchronicity of events can be explained consistently. Thus, it is probable that the cause of relative sea-level rise events is coseismic subsidence. Thus, we believe that four coseismic subsidence events occurred during the middle to late Holocene (events II and V, 600-100 yr BP; events I, III, and IV , 1,300-900 yr BP; event VI, 4,200-3,800 yr BP; events VII and IX, 5,600 to 4,700 yr BP) can be detected from the analyses of sediment cores drilled in the western Nobi Plain.

Activity within the Yoro Fault System Determined from Estimation Age of Coseismic Subsidence

Coseismic subsidence events recorded in the core data in the western Nobi Plain can be attributed to faulting within the Yoro fault system.

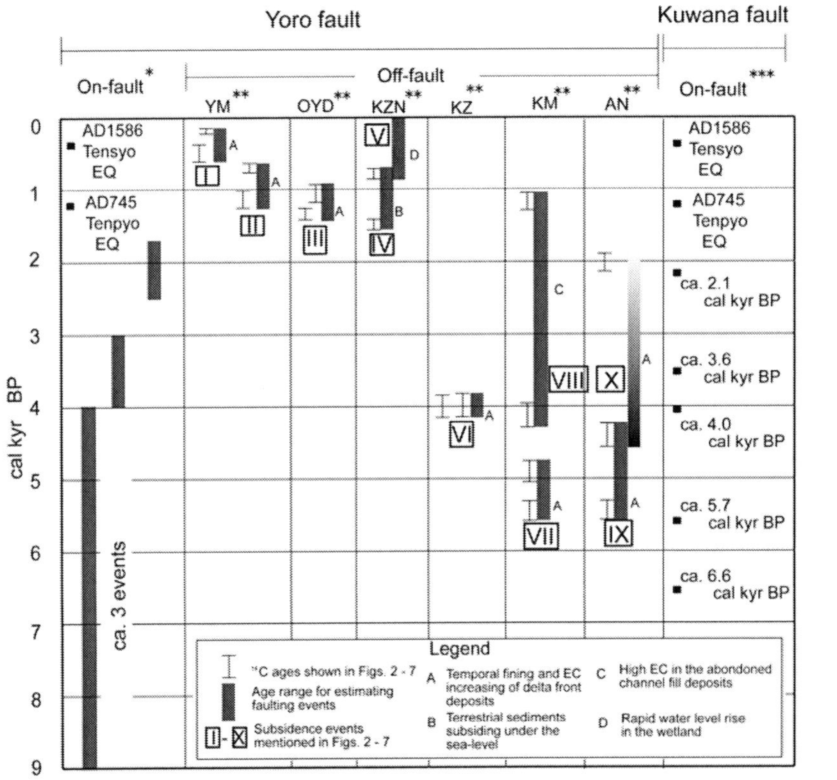

* Evidence for faulting events of Yoro fault identified by Sugai *et al.* (1999) and Sugai (2011)

** Evidence for coseismic subsidence identified by in this study

*** Evidence for faulting events of Kuwana fault identified by Sugai *et al.* (1998) and Naruhashi *et al.* (2004,2008)

Figure 11. Space–time diagram of the Nobi Plain near the Yoro fault system (modified from Niwa et al. (2012).

The detected coseismic subsidence events and on-fault paleoseismic evidence on the Yoro fault system are summarized in Figure 11. In addition to the Tensho (AD 1586) and the Tempyo (AD 745) earthquakes on the Yoro fault, the faulting events at the Yoro fault are estimated to have occurred, once at 2,500-1,700 yr BP, once at 4,000-3,000 yr BP and about thrice at 9,000-4,000 yr BP (Sugai et al., 1999; Sugai, 2011). The Kuwana fault is also estimated to have generated the Tensho earthquake in AD 1586 and the Tempyo earthquake in AD 745 (Sugai et al., 1998). Further, the faulting at the Kuwana fault is estimated to have occurred at ca. 6,600, 5,700, 4,000, 3,600, and 2,100 yr BP (Naruhashi et al., 2004, 2008).

The events II and V occurred at 600-100 yr BP can be correlated with the Tensho earthquake in AD 1586, and the events I, III, and IV occurred at 1,300-900 yr BP, can be connected to the Tenpyo earthquake in AD 745. Both the earthquakes are assumed to be generated both the Yoro and Kuwana faults (Figure 11). The event VI, occurred at 4,200 to

3,800 yr BP, can be correlated with the Yoro fault activity at 4,000-3,000 yr BP or one of the three activities of the Yoro fault at 9,000-4,000 yr BP (Figure 11). The timing of this event corresponds to the activity on the Kuwana fault about 4,000 yr BP (Figure 11).

Events VII and IX, occurred at 5,600-4,700 yr BP, can be correlated with the Yoro fault activity at one of the three activities of the Yoro fault at 9,000-4,000 yr BP (Figure 11). The timing of these events corresponds to activity on the Kuwana fault about 5,600 yr BP (Figure 11). Since the age range of events VIII and X is large, the correlation of these events with previously reported faulting events is not possible.

The timing of the four coseismic subsidence events estimated to have occurred at different times is consistent with that of the Yoro fault faulting. Thus, it is probable that the cause of coseismic subsidence is induced by the Yoro fault activity. Besides, events II and V (occurred at 600-100 yr BP), events I, III, and IV (at 1,300- 900 yr BP), event VI (at 4,200-3,800 yr BP), and events VII and IX (at 5,600-4,700 yr BP ago) can be correlated with the Kuwana fault activity. Thus, the coseismic subsidence events recorded in the core sediments in the western Nobi Plain which are interpreted to be the result of the activity on the Yoro fault corresponds to faulting events on the Kuwana fault, indicating the possibility that these two faults have been synchronously active during the past earthquakes. Based on the activity of these faults during the historical ages, Sugai et al. (1999) has indicated that the Yoro and Kuwana faults constitute a behavioral segment within the Yoro fault system. The results of our off-fault paleoseismological study are consistent with the previous interpretations that the Yoro and Kuwana faults have generated large earthquakes synchronously and comprised a behavioral segment within the Yoro fault system.

CONCLUSION

In this chapter, as an example of off-fault paleoseismological study, our trial to reconstruct paleoseismic history based on the detection and age estimation of coseismic subsidence recorded in the Holocene delta sequence (Niwa et al., 2009, 2010, 2012) was reviewed. Analyses of six sediment cores drilled in the Nobi plain on the footwall (downthrown) side of the Yoro fault system revealed the following.

We identified temporal facies changes which cannot be explained by normal delta progradation. Four episodes of temporal fining and increasing EC in delta front deposits were identified at 600-200, 1,300-900, 4,200-3,800 and 5,600-4,700 yr BP. We also identified a facies change from back swamp to delta front at 1,500-700 yr BP, The abandoned channel deposits with high EC were deposited at 4,300-1,000 yr BP. These findings suggest that there were several temporal relative sea-level rise events in the western part of the Nobi Plain during the past 6,000 years. Additionally, an abrupt change in sediments about 800 yr BP from back swamp peat deposits to inorganic silt was identified. This change was accompanied by a sudden decrease of C/N, suggesting water level rise in the wetland at that time.

These vertical changes in the core sediments which cannot be explained by normal delta progradation indicate that the temporal relative sea-level rise events occurred at 200-600, 600-1,300, 3,800-4,200 and 4,700-5,600 yr BP. These events of the relative sea level were probably caused by the coseismic subsidence, induced by activity on the Yoro fault. The radiocarbon ages are consistent with synchronous activity on the Yoro and Kuwana faults.

Our results of correlation of paleoseismic events are consistent with the previous interpretations that the Yoro and Kuwana faults comprise a behavioral segment within the Yoro fault system. Our results also show that the information of off-fault paleoseismmological study obtained from the Holocene coastal plain deposits can provide further evidence of faulting events, inferred from the on-fault paleoseismic studies.

ACKNOWLEDGMENTS

We thank staff of the Tono Geoscience Center of the Japan Atomic Energy Agency, especially Dr. Osamu Fujiwara (now at Active Fault and Earthquake Research Center, National Institute of Advanced Industrial Science and Technology), Dr. Eiji Sasao, Dr. Ken-ichi Yasue (now at Nuclear Safety Commission of Japan, Cabinet Office, Government of Japan) and Dr. Masayoshi Tajikara (now at Earthquake Research Center, Association for the Development of Earthquake Prediction—for measurements of ^{14}C-AMS ages and helpful comments. We also thank staff of the Natural Environmental Changes Laboratory at The University of Tokyo, especially Dr. Takashi Ogami (now at Chuo University) for fruitful discussions.

REFERENCES

Atwater, B.F., 1992. Geologic evidence for earthquakes during the past 2000 years along the Copalis River, southern coastal Washington. *Journal of Geophysical Research,* 97, B2, 1901–1919.

Awata, Y., Yoshida, F., 1991. Holocene movement of the Kuwana and Yokkaichi faults, northwest of Ise Bay, Central Japan. *Active Fault Research,* 9, 61– 68 (in Japanese).

Burbank, D. W., Anderson, R. S. Anderson., 2001. Paleoseismology. in Tectonic geomorphology. Burbank, D. W., Anderson, R. S. (Eds.), Blackwell Science, Malden, 105–129.

Hori, K., Koide, S., Sugiura, M., 2008. Holocene fluvial deposits of core ST1 and ST2 from the northern part of Nobi plain, central Japan. *The Quaternary Research,* (Japan) 47, 51–56 (in Japanese with English abstract).

Hsieh, M. L., Rau, R. J., 2009. Late Holocene coseismic uplift on the Hua-tung coast, eastern Taiwan: Evidence from mass mortality of intertidal organisms. *Tectonophysics*, 474, 595 – 609.

Iida, K., 1987. Tensyo earthquake. The University of Nagoya Press, Nagoya, 552 p (in Japanese).

Ikeda, Y., Imaizumi, T., Togo ,M., Hirakawa, K., Miyauchi, T., Sato, H., (Eds.), 2002. Atlas of Quaternary Thrust Faults in Japan. Univ. Tokyo Press, Tokyo, 260 pp (in Japanese).

Investigation Committees for land subsidence at the Tokai three prefectures, 2009. Situation of land subsidence in the Nobi plain, central Japan, in 2008. Investigation Committees for land subsidence at the Tokai three prefectures, 71 p (in Japanese).

Iseki, H., 1962. Basic issues for researches of alluvial plains. *Journal of the Faculty of Literature, Nagoya University History*, 24, 51 – 74 (in Japanese).

Ishiyama, T., Mueller, K., Sato, H., Togo, M., 2007. Coseismic fault-related fold model, growth structure, and the historic multisegment blind thrust earthquake on the basement-involved Yoro thrust, central Japan. *Journal of Geophysical Research,* 112,B03S07, doi:10.1029/2006JB004377.

Jankaew, K., Atwater, B. F., Sawai, Y., Choowong, M., Charoentitirat, T., Marin, M. E., Prendergast, A., 2008. Medieval forewarning of the 2004 Indian Ocean tsunami in Thailand. *Nature*, 455, 1228 – 1231.

Kaizuka, S., 1950. Fault topography of western part of Kuwana city, central Japan. *Geographical Review of Japan*, 22, 352 – 356 (in Japanese).

Kelsey, H. M., Nelson, A. R., Hemphill-Haley, E., Witter, R. C., 2005. Tsunami history of an Oregon coastal lake reveals a 4600 yr record of great earthquakes on the Cascadia subduction zone. *Geological Society America Bulletin,* 117, 1009 – 1032.

Knudsen, K. L., Witter, R. C., Garrison-Laney, C. E., Baldwin, J. N., Carver, G. A., 2002. Past earthquake-induced rapid subsidence along the northern San Andreas fault: A paleoseismological method for investigating strike-slip faults. *Bulletin of the Seismological Society of America*, 92, 2612 – 2636.

Kumon, F., Kawai, S., Iniuchi, Y., 2003. Climate changes between 25,000 and 6,000yrsBP deduced from TOC, TN, and fossil pollen analyses of a sediment core from Lake Nojiri, central Japan. *The Quaternary Research,* (japan) 42, 13 – 26 (in Japanese with English abstract).

Kuwahara, T., 1968. The Noobi basin and its fault block movements. *The Quaternary Research,* (Japan) 7, 235 – 247 (in Japanese with English abstract).

Langridge, R., Dissen, R. V., Rhoades, D., Villamor, P., Little, T., Litchfield, N., Clark, K., Clark, D., 2011. Five thousand years of surface ruptures on the Wellington fault, New Zealand: Implications for recurrence and fault segmentation. *Bulletin of the Seismological Society of America*, 101, 2088 – 2107.

Liu, J., Klinger, Y., Sieh, K., Rubin, C., 2004. Six similar sequential ruptures of the San Andreas fault, Carrizo Plain, California. *Geology,* 32, 649 – 652.

Liu, J., Klinger, Y., Sieh, K., Rubin, C., Seitz, G., 2006. Serial ruptures of the San Andreas fault, Carrizo Plain, California, revealed by three-dimensional excavations. *Journal of Geophysical Research,* 111, B02306, doi:10.1029/2004JB003601.

Machette, M. N., Personius, S. F., Nelson, A. R., 1992. Paleoseismology of the Wasatch Fault zone: a summary of recent investigation, interpretations, and conclusions. in Assessment of regional earthquake hazards and risk along the Wasatch Front, Utah Gori, P. L., Hays, W. W. eds, United States Geological Survey Professional Paper 1500, A1 – A30.

Masuda, F., Iwabuchi, Y., 2003. High-accuracy synchronism for seismic reflectors and ^{14}C ages: Holocene prodelta succession of the Kiso River, central Japan. *Marine Geology,* 199, 7 – 12.

Matsuda, T., Ota, Y., Okada, A., Shimizu, F., Togo, M., 1977. Aerial photo-interpretation of active faults —the individual difference and examples—. *Bulletin of Earthquake Research Institute*, 52, 461 – 496 (in Japanese with English abstract).

McCalpin, J., Nelson, A. R., 1996. Introduction to paleoseismology. In: McCalpin, J. (Eds.), Paleoseismology, Academic Press, San Diego, 1 – 28.

McCalpin, J. P., Nishenko, S. P., 1996. Holocene paleoseismicity, temporal clustering, and probabilities of future large (M > 7) earthquakes on the Wasatch fault zone, Utah. *Journal of Geophysical Research*, 101, B3, 6233 – 6253.

Nakada, M., Lambeck, K., 1988. The melting history of the late Pleistocene Antarctic ice sheet. *Nature,* 333, 36 – 40.

Nakada, M., Yonekura, N., Lambeck, K., 1991. Late Pleistocene and Holocene sea-level changes in Japan: Implications for tectonic histories and mantle rheology. *Palaeogeography Palaeoclimatology Palaeoecology*, 85, 107 – 122.

Nakada, M., Okuno, J., Yokoyama, Y., Nagaoka, S., Takano, S., Maeda, Y., 1998. Mid-Holocene underwater Jomon sites along the west coast of Kyushu, Japan and asthenospheric viscosity. *The Quaternary Research*, (Japan) 37, 315 – 327.

Naruhashi, R., Sugai, T., Fujiwara, O., Awata, Y., 2004. Recurrence interval of the Kuwana fault reconstructed from Holocene synfaulting shallow marine sediments on the western margin of the Nobi plain, central Japan. *The Quaternary Research*, (Japan) 43, 317 – 330 (in Japanese with English abstract).

Naruhashi, R., Sugai, T., Fujiwara, O., Awata, Y., 2008. Detecting vertical faulting event horizons from Holocene synfaulting in shallow marine sediments on the western margin of the Nobi plain, central Japan. *Bulletin of the Seismological Society of America,* 98, 1447 – 1457.

Nelson, A. R., Shennan, I., Long, A. J., 1996. Identifying coseismic subsidence in tidal-wetland stratigraphic sequences at the Cascadia subduction zone of western North America. *Journal of Geophysical Research*, 101, B3, 6115 – 6135.

Nelson, A. R., Asquith, A. C., Grant, W. C., 2004. Great earthquakes and tsunamis of the past 2000 years at the Salmon River Estuary, Central Oregon Coast, USA. *Bulletin of the Seismological Society of America,* 94, 1276 – 1292.

Nelson, A. R., Sawai, Y., Jennings, A. R., Bradley, L. A., Gerson, L., Sherrod, B. L., Sabean, J., Horton, B. P., 2008. Great-earthquake paleogeodesy and tsunamis of the past 2000 years at Alsea Bay, central Oregon coast, USA. *Quaternary Science Reviews*, 27, 747 – 768.

Nelson, A. R., Kashima, K., Bradley, L. A., 2009. Fragmentary evidence of great-earthquake subsidence during Holocene emergence, Valdivia Estuary, south central Chile. *Bulletin of the Seismological Society of America,* 99, 71 – 86.

Nichol, S. L., Goff, J. R., Devoy, R. J. N., Changue-Goff, C. Hayward, B., James, I., 2007. Lagoon subsidence and tsunami on the West Coast of New Zealand. *Sedimentary Geology*, 200, 248 – 262.

Niwa, Y., Sugai, T., Ogami, T., Tajikara, M., Yasue, K., Saito, T., Fujiwara, O., 2009. Evidence for late Holocene coseismic subsidence identified in drilling cores along the Yoro fault system, western margin of the Nobi plain, central Japan. *The Quaternary Research*, (Japan) 48, 339 – 349 (in Japanese with English abstract).

Niwa, Y., Sugai, T., Ogami, T., Tajikara, M., Yasue, K., Fujiwara, O., 2010. Holocene activity of the Yoro fault system, western margin of the Nobi Plain, central Japan, detected by rapid relative sea-level rise, recorded in 20 m-long drilling cores. *Journal of Geography*, 119, 668 – 682 (in Japanese with English abstract).

Niwa, Y., Sugai, T., Saegusa, Y., Ogami, T., Sasao, E., 2011. Use of electrical conductivity to analyze depositional environments: Example of a Holocene delta sequence on the Nobi Plain, central Japan. *Quaternary International*, 230, 78 – 86.

Niwa, Y., Sugai, T., Yasue, K., 2012. Activity of the Yoro fault system determined from coseismic subsidence events recorded in the Holocene delta sequence of the Nobi Plain, central Japan. *Bulletin of the Seismological Society of America,* 102, doi: 10.1785/0120110211.

Obermeier, S. F., 2009. Using liquefaction-induced and soft-sediment features for paleoseismic analysis. In: McCalpin, J. (Eds.), *Paleoseismology,* (second edition), Academic Press, San Diego, 497 – 564.

Ogami, T., 2009. The evolution of deltaic coastal sediment associated with sea-level changes since the last glacial at the Nobi Plain, Central Japan. Ph.D. Thesis, University of Tokyo, 1–123 (in Japanese with English abstract).

Ogami, T., Sugai, T., 2006. Evaluation of the activity of the Yokkaichi fault since the late Pleistocene. *The Quaternary Research*, (Japan) 45, 131 – 139 (in Japanese with English abstract).

Ogami, T., Sugai, T., Fujiwara, O., Yamaguchi, M., Sasao, E., 2009. The latest Pleistocene-Holocene sequence of the Kiso river delta based on analyses of sedimentary cores and ^{14}C datings. *Journal of Geography*, 118, 665 – 685 (in Japanese with English abstract).

Ohmori, H., 1983. Erosion rates and their relation to vegetation from the viewpoint of world-wide distribution. *Bulletin of the Department of Geography, University of Tokyo* 15, 77 – 91.

Okada, A., 1979. Active faults—Active faults and historical earthquakes in and around Aichi Prefecture, central Japan—Division of Earthquake in Disaster Prevention Council of Aichi Prefecture, 102 p (in Japanese).

Ota, Y., Sangawa, A., 1984. Active faults in the eastern foot area of the Suzuka range, central Japan. *Geographical Review of Japan*, 57, 237 – 262 (in Japanese with English abstract).

Ota, Y., Hull, A. G., Berryman, K. R., 1991. Coseismic uplift of Holocene marine terrace in the Pakarae River Area, Eastern North Island, New Zealand. *Quaternary research*, 35, 331 – 346.

Palyvos, N., Lemeiile, F., Sorel, D., Pantosti, D., Pavlopoulos, K., 2008. Geomorphic and biological indicators of paleoseismicity and Holocene uplift rate at a coastal normal fault footwall (western Corinth Gulf, Greece). *Geomorphology*, 96, 16 – 38.

Ran, Y., Chen, L., Chan, J., Wang, H., Chen, G., Yin, J., Shi, X., Li, C., Xu, X., 2010. Paleoseismic evidence and repeat time of large earthquakes at three sites along the Longmenshan fault zone. *Tectonophysics*, 491, 141 – 153.

Research Group for Active Faults of Japan, 1981. Active Faults in Japan—Sheet Maps and Inventories. Univ. Tokyo Press, Tokyo. 363 pp (in Japanese with English summary).

Research Group for Active Faults of Japan, 1991. Active Faults in Japan—Sheet Maps and Inventories (revised edition). Univ. Tokyo Press, Tokyo. 437 pp (in Japanese with English summary).

Reimer, P. J., Baillie, M. G. L., Bard, E., Bayliss, A., Beck, J. W., Blackwell, P. G., Ramsey, C. B., Buck, C. E., Burr, G. S., Edwards, R. L., Friendrich, M., Grootes, P. M., Guilderson, T. P., Hajdas, I., Heaton, T. J., Hogg, A. G., Hughen, K. A., Kaiser, K. F., Kromer, B., McCormac, F. G., Manning, S. W., Reimer, R. W., Richards, D. A., Southon, J. R., Talamo, S., Turney, C. S. M., van der plicht, J., Weyhenmeyer, C. E., 2009. Intcal09 and Marine09 radiocarbon age calibration curves, 0 – 50,000 years cal BP. *Radiocarbon*, 51, 1111 – 1150.

Rubin, C. M., Lindval, S. C., Rockwell, T. K., 1998. Evidence for large earthquakes in metropolitan Los Angeles. *Science,* 281, 398 – 402.

Saegusa, Y., Kashima, K., Yamamuro, M., Sasao, E., Ogami, T., Niwa, Y., Wakabayashi, T., Sugai, T., 2008. Reconstruction of Holocene environments by diatom, $\delta^{13}C$, C/N, EC, and TS analysis in Nobi plain, central Japan. Proceedings of the General Meeting of the Association of Japanese Geographers 74, 90 (in Japanese).

Saegusa, Y., Sugai, T., Ogami, T., Kashima, K., Sasao, E., 2011. Reconstruction of Holocene environmental changes in the Kiso-Ibi-Nagara compound river delta, Nobi plain, central Japan, by diatom analyses of drilling cores. *Quaternary International*, 230, 67 – 77.

Sato, H., 2008. Reconstruction of Holocene sea-level change along the coast of Harimanada in the eastern part of Seto Inland Sea, western Japan. *The Quaternary Research,* (Japan) 47, 247 – 259 (in Japanese with English abstract).

Sebrier, M., Chariri, A., Bles, J. L., 1997. Paleoseismicity in France: Fault trench studies in a region of moderate seismicity. *Journal of Geodynamics,* 24, 207 – 217.

Sherrod, B. L. 2001. Evidence for earthquake-induced subsidence about 1100 yr ago in coastal marshes of southern Puget Sound, Washington. *Geological Society America Bulletin,* 113, 1299 – 1311.

Shishikura, M., Echigo, T., Namegaya, Y., 2009. Evidence for coseismic and aseismic uplift in the last 1000 years in the focal area of a shallow thrust earthquake on the Noto Peninsula, west-central Japan. *Geophysical Research Letters,* 36, L02307, doi:10.1029/2008GL036252.

Stuiver, M., Reimer, P. J., 1993. Extended ^{14}C database and revised CALIB radiocarbon calibration program. *Radiocarbon,* 35, 215 – 230.

Sugai, T., 2011. Geomorphological and geological records indicate that activity on the Yoro fault system caused the AD 1586 Tensho earthquake. *Active fault Research*, 35, 15 – 28 (in Japanese with English abstract).

Sugai, T., Sugiyama, Y., 1998. Deep seismic reflection survey of concealed active structures in the Nobi Plain, central Japan. Geological Survey of Japan Interim Report EQ/98/1, 55 – 65 (in Japanese with English abstract).

Sugai, T., Sugiyama, Y., 1999. Basin dynamics and subsidence history of the Nobi Plain, central Japan revealed by 600 m deep core analyses and deep seismic reflection survey. Geological Survey of Japan Interim Report EQ/99/3, 77 – 87 (in Japanese with English abstract).

Sugai, T., Awata, Y., Shimokawa, K., 1998. Paleoseismological study of the Kuwana fault and the Yokkaichi fault in Mie Prefecture, central Japan. Geological Survey of Japan Interim Report EQ/98/1, 75 – 90 (in Japanese with English abstract).

Sugai, T., Fusejima, Y., Awata, Y., Azuma, T., Kariya, Y., Suzuki, Y., 1999. Late Holocene paleoseismicity of the Yoro fault system, central Japan – Geologic evidence of the last two historical large (M7.8 to 8) earthquakes of Tensyo Jishin in 1586 and Tenpyo Jishin in 745. Geological Survey of Japan Interim Report EQ/99/3, 89 – 102 (in Japanese with English abstract).

Takada, K., Atwater, B. F., 2004. Evidence for Liquefaction Identified in Peeled Slices of Holocene Deposits along the Lower Columbia River, Washington. *Bulletin of the Seismological Society of America*, 94, 550 – 575.

Takada, Y., Kondo, Y., Miyamura, M., 1979. Geology of the Tsushima district. Quadrangle Series, scale 1 : 50,000. Geological Survey of Japan, 56 p (in Japanese with English abstract).

Tuttle, M. P., 2001. The use of liquefaction features in paleoseismology: Lessons learned in the New Madrid seismic zone, central United States. *Journal of Seismology*, 5, 361 – 380.

Tuttle, M. P., Collier, J., Wolf, L. W., Lafferty, R. H., 1999. New evidence for a large earthquake in the New Madrid seismic zone between A.D. 1400 and 1670. *Geology*, 27, 771 – 774.

Tuttle, M. P., Schweig, E. S., Sims, J. D., Lafferty, R. H., Wolf, L. W., Haynes, M. L., 2002. The Earthquake Potential of the New Madrid Seismic Zone. *Bulletin of the Seismological Society of America*, 92, 2080 – 2089.

Tuttle, M. P., Dyer-Williams, K., Schweig, E. S., Prentice, C. S., Moya, J. C., Tucker, K. B., 2005. Liquefaction induced by historic and prehistoric earthquakes in western Puerto Rico, the Virgin Islands, and offshore areas. *Geological Society of America Special Paper*, 385, 263 – 276.

Umitsu, M., 1992. Holocene deltaic sequence of the Kiso river delta, central Japan. *Journal of Sedimentological Society of Japan*, 36, 47–56 (in Japanese with English abstract).

Usami, T., 1996. Materials for Comprehensive List of Destructive earthquakes in Japan, Revised and Enlarged Edition. University of Tokyo Press, 493 pp (in Japanese).

Williams, H., Hutchinson, I., 2000. Stratigraphic and microfossil evidence for late Holocene tsunamis at Swantown marsh, Whidbey Island, Washington. *Quaternary Research*, 54, 218 – 227.

Yamaguchi, M., Sugai, T., Ota, T., Ohmori, H., Sugiyama, Y., Fujiwara, O., 2001. The estimation of sedimentary environment based on synthetic analysis of boring core in the Nobi Plain, Central Japan. Proceedings of the General Meeting of the Association of Japanese Geographers 60, 90 (in Japanese).

Yamaguchi, M., Sugai, T., Fujiwara, O., Ohmori, H., Kamataki, T., Sugiyama, Y., 2003. Depositional process of the Holocene Nobi Plain, central Japan, reconstructed from drilling core analysis. *The Quaternary Research,* (Japan) 42, 335 – 346 (in Japanese with English abstract).

Yamaguchi, M., Sugai, T., Fujiwara, O., Ohmori, H., Kamataki, T., Sugiyama, Y., 2005. Depositional process and landform of the Kiso river delta, reconstructed from grain size distributions, and accumulation rates of sediment cores. *The Quaternary Research*, (Japan) 44, 37 – 44 (in Japanese with English abstract).

Yamaguchi, M., Sugai, T., Ogami, T., Fujiwara, O., Ohmori, H., 2006a. Three-dimensional structures of latest Pleistocene to Holocene sequence at Nobi plain, central Japan. *Journal of Geography*, 115, 41 – 50 (in Japanese with English abstract).

Yamaguchi, M., Sugai, T., Fujiwara, O., Ogami, T., Ohmori, H., 2006b. Stratigraphy of upper Holocene deposits and landform evolution in the Kiso river delta system, central Japan. *The Quaternary Research,* (Japan) 45, 451 – 462 (in Japanese with English abstract).

Yamazaki, H., Shimokawa, K., Mizuno, K., Tanaka, T., 2002. Off-fault paleoseismology in Japan: With special reference to the Fujikawa-kako fault zone, central Japan. *Geographical Reports of Tokyo Metropolitan University,* 37, 1 – 14.

Yokoyama, T. Sato, M., 1987. Analysis of paleo-environments based on electric conductivity of stirred clayey sediments into water: on two drill cores of Osaka Group in Senriyama Hills and Biwako Group at bottom of Lake Biwa, Japan. *Journal of the Geological Society of Japan*, 93, 667 – 679 (in Japanese with English abstract).

In: Holocene
Editor: Bahadur Singh Kotlia

ISBN: 978-1-62257-722-4

Chapter 16

EMERGENCE OF GRACILE HUMAN ANATOMY DURING THE HOLOCENE IN ASIA; CONVERGENT ACCLIMATION AND/OR PHYLOGENETIC TREND? EXAMPLE OF OCCLUSION

Anne Dambricourt Malassé*

UMR 7194 CNRS, Department of Prehistory, National Museum of Natural History, Paris, France

ABSTRACT

The dento-cranial architecture of current human species, common to all ethnic groups, appeared during the Holocene. The first thousand years are an important transition period in the long process of hominization from cultural, behavioral and anatomical point of views. One peculiarity of this new grade in human evolution is the occlusal balance between both maxillar and mandibular dental archs. Orthodontics, osteopaths, posturologists show that this balance is closely linked to the skull base and the erected posture of cerebellum and spine. However, the evolutionary origin of such balance is still unknown. A first comparison of Late Pleistocene and Early Holocene skulls in their cultural context opens new perspectives to clarify the factors of this recent human anatomical evolution, between climate, behavior and phylogenetic trends.

Keywords: Human evolution, Dental occlusion, Gracilization, Pleistocene, Holocene, Multiregional continuity

INTRODUCTION

The current chapter deals with the hominization processes and their periods of the transition. The recent anthropological finds in continental and insular Asia, covering a period

* Email: iphadm@mnhn.fr.

of transition (Kaifu and Fujita, 2012; Curnoe et al., 2012), open new perspectives on the origin of modern human anatomy. Human groups, widely dispersed across five continents, share common craniofacial characters first known with proto and cromagnoid forms in Europe (Cromagnon, Hoedic, Teviec; see Figure 1.), North Africa (Taforalt, Afalou-Bou-Rhummel), Southeast Asia (Wajak) and more recently in Mongolia (Salkhit, Coppens et al., 2008), South China (Zhirendong, Liu et al., 2010) and probably in India (Chamyal et al., 2011; Beck et al., 2012; Dambricourt Malassé et al., 2012) and even in Northwest America (Chatters, 2000; Barker et al., 2000).

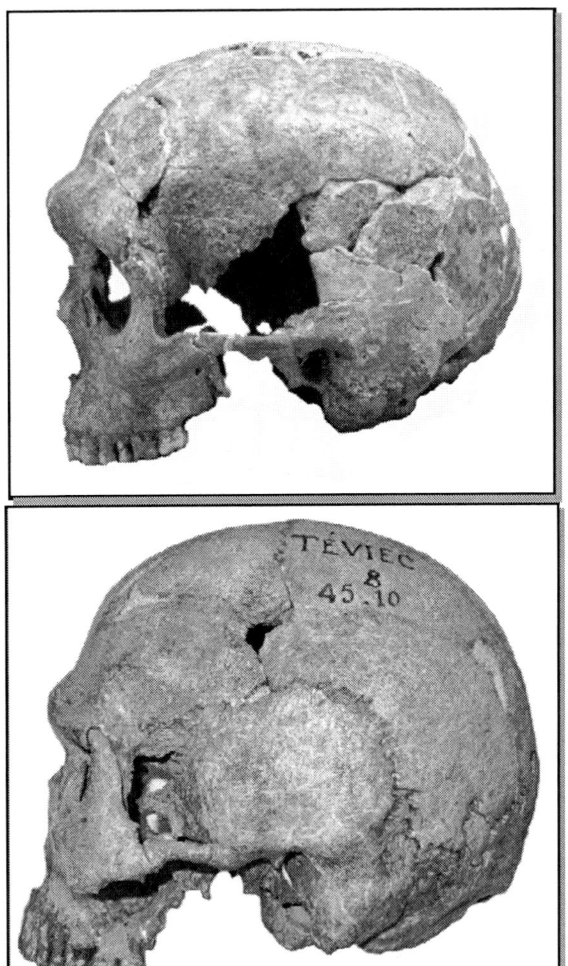

Coll. IPH, after Dambricourt Malassé, 2011b.

Figure 1. Proto-cromagnoid Qafzeh 6, 95 ka, and late cromagnoid Teviec, 9 ka.

The Craniofacial changes that characterize the current anatomy concern the relative proportions of the face with the neurocranium, the gracilization of bony exocranial superstructures, a change in the incisor-canine occlusion and volumetric development of the cerebellum, proportionally larger than the brain. According to (Weaver, 2005), the Middle and Late Pleistocene humans including Neanderthals and Cro-Magnon 1, have the largest cerebral hemispheres relative to cerebellum volume of any primates, including earlier and Holocene humans. In recent humans, however, the pattern is reversed, the cerebellum is larger with respect to the rest of the brain (and conversely, the cerebral hemispheres are smaller with respect to the cerebellum) than in Late Pleistocene humans. At this organizational level of neural development and occlusal balance, it is reasonable to question the intrinsic causes of such an emergence. Indeed, the cerebellum is involved in the neocorticalization and its cognitive functions are now well recognized (Bellebaum et al., 2012). Such a "neo-cerebellisation" reveals a general evolutionary trend of the central nervous system. According to Ferembach (1986), Epipaleolithic and Mesolithic are of great interest, especially for understanding the human evolution.

Dental Occlusion and Verticalization of the Central Nervous System

The anatomical changes, less dramatic in appearance than between Ardipithecus and Australopithecus, nevertheless reveal an evolutionary process in the continuity of the cerebrospinal verticalization. This phenomenon emerges during the last weeks of the embryonic period, 7 weeks after fertilization in the modern human (Dambricourt Malassé, 1988, 1993, 2006, 2009, 2011a). Indeed, the cephalic skeleton of the mammal embryo is limited to the future basicranium, flat and cartilaginous. This flattened shape is due to the antero-posterior axis of the vertebrates given by an embryonic longitudinal structure, the chord. The apex is just below the hypophysis. This chord defines the future quadrupedal organization of the axial skeleton where the pelvis is behind the head and not below as in hominids. The lowering of the cerebellum, which breaks this plan just at the chordal apex, occurs at the end of the embryonic development only in the Simian clade (Dambricourt Malassé, ib. 2011b). This evolutionary event, unique in mammal phylogeny, was of great importance because for the first time, the vertical dimension emerged from the embryonic antero-posterior axis. This started 39 Ma ago (first simiiform) and increased two times with great ape's pattern then with hominids. In such a threshold (punctuated equilibria), the quadrupedal equilibrium is no longer possible and the increasing instability of the erected axial skeleton requires more time to ensure the psychomotor equilibrium and to develop a new and more complex neural network between the cerebellum and brain. Because the bones of the basicranium are in connection with the maxilla and mandible, the links with erected posture and occlusion are simultaneous. Moreover, the cerebellum is erected more when the adult face is close to the neurocranium. The space between the cerebellum and the vocal tract is then reduced (Figure 2). A number of works of this complex morphodynamics are published (Dambricourt and Deshayes, 1992; Deshayes, 2006; Dambricourt Malassé, 2011a; Hadjouis, 2012).

During the hominid evolution, the occlusion moved from "promandibular canine balance" (the lower canine in front of the upper) common to Great Ape's species, Australopithecines and the most archaic species of the genus Homo (*habilis*, *georgicus*, *ergaster*) to labidonty with *Homo antecessor* and the ancestors of *Homo florensiensis* (canines and incisors in edge-to-edge bite), while cephalization followed different processes in Europe (*Homo heidelbergensis*), Africa (*Homo rhodensiensis*) and Southeast Asia (*Homo *florensiensis*). Among these, the multiregional evolutionary trends, some of them changed around 200,000 years BP with a new lowering of the cerebellum and a spatial reduction between the cerebellar fossa and the adult vocal tract. This evolutionary grade corresponds to the African robust *Homo sapiens* (Omo Kibish 1, Bouri), and later, to the near eastern proto-cromagnoid without changing labidonty. Nevertheless, the dental growth is changed, revealing an ossified triangular notch acquired with *Homo* pattern and becoming the chin with the drop of cromagnoid dental arch (Figure 3).

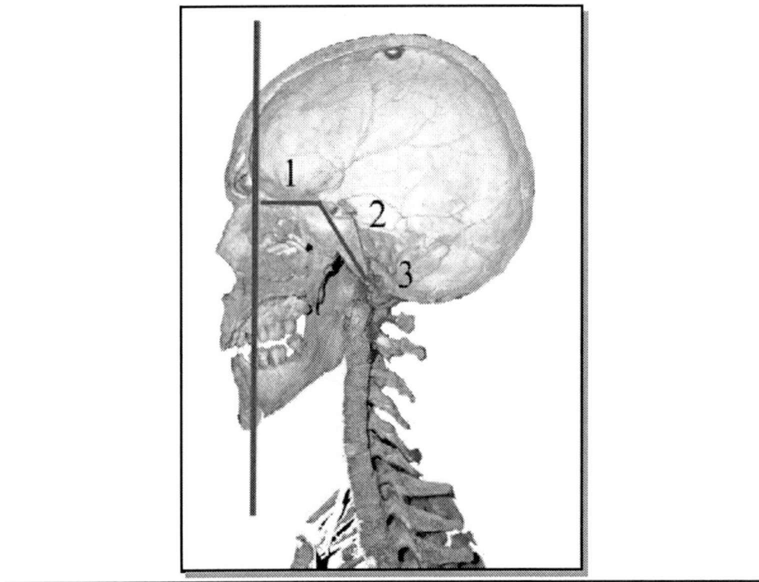

After Dambricourt Malassé, 2011b.

Figure 2. Sagittal section of basicranium and face, *Homo sapiens* and Great Ape (Gorilla). 1, superior cerebral fossa; 2, middle cerebral fossa; 3, cerebellar fossa.

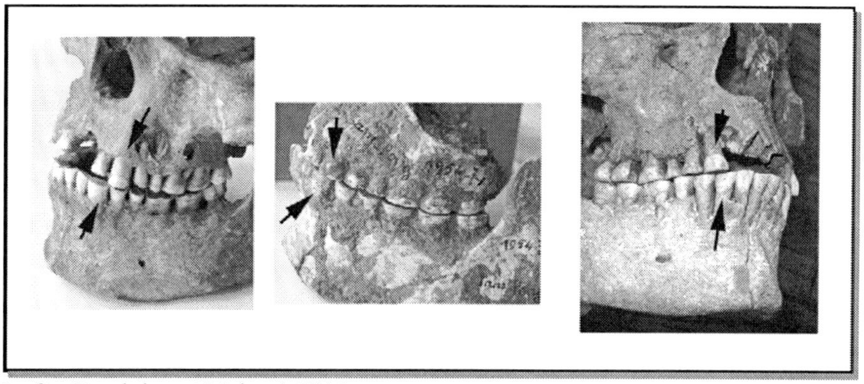

Coll. IPH, after Dambricourt Malassé, 2011a.

Figure 3. Labidonty. Epipaleolithic, Afalou (Algeria); Mesolithic, Tam Pong (Laos); Mesolithic Teviec (France). Black arrow shows canines in edge-to-edge occlusion.

The chin never exists in the previous stage of less cerebrospinal erected posture, despite the degree of cephalization, high of the vault and endocranial capacity. Becoming *sapiens* is much more complex than a gradual increasing complexity of the brain. Again, the cephalo-caudal embryonic organization is concerned; proto-cromagnoid or *sapiens* embryonic pattern is different from *Homo neanderthalensis*, late Asian *Homo erectus* and *Homo florensiensis*. However, the Asian *Homo erectus* was not necessarily genetically devoid of such evolutionary properties. Finally, the labidonty disappears during the Holocene, replaced by our psalidontic occlusion, the mandibular canine-incisor arch which is located behind its maxillary homolog (Figure 4).

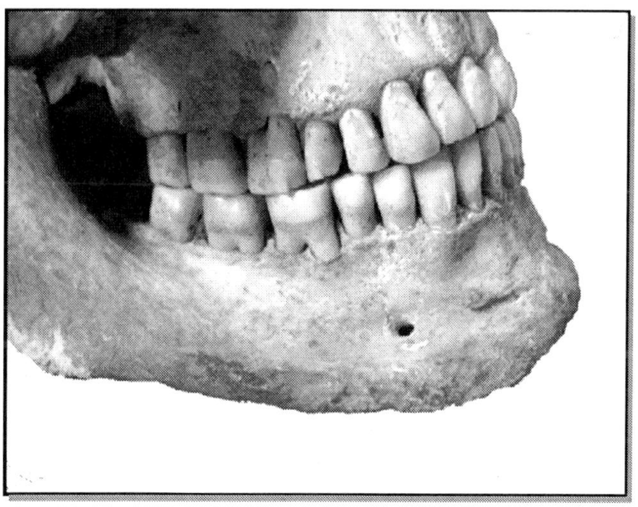

Figure 4. Psalidonty.

Where, when and how was the psalidonty manifested? Was this phenomenon unique (monocentric) or global? An adaptation to the Holocene Climate Optimum (HCO) (9,000-5,000 years BP), was it a consequence of the productive economy (livestock, agriculture) or the evolutionary process of neural networks complexification? The Indian and Chinese skulls, dated to the Early Holocene and anatomically modern and robust, have inherited specific traits of the oldest Eurasian *Homo* species (*Homo georgicus* 1.8 Ma BP), that is absent on their African contemporaries and still visible on all later Asian *Homo erectus* (Dambricourt Malassé, 2008; Chamyal et al., 2011). Only the phylogenetic continuity can explain this peculiarity and it is reasonable to predict that among Asian lineages, some of them were also concerned by a "cromagnoid" grade (Dambricourt et al., 2012).

ORIGIN OF PSALIDONTY

The Neolithic man with new food including milk is often referred as to the factor of general gracilization (Ferembach, 1986; Sardi et al., 2004) but this hypothesis presupposes that behavior can change the cromagnoid genetic growth pattern. The comparative studies relativize this hypothesis (Schwidetzky, 1989; Grupe, 1989). A decrease in the size of incisor-canine crowns and roots can possibly explain the drop of the alveolar arch but not the position back from the maxillary homolog. According to the morphodynamics, such a drop presupposes a change in the genetic pattern of neural growth and basi-cranial flexure with epigenetic changes such as delayed psychomotor development, decreased endocrine function and gracilization of bone tissues and muscles. A preliminary study has identified in time and space, labidontic peoples in their cultural environments and has compared the results with the oldest psalidontics. We present a first overview of occlusions of the Late Pleistocene and Early Holocene with special attention to Asia (India, China, and Southeast Asia).

MATERIALS AND METHODS

The mandible is correctly articulated when the lower M1 exceeds upper M1 of one cusp (Angle's dental classification, Figure 4). The characteristic of labidonty is the strong apical wear of incisors and canines. However, an isolated mandible with abraded teeth is not necessarily labidontic, it may have been abraded by tanning (Plenot and Gessain, 1982). Similarly, some individuals are no longer in Angle class 1, but in promandibular imbalance (class 3), which also generates significant wear of incisor and canine apex. Until Mesolithic, neurocranium associated with mandible are rare and infantile stages are exceptional. The list of individuals is therefore not representative of inventoried skulls.

However, given the low number of complete cranium in the Upper Paleolithic and the Mesolithic, the observations include skulls without mandible taking into account the wear of maxillary teeth. The skulls and casts come from the Institute of Human Paleontology, Paris (IPH), a few from India (Orsang in Baroda University, Tekka Lakata from Karnataka in Deccan College) and from China (Wanrengang with Ji Xuping, Yunnan Institute of Archaeology).

A total of 136 skulls include Epipaleolithic populations of Afalou (n=50) and Taforalt (n=27) in North Africa, Mesolithic with Hoëdic (n=11) and Teviec (n=7) on the Atlantic coast in France, and other skulls described by authors in this work. In Mainland Southeast Asia (Myanmar, Thailand, Laos, Cambodia, Vietnam), the Epipaleolithic and Mesolithic equivalences are Hoabhinien and Bacsonien (Saurin and Carbonnel, 1974).

In the Southeast insular Asia (Malaysia, Indonesia, Philippines), these are Toalian (Sulawesi) and Sampungian (East Java) traditions (Forestier and Edoumba, 2000). The oldest traces of horticultural activity are known from the Spirit Cave, Thailand, in the Hoabinhian level I, dated to 9,180 yr BP (Gorman, 1969, 1972).

CONCLUSION

Prehistory recognizes at least three independent centers of Neolithic, Middle East, Asia and Sub-Saharan Africa with a noticeable implanting in the Upper Paleolithic in Europe, where food storage and the associated settlements are already evident between 28 and 20 ka BP (Kozlowski, 2001).

The majority of Epipaleolithic and Mesolithic skulls are labidontic (Table 1). However, some individuals associated with Upper Paleolithic are very probably psalidontic, e.g., Gamble's Cave in Kenya, Le Placard (France) and Grimaldi (Monaco) in Europe. During Mesolithic, the psalidonty is observed in North Africa (with the child of Tebessa), two types coexisting in Ofnet (Germany) and Kostenki (Russia) in an environment more recent than was previously expected (Henry-Gambier, 2008).

The child of La Genière (France) may be Neolithic. Conversely, in Vietnam, a Neolithic skull at Lang Son is labidontic (Verneau, 1909), like Song Gentong II (Java), reported to be a lower Neolithic (Detroit, 2002). The Niah Cave, Borneo, has yielded several burials from the Meso-Neolithic periods. Manser (2008) describes homogeneity in the characters demonstrating a genetic continuity. In West Malaysia, the Gua Cha skeletons allow comparisons between Hoabinhian and Neolithic people. Bulbeck (2000) has compared the dental morphology to verify the phyletic link (sondadont, M1 has two roots), or replacement (sinodont, lower M1 has three roots). The occlusions still remain to be compared.

The transition from the predator to the productive economy was progressive, diachronic and not a monocentric "revolution." The transition from labidonty to psalidonty seems to be correlated, but the hypothesis of a dietary epiphenomenon does not explain the genetic heritability of such a global evolution in separated populations, such as North Africa and Insulindia.

The reversal of relative proportions between the cerebellum and the brain, as well as the proliferation of dental agenesis, continue to question anthropologists on the real modalities of hominization processes, closely linked to the verticalization of the cerebellum and its neocorticalization.

ACKNOWLEDGMENTS

I am grateful to Prof. B.S. Kotlia for inviting me to submit this chapter. I thank Dr. Ousmane Chérif Touré for assisting in the English translation.

Table 1. ab, abraded; L , labidonty; low ab, low abraded; no ab., no abraded; mand, mandible; P, probable; Ps , psalidonty; strong ab., strong abraded

Asia	Country	Occlusion	Cultural Association	Age ka	Source
Ryonggok 1	North Korea	No mand	burial	46-48	IPH cast Bae 2011
Ryonggok 2	North Korea	No mand	burial	46-48	IPH cast Bae 2011
Minatogawa 1	Japan	abraded	no industry	18	Suzuki
CKT 1 Up Cave	China	L	Upper pal.	29	IPH cast
Red Deer Cave	China Yunnan	L		14-11	Curnoe
Liukiang	China wangsi	no mand	no industry	20 – 30	Wu Liu
Wanrengang	China Yunnan	L	Hoabhinian ?		original
Lang Cuom 3	Vietnam	abraded	Bacsonian		IPH cast
Lang Cuom 7	Vietnam	abraded C	Bacsonian		IPH cast
Hang Cho	Vietnam	suspect Ps	Hoabinhian		Matsumara
Tam-Hang	Laos	abraded	Hoabinhian		IPH cast
Pho-Binh-Gia 1	Vietnam	L	Neolithic		Verneau
Mai Da Nuoc	Vietnam	Ps	Hoabhinian	10	Nguyen
MohKhiew -1	Thaïland	L	Hoabhinian	27 - 11	Detroit
MohKhiew -2	Thaïland	L	Hoabhinian	<11	Detroit
Niah Cave	Borneo	?	Meso-Neo		Manser
Gua Cha	Malaysia	?	Hoabinhian		Bulbeck
Song Gentong 2	Java	L (?)	Low Neolithic		Detroit
Song Terus 1	Java	L	Microlith		Detroit
Song Keplek 5	Java	L	Microlith	8-5	Detroit
Australia					
Nacurrie 1		L		11	Brown
Coobool Creek		L		14, >20	Brown
North America					

Asia	**Country**	**Occlusion**	**Cultural Association**	**Age ka**	**Source**
Kennewick Man		L	No industry	8.4	Chatters
Spirit Cave		L	mummy	9-10	Barker
Africa					
Fish Hoek	South Africa	L	Upper Paleo.	12	IPH cast
Gamble's Cave	Kenya	Ps	Upper Paleo.		IPH cast
Afalou n = 50	Algeria	L	Epipaleolithic		IPH
Taforalt n = 27	Marocco	L	Epipaleolithic		IPH
Mechta el Arbi 5		Low ab.	Epipaleolithic		IPH
Tebessa 6 years		L	Up Capsian		IPH
Europe					
Isturitz III	France	L	Aurignacian		IPH cast
Grimaldi	Monaco	Ps prob	Aurignacian	26	IPH cast
Predmost ♀	Moravia	L	Upper Pal.	35-25	IPH cast
Predmost ♂	Moravia	L	Upper Pal.	35-25	IPH cast
Brno III	Czech republic	L	Up Pal uncert.		IPH cast
Dolni Vestonice	Moravia	L	Gravettian		IPH cast
Ofnet K 1802	Germany	strong ab.	Tardenoisian		IPH cast
Ofnet K 1818	Germany	no ab.	Tardenoisian		IPH cast
Kostenki XIV	Russia	Ps	Up Pal uncert		IPH cast
Saint Germain	France	L	Magdalenian		IPH cast
Les Hoteaux	France	L	Magdalenian		Vallois
Le Placard	France	Low ab.	Solutrean		IPH cast
Hoedic n = 11	France	L	Mesolithic		IPH
Teviec n = 7	France	L	Mesolithic		IPH
La Genière (M1)	France	Class 3	Uncertain		IPH cast

REFERENCES

[1] Bae, C.J., Bae, K. 2011. The nature of the Early to Late Paleolithic transition in Korea, Current perspectives. *Quaternary International* 20, 1-10, dx.doi.org/10.1016/ j.quaint.2011.08.044.

[2] Barker, P., Cynthia, E., Damadio, S. 2000. Determination of Cultural Affiliation of Ancient Human Remains from Spirit Cave, Nevada. Bureau of Land Management Nevada State Office p. 39.

[3] Beck, L., Cuif, J.-P., Pichon, L., Vaubaillon, S., Dambricourt Malassé, A., Abel, R.L. 2012. Checking collagen preservation in a bone fragment of the potentially Oldest Modern Indian by non-destructive studies. *Nuclear Instruments and Methods in Physics Research section B* 273, 203-207.

[4] Bellebaum, C., Daum, I., Suchan, B. 2012. Mechanisms of cerebellar contributions to cognition in humans. *Cognitive Science* 3 (2), 171–184.

[5] Brothwell, D. 1960. Upper Pleistocene human skull from Niah Caves, Sarawak. *The Sarawak Museum Journal* 9, 323-349.

[6] Brown, T., Townsend, G.-C, Richards, L.-C, Burgess, V.B. 1990. Concepts of occlusion, Australian evidence. *American Journal of Physical* Anthropology 82, (3), 247–256.

[7] Brown, P. 1992a. Recent human evolution in East Asia and Australasia. *Philosophical Transactions of the Royal Society London, Series B* 337,235-242.

[8] Brown, P. 1992b. Post-Pleistocene change in Australian Aboriginal tooth size, dental reduction or relative expansion? In: T. Brown and S. Molnar (Eds.), Human craniofacial variation in Pacific Populations. Adelaide, Anthropology and Genetics Laboratory, University of Adelaide, 33-52.

[9] Brown, P. 2001. Chinese Middle Pleistocene hominids and modern human origins in East Asia. In: L. Barham and K. Robson Brown (Eds.) Human Roots - Africa and Asia in the Middle Pleistocene. Bristol, Western Academic & Specialist Publisher, 135-145.

[10] Bulbeck, P. 2000. Dental morphology at Gua Cha, West Malaysia, and the implications for "sundadonty." *Bulletin of the Indo-Pacific Prehistory Association* 19, 17-41.

[11] Bulbeck, D. 2001. The Gua Cha burials. Concordance, chronology, demography. Report, Duckworth Laboratory, Cambridge University, England.

[12] Chamyal, L., Dambricourt Malassé, A., Maurya, D.M., Raj, R., Bandhari. S., Pant, S.K., Gaillard, C. 2011. Discovery of a robust *Homo sapiens* in India (Orsang, Lower Narmada Basin, Gujrat). Possible continuity with Indian *Homo erectus*. *Acta Anthropologia Sinica* 2, 167-191.

[13] Chatters, J.C. 2000. The Recovery and First Analysis of an Early Holocene Human. *American Antiquity* 65 (2), 291–316.

[14] Coppens, Y., Tseveendorj, D., Demeter, F., Turbat, T., Giscard, P-H. 2008. Discovery of an archaic Homo sapiens skullcap in Northeast Mongolia. Comptes Rendus de l'Académie des Sciences, *Palévol* 7 (1) , 51-60.

[15] Cuong, N.L. 1987. An early Hoabinhian skull from Vietnam. *Bulletin of the Indo-Pacific Prehistory Association* 7, 30-35.

[16] Curnoe, D., Ji, X., Herries, A.I.R., Bai, K., Taçon, P.S.C., Bao, Z., Fink, D., Zhu, Y., Hellstrom, J., Luo, Y., Cassis, G., Su, B., Wroe, S., Hong, S., Parr, W.C.H., Huang, S.,

Rogers, N. 2012. Human Remains from the Pleistocene-Holocene Transition of Southwest China Suggest a Complex Evolutionary History for East Asians, *PLoS ONE* 7(3), e31918 [J. Alroy/J. Alroy].

[17] Dambricourt Malassé, A. 1988. Hominisation et foetalisation. *Comptes Rendus de l'Académie des Sciences* 307 (II), 199-204.

[18] Dambricourt Malassé, A., Deshayes M.J. 1992. Modeling of the craniofacial architecture during ontogenesis and phylogenesis. In: Berthoz, Graf, Vidal (Eds), The Head-Neck sensory motor system. New York-Oxford University Press, 37-46.

[19] Dambricourt Malassé, A. 1993. Continuity and discontinuity during modalities of hominization. In: Lars Wederlin, Jean Chaline (Eds), Modes and Tempos in Evolution in the Quaternary. *Quaternary International* 19, 85-100.

[20] Dambricourt Malassé, A. 2006. Evolution du chondrocrâne et de la face des grands anthropoïdes miocènes jusqu'à *Homo sapiens*, continuités et discontinuités. In : Climats, cultures et Sociétés aux temps préhistoriques. De l'apparition des hominidés jusqu'au Néolithique. Comptes Rendus de l'Académie des Sciences, *Palévol* 5, 109-117.

[21] Dambricourt Malassé, A. 2008. The human settlement in Eurasia, the mountainous Central Asia and the Sub-Himalayan piedmonts from Plio-Pleistocene to Holocene, origins, human evolution and migrations (in French). *L'Anthropologie* 112 (3), 370-403.

[22] Dambricourt Malassé, A. 2009. Embryogeny and Human phylogeny. Chapter 8. In: A.R. Sankhyan, Asian Perspectives on Human Evolution, New Delhi, Serial Publications (Ed), 103-121.

[23] Dambricourt Malassé, A. 2010. The relationships between occlusion and posture in the hominid lineage, implications for the transition between Mesolithic and Neolithic populations. *International Journal of Modern Anthropology* 3, 12-63.

[24] Dambricourt Malassé, A. 2011. The human lineage, a macro-evolutionary process acting during embryogeny with emergent macro-evolutionary implications. In: *Biological Evolution, Facts and Theories, Analecta Gregoriana* 312, 243-274.

[25] Dambricourt Malassé, A. 2011. The permanent bipedal balance, embryonic origin, morphogenesis, occluso-postural balances, consequences for the psychomotor and behavioral evolution of hominids (in French). Accreditation to Direct Research, Biomechanics and Bioengineery, University of Compiègne.

[26] Dambricourt Malassé, A,. Chamyal, L., Raj, R., Maurya, M., Juyal, N., S., Bhandari, R., Pant, K., Shah, S. 2012. Orsang Man , a robust *Homo sapiens* in Central India with Asian *Homo erectus* features. Abstract. International Union of Anthropological and *Ethnological Sciences*, 2013, Manchester.

[27] Deshayes, M. J. 2006. Les déformations crâniennes asymétriques et leur retentissement dento-facial et occlusal. *Orthodontie Française* 77 (1), 87-99.

[28] Détroit, F. 2002. Origine et évolution des *Homo sapiens* en Asie du Sud-Est, Descriptions et analyses morphométriques de nouveaux fossiles. PhD Dissertation, Muséum national d'Histoire naturelle, Paris.

[29] Ferembach, D. 1986. Les hommes de l'Holocène. In: D. Ferembach., C., Susanne, M.C. Chamla (Eds), L'homme, son évolution, sa diversité, CNRS-Doin, 224-231.

[30] Forestier, H., Patole-Edoumba, E. 2000. Les industries lithiques du Paléolithique tardif et du début de l'Holocène en Insulinde. *Aséanie* 6 (6), 13-56.

[31] Grupe, G. 1989. Grazilisierung und Ernährung - Ursache oder Wirkung? *Homo* 40 (1-2), 58-64.

[32] Gorman, C.F. 1969. Hoabinhian, A Pebble-tool Complex with Early Plant Associations in Southeast Asia, *Science* 163, 671-673.

[33] Gorman, C.F. 1972. Excavations at Spirit Cave, North Thailand, *Asian Perspectives* (1970) 13, 79-107.

[34] Hadjouis, D. 2012. Les fonctions masticatrices et para-masticatrices des hommes fossiles d'Algérie. De la labidontie à la psalidontie. Libyca, Alger.

[35] Henry-Gambier, D. 2008. Comportement des populations d'Europe au Gravettien, pratiques funéraires et interprétations. *Paleo* 20, 399-438.

[36] Kaifu, Y., Fujita, M. 2012. Fossil record of early modern humans in East Asia. *Quaternary International* 248, 2-11.

[37] Kayser, O. 1984. Autour du mésolithique en Bretagne. *Revue archéologique de l'Ouest* 1 (1), 7-13.

[38] Kozlowski, J. 2001. Les premiers hommes modernes et les premiers agriculteurs en Europe, voies de diffusion et interactions entre populations. *Etudes et Recherches Archéologiques de l'Université de Liège* 99, 9-34.

[39] Manser, J.M. 2008. Morphological analysis of the human burial series at Niah Cave, implications for late Pleistocene-Holocene Southeast Asian human evolution. *Borneo Research Bulletin* 39 : 346-347.

[40] Matsu'ura S., Kondo, M. 2011. Relative chronology of the Minatogawa and the Upper Minatogawa series of human remains from Okinawa Island, Japan. *Anthropological Science* 119, 173-182.

[41] Matsumura, H., Yoneda, M., Dodo, Y., Oxenham, M.F., Cuong, N.L., Thuy, N.K., Dung, L.M., Long, V.T., Yamagata, M., Sawada, J., Shinoda, K., Takigawa, W. 2008. Terminal Pleistocene human skeleton from Hang Cho Cave, northern Vietnam, implications for the biological affinities of Hoabinhian people *Anthropological Science* 116 (3), 201-217.

[42] Plenot, H.R., Gessain, R. 1982. La dentition des Ammassalimiut (Eskimo de la côte Est du Groenland). *Bulletins et Mémoires de la Société d'Anthropologie de Paris* 9 (9-4), 321-332.

[43] Sardi, M.L., Ramírez-Rozzi, F., Pucciarelli, H.M. 2004. Morphologie crânio-faciale des populations de l'Europe et du Maghreb depuis le Paléolithique Supérieur jusqu'à présent. *Antropo* 7, 145-153.

[44] Saurin, E., Carbonnel, J.P. 1974. Evolution préhistorique de la péninsule indochinoise d'après les données récentes. *Paléorient* 2, 133-165.

[45] Schwidetzky, I. 1989. Some Comments on the Problem of Gracilization. *Bulletin de la Société Royale Belge d'Anthropologie et de Préhistoire* 100, 25-31.

[46] Suzuki, H., 1982. Skulls of the Minatogawa Man. Chapter 2. In: The Minatogawa Man. The Upper Pleistocene Man from the Island of Okinawa, Suzuki H. and Hanihar K. (Eds). The University Museum. The University of Tokyo Bulletin 19.

[47] Vallois, H. 1972. Le crâne magdalénien des Hôteaux. *Bulletin et Mémoire Société d'Anthropologie de Paris* 9 (XII), 7-25.

[48] Verneau, R. 1909. Les crânes humains du gisement préhistorique de Pho-Binh-Gia (Tonkin). *L'Anthropologie* 20, 545-559.

[49] Weaver, A.H. 2005. Reciprocal evolution of the cerebellum and neocortex in fossil humans. *Proceedings of the National Academy of Sciences USA* 102 (10), 3576-3580.

[50] Wu, Liu., Xiujie, Wu., S-L, Wang., Haijun, Li. 2010. Le bassin de l'homme de Pléistocène supérieur du Liujang, Sud de la Chine, taille du corps, forme du corps, encéphalisation. *L'Anthropologie* 114, 543-563.

In: Holocene
Editor: Bahadur Singh Kotlia

ISBN: 978-1-62257-722-4

Chapter 17

LATE HOLOCENE DESERT-INDUCED STRESS AND HUMAN MIGRATIONS THROUGH THE SAHARA: THE CASE OF THE GARAMANTES

***Efthymia Nikita*[1,*], *David Mattingly*[2] *and Marta Mirazón Lahr*[1]**
[1]Leverhulme Centre for Human Evolutionary Studies, University of Cambridge, UK
[2]School of Archaeology and Ancient History,
University of Leicester, UK

ABSTRACT

The current chapter focuses on the Garamantes, a population that flourished in the Late Holocene in the harsh environment of the Sahara Desert in southwestern Libya. The Garamantes are associated with important innovations, such as the development of urbanism, trans-Saharan trade, irrigated agriculture and metallurgy. They created a hierarchical and probably slave-using society and substantially increased their population. Therefore, the material evidence of this civilization seems to indicate that the hyper-arid environment of the Sahara Desert did not impose any particular stresses on the population and neither did it restrict population movements.

The analyses of musculoskeletal stress markers and biomechanical properties point to the fact that either the stress levels imposed on the Garamantes were not particularly high and/or the Garamantes were particularly successful at coping with their environment. Furthermore, the North African populations were in contact with each other during the Late Holocene exchanging goods but also exhibiting some degree of gene flow. However, the Garamantes, being located in the centre of the Sahara, demonstrate greater biological distance in relation to their neighbours. These conclusions have important anthropological evolutionary implications relating to human adaptation to extreme, hyper-arid, environments.

Keywords: Garamantes, Sahara Desert, Late Holocene, human adaptation

---

* E Mail: efi.nikita@gmail.com.

INTRODUCTION

The Garamantian civilization developed in the core of the Sahara Desert in what is today the Fazzān basin in southwestern Libya. Despite existing in a harsh natural environment, the population is associated with very important innovations in the Sahara Desert and developed contacts with many other North African groups. The aim of the current study is to assess biologically the stress levels imposed on the Garamantes by the desertic environment, as well as to explore the extent of contacts among North African groups in the Late Holocene with an emphasis on the role of the Garamantes, being the population located at the centre of the Sahara Desert and controlling a significant portion of ancient trans-Saharan trade.

Several methods were employed for the above purposes. The examination of musculoskeletal stress markers and biomechanical reconstructions were used to explore how strenuous the daily activities necessary for survival in the desert were, while three-dimensional morphometric and non-metric trait analyses were performed in order to examine the biological affinities of the various Late Holocene North African groups, offering an insight to population movements across the Sahara.

HISTORICAL, CHRONOLOGICAL, ENVIRONMENTAL AND ARCHAEOLOGICAL BACKGROUND OF THE STUDY REGION

Topographic and Geological Information about the Sahara and Fazzān

The Sahara is the largest desert on the planet, covering an area of 8 to 9 million km^2 and spreading across Morocco, Mauritania, Algeria, Tunisia, Niger, Libya, Chad, Mali, Egypt and Sudan (Holl, 2004). Its physiographic features include swampy depressions, flat stony surfaces, sand seas, plateaus and even some mountain areas (Camps, 1982).

Fazzān (Figure 1) is located in southwestern Libya. It is a large basin separated by the Messak Settafet, a Nubian Sandstone escarpment into two sub-basins. The landscape in the area comprises great sand seas (Dahān Ubārī and Dahān Murzuq), gravel and boulder strewn wastelands and hyper-arid rock plateaux (Cremaschi and Di Lernia, 1996). The core of the Garamantian territory was the Wādī al-Ajāl. This is a valley-like depression in Fazzān containing parallel groups of oases. Among these, three main oasis belts can be distinguished; Wādī ash-Shatī in the North, Wādī al'Ajāl in the centre and Murzuq oasis and Wādī Barjūj in the South (Wilson, 2003).

Climatic Conditions in the Sahara and Fazzān

At present, the Sahara Desert is characterised by extreme shortage of water, substantial diurnal range of temperature, occasional strong winds and negligible rain-fall (Oberlander, 1994; Parsons and Abrahams, 1994; Brooks et al., 2003). In modern Fazzān, the conditions are similar to those of the rest of the Sahara, that is, the climate is hyper-arid; during summer the temperatures reach 50°C, while the winters are dry and warm.

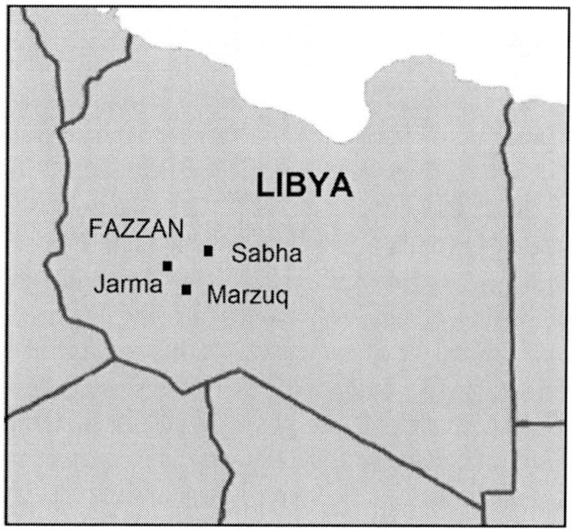

Figure 1. The location of Fazzān.

It should be pointed out that the conditions have not always been such. In the early Holocene, the climatic conditions improved, expressed as greater rainfall and lake development (Holl, 2004; Brooks et al., 2005; Barker, 2006; Kuper, 2006; Kuper and Kröpelin, 2006; Cremaschi and Zerboni, 2009).

In the Late Holocene, from 3,000 BC, desiccation was established and the large mammalian fauna disappeared (Cremaschi, 2002; Di Lernia and Palombini, 2002; Brooks et al., 2003, 2005; Kuper, 2006). By 1,000 BC the Saharan and Fazannese climate was as of today (Mattingly, 2000). The dropping of the water table in the region is ongoing and has been further intensified in the past few decades due to the extensive use of the water resources as well as oil extraction (Brooks et al., 2001).

Pre-Garamantian Fazzān

The Hominins were present in Fazzān since the Lower Pleistocene, but occupation appears to have been sporadic and dependant upon periodic wet intervals (White and Mattingly, 2006; Lahr et al., 2008). The climatic conditions from the Late Middle Palaeolithic to the Early Holocene made human subsistence in the Sahara impossible (Holl, 2004; White and Mattingly, 2006). Recolonisation of Fazzān occurred in the early Holocene, when the conditions became wetter (Holl, 2004).

The settlements were located in the valleys, where small lakes and springs were found (Barich, 1998; Mattingly, 2000, 2005; Garcea, 2006). As the environment gradually became more arid from the Middle Holocene onwards, the semi-permanent settlements developed nearby the water resulted in population growth and increased socio-cultural complexity (Barich, 2002; Brooks, 2006; Garcea, 2006; Cremaschi and Zerboni, 2009).

The Garamantes

The Garamantes were the founders of the first Libyan state and flourished between approximately 900 BC and 500 AD (Mattingly, 2006). Based on archaeological evidence and palaeoanthropological analyses, it is found that they descended from the local Neolithic populations (Mattingly, 2000; Di Lernia et al., 2002). The Garamantian society went through approximately 1,500 years of evolution, which are divided into four sub-periods; In the Early phase (1,000-500 BC), a few settlements were built on defensible posts. However, a well-developed irrigated cultivation system was already present. During the Proto-Urban phase (500-1 BC) Garama, the capital of the state, was built, the fortified hilltop villages were abandoned and settlements at the bottom of the valleys were developed as agricultural exploitation was intensified. In the Classic phase (1-400 AD), Garama was adorned with public buildings as a result of contact with the Roman Empire. Other towns and villages lying in the centre of depressions, where agriculture was practised, also made their appearance. Finally, the Late phase (400-700 AD) placed an emphasis again on defensive structures, which may suggest a collapse of the centralised state. In this phase, the agriculture was restricted around wells and the trading activity was limited (Liverani, 2004; Mattingly, 2006).

Basic Osteological Assessments

Number of Individuals

For each tomb, the most likely number of individuals, *N,* was estimated following two equations (Nikita, in press a):

$$N = \frac{M_1^2 + M_2^2 + \cdots + M_n^2}{4P} \tag{1}$$

and

$$N = \frac{R_1L_1 + R_2L_2 + \cdots + R_nL_n}{P} \tag{2}$$

where P is the number of pairs between left and right elements, $M_i = R_i + L_i$, and R_i and L_i are the numbers of right and left elements per bone type. These equations are extensions of the so-called Lincoln's expression (Adams and Konigsberg, 2004). The obtained results were further corroborated by means of a computer simulation (Nikita and Lahr, 2011)

Assessment of Sex

In the current study, only the adult individuals were examined. The pelvis was used as the most valid skeletal sex indicator applying Phenice's (1969) method. In addition, the shape of the sciatic notch and the presence of a pre-auricular sulcus were examined. The crania were also used for sexual discrimination. The anatomical regions of greater importance included the mastoid process, the upper orbital margin, the supraorbital crest, the glabella and the external occipital protuberance (Buikstra and Ubelaker, 1994).

Comparative Material

North African populations, chronologically close to the Garamantes as well as Egyptians from the 4th and 5th millennia BC, were incorporated in the analyses (Table 1). The location of the populations under study is given in Figure 2. The pairwise straight-linear geographic and temporal distances among populations are given in Table 2. The sample size for each of these groups varied according to the analyses performed and the elements employed in each of them. Thus, it is given separately in the relevant sections.

The Badarian culture flourished on the East bank of the Nile River and is associated with the earliest evidence of agriculture in Upper Egypt. The culture appears to have originated from multiple sources, with the western Desert being the most prominent (Hendrickx and Vermeersch, 2000; Keita, 2005). The Badarian settlements were occupied seasonally (Wengrow, 2006), and the main subsistence activities included agriculture, livestock keeping and fishing (Hendrickx and Vermeersch, 2000; Midant-Reynes, 2000a).

The Naqada culture appeared on the West bank of the Nile in Upper Egypt. The Naqada I period (4,400–3,500 BC) was similar to the Badarian, but with greater emphasis on food production, larger burial sites and greater variation in the types of grave goods, that seems to indicate social inequality (Midant-Reynes, 2000b). In Naqada II (3,500–3,200 BC), the culture expanded geographically, and more elaborate tombs and complex funerary rituals appeared (Midant-Reynes, 2000a, 2000b). This period also witnessed increased trade, craft specialization and urbanization (Friedman, 1996).

Naqada III (3,200–3,000 BC) was characterized by even more elaborate grave goods, indicating more pronounced social differences (Kemp, 1991). The economy was based on agriculture and stock breeding (McArdle, 1982). Hunting was marginal, but fishing played an important role (Hassan, 1988).

Table 1. Comparative samples

Population	Symbol	Date	Country	Collection
Garamantes	GAR	900BC-500AD	Libya	Museum of Jarma (Libya)
Carthage	CAR	751BC-435AD	Tunisia	Musée de l'Homme (Paris)
Algerians (Ain Beida, Skikda, Setif, Bijaia, Khenchela)	ALG	1500 BC	Algeria	Musée de l'Homme (Paris)
Alexandria	ALE	323BC-330AD	Egypt	Musée de l'Homme
Gizeh	GIZ	664-343 BC	Egypt	Duckworth Lab. (Cambridge)
el-Badari	BAD	4400-4000 BC	Egypt	Duckworth Lab. (Cambridge)
Naqada	NAQ	4000-3200 BC	Egypt	Duckworth Lab. (Cambridge)
Soleb	SOL	1575-1380 BC	Sudan	Musée de l'Homme (Paris)
Kerma	KER	2000-1550BC	Sudan	Duckworth Lab. (Cambridge)
Jebel Moya	JM	3000-100 BC	Sudan	Duckworth Lab. (Cambridge)

Kerma was the capital of the Kingdom of Kush in Upper Nubia (Hafsaas-Tsakos, 2009). The Kerma culture represents an indigenous society that was gradually influenced by the Egyptian culture, although its Nubian features remained dominant (Phillipson, 2005). This culture controlled the trade networks between Egypt, central Africa and the East, while the favorable position of the city on a wide floodplain enabled productive agriculture and access to local mineral resources (Trigger, 1976; Adams, 1977).

Skeletons from Soleb (Nubia), dated to the 18th Pharaonic Dynasty, were also analyzed (Trigger, 1976). During this period, the state of Kerma had been destroyed and Nubia was under Egyptian control (Adams, 1977). The origin of the Nubian samples is unclear. The individuals may have been locals that possessed a sub-Saharan component (Carlson and Van Gerven, 1979), or they may have been admixed migrants from elsewhere in North Africa (Turner and Markowitz, 1990).

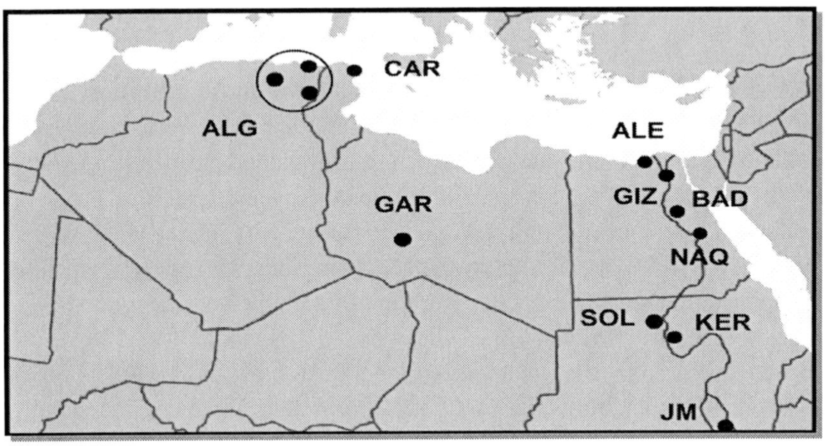

Figure 2. Location of the populations under study.

Table 2. Temporal (in years below the diagonal) and straight-linear geographic (in km above the diagonal) distances among populations under study

	GAR	CAR	ALG	ALE	GIZ	BAD	NAQ	SOL	KER	JM
GAR	0	1190	1235	1730	1810	1830	2000	1860	1920	2460
CAR	44	0	260	1950	2110	2250	2460	2690	2770	3470
ALG	1300	1340	0	2180	2350	2490	2690	2850	2900	3630
ALE	205	160	1504	0	180	485	650	1185	1280	1990
GIZ	300	345	1000	505	0	330	480	1070	1150	1840
BAD	4000	4045	2700	4205	3700	0	180	750	820	1520
NAQ	3400	3445	2100	3605	3100	600	0	650	720	1375
SOL	1280	1325	25	1485	980	2720	2120	0	90	840
KER	1575	1620	275	1780	1275	2425	1825	295	0	750
JM	1250	1290	50	1455	950	2750	2150	30	325	0

Jebel Moya is located south-southeast of Khartoum, Sudan (Mukherjee et al., 1955). The permanent water sources made the location favorable to occupation (Williams et al., 1982), although the site of Jebel Moya was most likely the meeting place of mobile groups, who subsisted on livestock keeping. After 500 BC, the settlement became permanent and flourished, while since the 8th century, imported goods started appearing and probably related to a social elite (Gerharz, 1994).

Gizeh is located near the south entrance of the Delta and is one of the northernmost cemeteries of the Memphite Necropolis. The material used in the present study was excavated in a cemetery south of the Gizeh pyramids and it belongs to the 26th–30th Dynasty (Petrie, 1907), a period characterized by intermittent Persian occupation in the area (Lloyd, 2000).

Alexandria was founded in 332 BC by Alexander the Great. From its port, the traders sailed throughout the Mediterranean to India and the East African coast (Phillipson, 2005). The city also developed significant agricultural production and exploited natural resources (Haas, 1997). In the Roman period, the social differentiation resulted in ethnic tensions and revolts (Alston, 1997).

The Phoenician civilization established commercial outposts throughout the Mediterranean, the most prominent being the Carthage, founded at 751 BC. Besides trading activities, the Phoenician prosperity was due to the cereal cultivation of the Tunisian plains (Phillipson, 2005). The decline began in 539 BC when Cyrus the Great conquered Phoenicia. However, Carthage continued to flourish until it was destroyed by Rome in 146 BC (Aubet, 2001).

Finally, protohistoric Algerians (1500 BC) were also studied (Keita, 1988). At that time, the Maghreb was occupied by Berber-speaking populations. The origin of these groups is uncertain, but they were most likely autochthonous to the region (Camps, 1974).

How Stressful Was the Desert?

Archaeological Data and Hypotheses Tested

As mentioned above, the Garamantian civilization developed at a time when the desertification of the Sahara had reached levels comparable to the present. Despite this harsh environment, the population flourished financially and culturally. The study of Garamantian activity markers can contribute to the evaluation of the daily stresses imposed by the desert environment and the effectiveness of the population in coping with the environmental pressures.

The principal Garamantian subsistence activity was agriculture based on subterranean channels, and later on wells (Wilson, 2003). These structures must have required substantial labor for their construction and maintenance, particularly on behalf of the male subjects. Moreover, the field preparation and tending, planting and harvesting involved numerous activities, more or less strenuous. The advent of farming is generally associated with reduced bilateral asymmetry (Ruff, 1987, 2000; Bridges, 1989; Marchi et al., 2006) and sexual dimorphism (Murdock and Provost, 1973; Ruff, 1987). Therefore, our expectation would be to find low levels of sexual dimorphism and bilateral asymmetry in the upper and lower limbs of the Garamantes.

Keeping livestock was also important to the Garamantes (Mattingly, 2003a), although it probably engaged a specialized sub-set of the population (Chang and Koster, 1986). The uneven terrain in Fazzān should have imposed great stresses on the lower limbs even when individuals covered small distances seeking pasture (Marchi et al., 2006; Pearson et al., 2006;

Sparacello and Marchi, 2008). The general pattern in agro-pastoralist societies is that males are primarily involved in the main production activities, whereas females perform more sedentary tasks (Ehrenberg, 1989). Thus, we would expect Garamantian males to be more involved in livestock keeping and consequently more mobile than females, a pattern which should manifest itself as greater robusticity in the lower limbs of male subjects.

The Garamantes also practiced metallurgy, worked with semi-precious stones and beads and produced pottery (Mattingly, 2003a). These tasks must have influenced the morphology of their limbs, but since they would have engaged a small part of the population, it would be unlikely to identify clear signs of these in the sample under examination. Trans-Saharan trade was also of primary importance to the population and extended across sub-Saharan Africa, Egypt and the North African Mediterranean littoral. The trade required long-distance travelling (Mattingly, 2003a), but as these trips took place on camels, they should also not have led to increased lower limb robusticity.

The construction of private houses and public buildings must have imposed strain on the skeleton of male individuals too. Indeed, the monumental architecture appeared in the Classic phase, while strong fortifications were built in the Early and Late phases (Mattingly, 2003b). Finally, the Garamantes may have engaged in hostile encounters with their neighbors especially in the Early and Late phases (Mattingly, 2003c). These tasks should result in males exhibiting more robust upper and lower limbs than females, while warfare may also increase the levels of bilateral asymmetry among males as a result of the unilateral use of weapons (Sparacello et al., 2011).

Based on the above data, the main hypotheses that can be formulated according to the variable Garamantian activities are the following. The agriculture and livestock keeping resulted in different patterns of robusticity and dimorphism. A large portion of the population should have engaged in agricultural production, whereas herding must have involved a subset of males. As a result, we would expect low levels of sexual dimorphism and low levels of mobility in the majority of the individuals under study. Nevertheless, the engagement of males in warfare and construction works should manifest as increased robusticity and greater bilateral asymmetry among the male subjects. Moreover, the harsh Central Saharan environment should have required more intense daily work on behalf of the Garamantes, resulting in greater upper and lower limb robusticity compared to the other North African populations occupying better watered environments. These hypotheses are tested by examining the musculoskeletal stress markers and biomechanical reconstructions (Nikita et al., 2011).

Methods and Materials

The populations whose geometric properties and musculoskeletal stress markers were used included, besides the Garamantes, the Jebel Moya, the Kerma and the el-Badari. The exact sample sizes per population and sex are given in the relevant tables. The remaining comparative groups were excluded as they did not preserve post-cranial elements. In the examined groups, all elements displaying poor preservation or pathological conditions that could affect bone shape were eliminated from the study.

Geometric Properties

Physical activity affects long bone diaphyses through structural modelling and remodelling (Weiss, 2003; Lieberman et al., 2004; Ruff et al., 2006; Shaw and Stock, 2009a, 2009b). In the current study, the cross-sectional morphology of clavicles, humeri, ulnae, radii, femora and tibiae was assessed using a periosteal mould based technique, which excludes the medullary cavity (Stock and Shaw, 2007; Sparacello and Pearson 2010). The moulds were taken at the mid-shaft of all elements except for the humeri, for which the 35% distance from the distal end was used in order to avoid the deltoid muscle attachment. Each mould was digitized and cross-sectional geometric properties were estimated using the software package Image J. The estimated cross-sectional geometric properties included *TA* and the ratios I_x/I_y and I_{max}/I_{min}. The *TA* is the total sub-periosteal area, that is, the total area encompassed by the outer perimeter of the bone section, while I_x, I_y, I_{max}, I_{min} are second moments of area and measure how the material is distributed, either using orthogonal axes defined by arbitrary reference planes (I_x, I_y) or orthogonal principal axes (I_{max}, I_{min}). Thus, the second moments of area represent indicators of cross-sectional shape (Lieberman et al., 2004; Ruff, 2008 and references therein).

Since skeletal robusticity is the strength/rigidity of a structure in relation to body size, *TA* was standardized according to body mass. The body mass was calculated as the average of the following three equations (Auerbach and Ruff, 2004):

$$BM = 2.239 \times FHD - 39.9 \text{ (McHenry, 1992)} \quad (3)$$

$$BM = 2.268 \times FHD - 36.5 \text{ (Grine et al., 1995)} \quad (4)$$

$$BM = 2.741 \times FHD - 54.9 \text{ (males); } BM=2.426 \times FHD-35.1 \text{ (females) (Ruff et al., 1991)} \quad (5)$$

where BM is the body mass (in kg) and FHD is the supero-inferior femoral head diameter (in mm). The bilateral asymmetry was calculated as:

$$a = 100 \times (X_{max}-X_{min}) / X_{min} \quad (6)$$

where X stands for any biomechanical property. The percent sexual dimorphism was determined by:

$$d = 100 \times (X_{male} - X_{female}) / X_{female} \quad (7)$$

Statistical analyses were performed in order to explore inter-population differences and intra-population sexual dimorphism and bilateral asymmetry. These analyses included Univariate Analysis of Variance (ANOVA), independent t-tests, Wilcoxon signed-rank and Mann-Whitney tests.

Musculoskeletal Stress Markers

Musculoskeletal stress markers (MSMs) are non-pathological markings at muscle, ligamentous or tendinous attachment sites, onto the bone surface. When these sites are subjected to stress, there is new bone formation so that the bone withstands the increased strain. However, the continuous stress applied on the insertion sites can lead to bone resorption (Knüsel, 2000 and references therein).

The musculoskeletal stress markers examined are given in Table 3. Each marker was recorded separately on each side of the body following the method proposed by Hawkey and Merbs (1995). However, only two dimensions were used, robusticity and stress lesion, and these were scored on a unified scale from 0 to 6. The arithmetic mean was used as the measure of mean tendency for each site (Molnar, 2006). The sexual dimorphism was examined using a two-sample Wilcoxon rank sum (Mann-Whitney) test. For bilateral asymmetry, a Wilcoxon matched-pairs signed-ranks test was applied. Finally, inter-population differences were assessed using a two-sample Wilcoxon rank sum (Mann-Whitney) test.

For assessing bilateral asymmetry the formula:

$$a = (\text{left MSM score}/\text{right MSM score}) \times 100 \quad (8)$$

was used (Hall, 1982). A result of less than 100 indicates right dominance and a score of more than 100 indicates left dominance (Eshed et al., 2004). The sexual dimorphism was calculated as (Hall, 1982):

$$d = 100 \times (\text{mean MSM score for males-mean MSM score for females})/(\text{mean MSM score for males}) \quad (9)$$

Table 3. List of the musculoskeletal stress markers (MSMs) studied

Clavicle	Humerus	Radius	Ulna	Femur	Tibia
pectoralis major (o)	deltoid (i)	biceps brachii	triceps brachii	gluteus maximus	soleus
trapezius	latissimus dorsi	supinator		gluteus medius	popliteus (o)
deltoid (o)	pectoralis major (i)			adductor magnus	semimembranosus
	teres major			vastus medialis	tibialis posterior
				piriformis	tibialis anterior
				gluteus minimus	
				obturator externus	
				popliteus (i)	
				vastus lateralis	
				gastrocnemius	
				Iliacus	
				Pectineus	

o= origin, i= insertion.

Table 4. Inter-population comparisons of cross-sectional geometric properties among males (upper limbs)

		Garamantes	Jebel Moya	Kerma	Badari	Significance
Bone	Property	Mean ± se	Mean ± se	Mean ± se	Mean ± se	
right side						
Clavicle GAR = 9, JM = 7 KER = 14, BAD = 3	*TA*	151±7	176±14	154±5	129±2	
	I_x/I_y	0.65±0.08	0.50±0.05	0.66±0.05	0.73±0.04	
	I_{max}/I_{min}	1.93±0.16	2.31±0.23	1.83±0.09	1.46±0.09	
Humerus GAR = 8, JM =15 KER = 26, BAD = 5	*TA*	420±17	491±19	442±9	366±16	GAR vs. JM*
	I_x/I_y	1.08±0.05	0.97±0.03	1.09±0.03	1.28±0.07	
	I_{max}/I_{min}	1.30±0.07	1.18±0.03	1.21±0.03	1.43±0.06	
Radius GAR = 7, JM=10 KER = 20, BAD = 4	*TA*	183±5	222±11	187±5	156±6	GAR vs. JM*
	I_x/I_y	0.67±0.04	0.77±0.27	0.68±0.22	0.68±0.28	
	I_{max}/I_{min}	1.55±0.08	1.39±0.04	1.56±0.05	1.70±0.20	
Ulna GAR = 7, JM = 8 KER = 9, BAD = 2	*TA*	205±6	245±11	204±6	169±1	GAR vs. JM* / GAR vs. BAD**
	I_x/I_y	0.92±0.08	1.24±0.05	0.78±0.03	0.74±0.11	GAR vs. JM**
	I_{max}/I_{min}	1.61±0.22	1.47±0.07	1.43±0.06	1.76±0.38	
left side						
Clavicle GAR = 8, JM = 8 KER = 19, BAD = 2	*TA*	145±8	189±9	151±5	135±2	GAR vs. JM***
	I_x/I_y	0.65±0.04	0.64±0.06	0.65±0.04	0.65±0.04	
	I_{max}/I_{min}	1.75±0.18	1.98±0.11	1.70±0.08	1.57±0.12	
Humerus GAR = 6, JM =14 K = 23, BAD = 5	*TA*	395±16	458±18	410±9	360±18	
	I_x/I_y	1.09±0.04	1.06±0.03	1.06±0.03	1.24±0.34	
	I_{max}/I_{min}	1.27±0.09	1.17±0.21	1.23±0.03	1.31±0.05	
Radius GAR = 7, JM = 9 KER = 21, BAD = 4	*TA*	178±6	215±10	185±5	164±6	GAR vs. JM*
	I_x/I_y	0.76±0.04	0.71±0.04	0.77±0.03	0.71±0.03	
	I_{max}/I_{min}	1.47±0.12	1.51±0.08	1.50±0.05	1.52±0.04	
Ulna GAR = 8, JM = 5 KER = 11, BAD = 3	*TA*	201±8	242±16	196±4	185±10	
	I_x/I_y	1.02±0.08	1.27±0.09	0.66±0.05	0.84±0.01	GAR vs. KER**
	I_{max}/I_{min}	1.49±0.18	1.41±0.08	1.66±0.12	1.46±0.12	

*significant at P < 0.05, **significant at P < 0.01, ***significant at P < 0.001.

Table 5. Inter-population comparisons of cross-sectional geometric properties among males (lower limbs)

		Garamantes	Jebel Moya	Kerma	Badari	Significance
Bone	Property	Mean ± se	Mean ± se	Mean ± se	Mean ± se	
right side						
Femur GAR = 7, JM =31 KER = 18, BAD = 4	*TA*	869±32	962±11	871±19	754.5±39	GAR vs. JM*
	I_x/I_y	1.42±0.15	1.32±0.05	1.29±0.05	1.36±0.18	
	I_{max}/I_{min}	1.65±0.16	1.52±0.04	1.37±0.04	1.43±0.18	

Table 5. (Continued)

Bone	Property	Garamantes Mean ± se	Jebel Moya Mean ± se	Kerma Mean ± se	Badari Mean ± se	Significance
right side						
Tibia GAR = 7, JM=14 KER = 17, BAD = 4	TA	751±36.1	850±17.9	760±21.8	681±22	
	I_x/I_y	2.12±0.12	1.89±0.09	1.88±0.08	2.14±0.12	
	I_{max}/I_{min}	2.28±0.13	1.93±0.09	1.99±0.08	2.17±0.11	
left side						
Femur GAR = 6, JM=26 KER = 19, BAD = 2	*TA*	915±25	962±13	866±11	752±103	GAR vs. BAD*
	I_x/I_y	1.37±0.12	1.35±0.05	1.15±0.05	1.46±0.21	
	I_{max}/I_{min}	1.52±0.11	1.43±0.05	1.30±0.04	1.53±0.16	
Tibia GAR = 6, JM= 12 KER = 16, BAD = 4	*TA*	730±20	821±18	731±28	638±32	
	I_x/I_y	2.25±0.20	2.06±0.14	1.99±0.09	2.22±0.22	
	I_{max}/I_{min}	2.38±0.23	2.21±0.12	2.05±0.08	2.28±0.21	

*significant at $P < 0.05$.

RESULTS

Cross-sectional Geometry

Tables 4-7 present the inter-population comparisons in I_x/I_y, I_{max}/I_{min} and standardized *TA* for the upper and lower limbs. As can be observed, most significant inter-population differences are found in the upper limbs. Moreover, the upper limb differences relate to both diaphyseal shape and rigidity, whereas the lower limb differences relate only to rigidity. More specifically, in the upper limb, statistically significant differences were observed in the ulnae of male Garamantes which were stronger than those of the males from el-Badari and Kerma (Table 4). In contrast, the Jebel Moya were significantly more robust than the Garamantes in several upper limb elements. The lower limbs of the male Garamantes showed very few significant differences to all comparative populations (Table 5). Among females, the Garamantian upper limbs exhibited stronger radii than the el-Badari, stronger ulnae and more gracile humeri than the Kerma. Compared to the Jebel Moya females, the Garamantes showed stronger humeri, while the Jebel Moya displayed stronger clavicles and ulnae (Table 6). In the lower limbs, the Jebel Moya showed stronger tibiae than the Garamantes (Table 7).

The above observations are visualized in the box-plots of Figures 3-6, which present the values of the cross-sectional properties concerning the lower limbs in order to illustrate more clearly the effect of the Saharan terrain on bone morphology. In particular, Figures 3-4 depict inter-population variation in I_x/I_y. It can be seen that all populations overall exhibit very similar values in diaphyseal shape for both sexes. In addition, Figure 4 shows that despite the overall similarity, el-Badari tibial values are greater than among the remaining groups, although this was not significant. Figure 5 shows that Jebel Moya males had stronger lower limbs in terms of *TA* than the other North African populations, while the el-Badari were the most gracile. The Garamantes and Kerma were very similar to each other. Finally, in Figure

6, we observe that the female Jebel Moya exhibit stronger tibiae compared to other groups, whereas, the remaining populations are more or less similar.

From the above data, it appears that the significant differences in robusticity and diaphyseal shape between the Garamantes on the one hand and the el-Badari and Kerma on the other were few. The mean *TA*, I_x/I_y, and I_{max}/I_{min} values among populations were quite similar in both sexes and showed no clear pattern. In contrast, compared to the Jebel Moya, the Garamantian males and females showed lower values of I_x/I_y, and I_{max}/I_{min}, particularly in the upper limbs. In addition, the mean *TA* values in Jebel Moya were higher than those in the Garamantes in the upper and lower limbs for both sexes, suggesting greater strength in the Jebel Moya. Table 8 shows that the Garamantes exhibited low levels of bilateral asymmetry. The males showed systematically greater asymmetry in *TA* than females, although this asymmetry reached statistical significance only for the clavicles. The Jebel Moya and el-Badari also showed low asymmetry and in only very few properties was a significant bilateral difference found. In contrast, several properties expressing diaphyseal shape and rigidity appeared significantly asymmetric in the Kerma sample. The right side was dominant in almost all the cases where significant differences were found. The examination of inter-population differences in bilateral asymmetry gave significant results only in *TA* for male humeri between the Garamantes and Jebel Moya ($p = 0.039$), and in I_x/I_y for female radii between the Garamantes and Kerma ($p = 0.039$). Therefore, we observe very low levels of bilateral asymmetry in the Garamantes, as we do among the el-Badari and Jebel Moya, in contrast to the pattern found in the Kerma in which numerous properties were significantly asymmetric in both the sexes. It is interesting to note that the Garamantes exhibited the least asymmetry compared to other North African populations and, although Garamantian males showed greater asymmetry in *TA* than females, this was not statistically significant.

Table 6. Inter-population comparisons of cross-sectional geometric properties among females (upper limbs)

		Garamantes	Jebel Moya	Kerma	Badari	Significance
Bone	Property	Mean ± se	Mean± se	Mean± se	Mean± se	
right side						
Clavicle	*TA*	138±6	169±6	129±4	139±18	GAR vs. JM*
GAR = 15, JM= 6,	I_x/I_y	0.71±0.04	0.58±0.04	0.64±0.04	0.64±0.06	
KER = 20, BAD = 3	I_{max}/I_{min}	1.64±0.08	1.92±0.19	1.78±0.08	1.70±0.17	
Humerus	*TA*	385±9	412±20	362±7	391±28	
GAR = 14, JM= 9,	I_x/I_y	1.02±0.06	1.03±0.04	1.17±0.04	1.18±0.085	GAR vs. KER*
KER = 32, BAD = 4	I_{max}/I_{min}	1.39±0.05	1.18±0.04	1.32±0.03	1.32±0.05	GAR vs. JM*
Radius	*TA*	167±6	186±10	162±3	169±8	
GAR = 14, JM= 6,	I_x/I_y	0.74±0.03	0.68±0.05	0.66±0.02	0.56±0.04	
KER = 19, BAD = 3	I_{max}/I_{min}	1.50±0.07	1.60±0.13	1.63±0.06	1.87±0.10	
Ulna	*TA*	190±6	196±61	178±4	207±5	
GAR = 16, JM= 2,	I_x/I_y	0.96±0.06	1.15±0.01	0.75±0.04	0.83±0.09	GAR vs. KER*
KER = 15, BAD = 3	I_{max}/I_{min}	1.53±0.08	1.27±0.02	1.62±0.07	1.70±0.29	
left side						
Clavicle	*TA*	144±7	156±13	124±4	144±14	
GAR = 12, JM= 5,	I_x/I_y	0.74±0.06	0.63±0.03	0.59±0.03	0.67±0.10	
KER = 17, BAD = 3	I_{max}/I_{min}	1.71±0.11	1.80±0.13	1.85±0.09	1.58±0.20	

Table 6. (Continued)

		Garamantes	Jebel Moya	Kerma	Badari	Significance
Bone	Property	Mean ± se	Mean± se	Mean± se	Mean± se	
Humerus GAR = 14, JM= 8, KER = 25, BAD = 4	*TA*	387±12	404±23	355±8	391±31	
	I_x/I_y	1.01±0.04	1.14±0.06	1.17±0.03	1.10±0.08	GAR vs. KER*
	I_{max}/I_{min}	1.31±0.04	1.23±0.04	1.33±0.03	1.25±0.04	
Radius GAR = 15, JM= 9,KER = 19, BAD = 3	*TA*	163±4	202±15	154±3	175±3	
	I_x/I_y	0.72±0.03	0.73±0.04	0.74±0.02	0.57±0.01	GAR vs. BAD**
	I_{max}/I_{min}	1.49±0.06	1.48±0.09	1.54±0.04	1.80±0.02	
Ulna GAR = 12, JM= 4, KER = 13, BAD = 3	*TA*	186±7	211±14	176±5	195±4	
	I_x/I_y	1.05±0.10	1.48±0.08	0.65±0.04	0.73±0.09	GAR vs. JM* / GAR vs. KER**
	I_{max}/I_{min}	1.66±0.11	1.83±0.15	1.74±0.12	1.61±0.14	

*significant at P < 0.05, **significant at P < 0.01, ***significant at P < 0.001.

Table 7. Inter-population comparisons of cross-sectional geometric properties among females (lower limbs)

		Garamantes	Jebel Moya	Kerma	Badari	Significance
Bone	Property	Mean ± se	Mean ± se	Mean ± se	Mean ± se	
right side						
Femur GAR = 6, JM=14, KER= 24, BAD=2	*TA*	772±40	828±18	773±12	833±43	
	I_x/I_y	1.17±0.14	1.28±0.05	1.14±0.04	1.25±0.02	
	I_{max}/I_{min}	1.41±0.13	1.39±0.05	1.29±0.02	1.28±0.04	
Tibia GAR = 15, JM= 8, KER=20, BAD= 2	*TA*	657±20	760±21	628±13	662±31	GAR vs. JM**
	I_x/I_y	1.92±0.11	1.97±0.13	2.24±0.13	2.40±0.28	
	I_{max}/I_{min}	2.05±0.12	1.98±0.13	2.27±0.13	2.49±0.22	
left side						
Femur GAR= 14, JM=15, KER=19, BAD= 4	*TA*	815±20	846±11	784±14	881±69	
	I_x/I_y	1.12±0.08	1.19±0.06	1.06±0.05	1.09±0.04	
	I_{max}/I_{min}	1.41±0.06	1.33±0.05	1.30±0.04	1.40±0.17	
Tibia GAR= 6, JM = 7 KER =17, BAD=2	*TA*	653±19	762±24	634±18	629±12	GAR vs. JM*
	I_x/I_y	1.94±0.07	1.81±0.10	2.21±0.15	2.42±0.24	
	I_{max}/I_{min}	2.06±0.08	1.85±0.10	2.26±0.15	2.52±0.16	

* significant at P < 0.05, **significant at P < 0.01.

As for sexual dimorphism (Table 9) among the Garamantes, only the male lower limbs were significantly more robust than the females, although no significant difference was traced in terms of diaphyseal shape. Overall, the el-Badari and Jebel Moya showed very low levels of dimorphism, while in the Kerma, several properties and elements were significantly different between males and females.

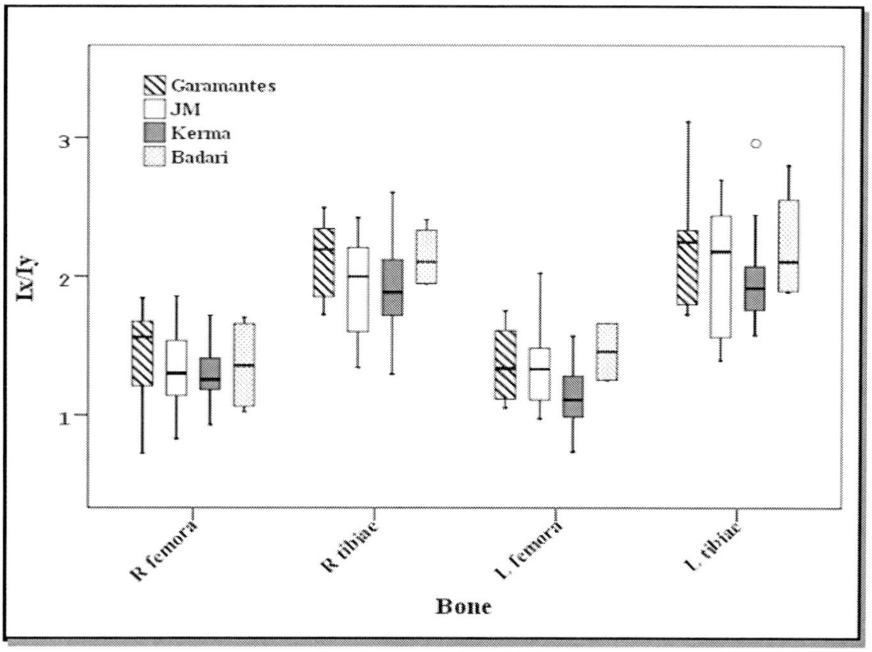

Figure 3. Inter-population comparisons for I_x/I_y (males, lower limbs).

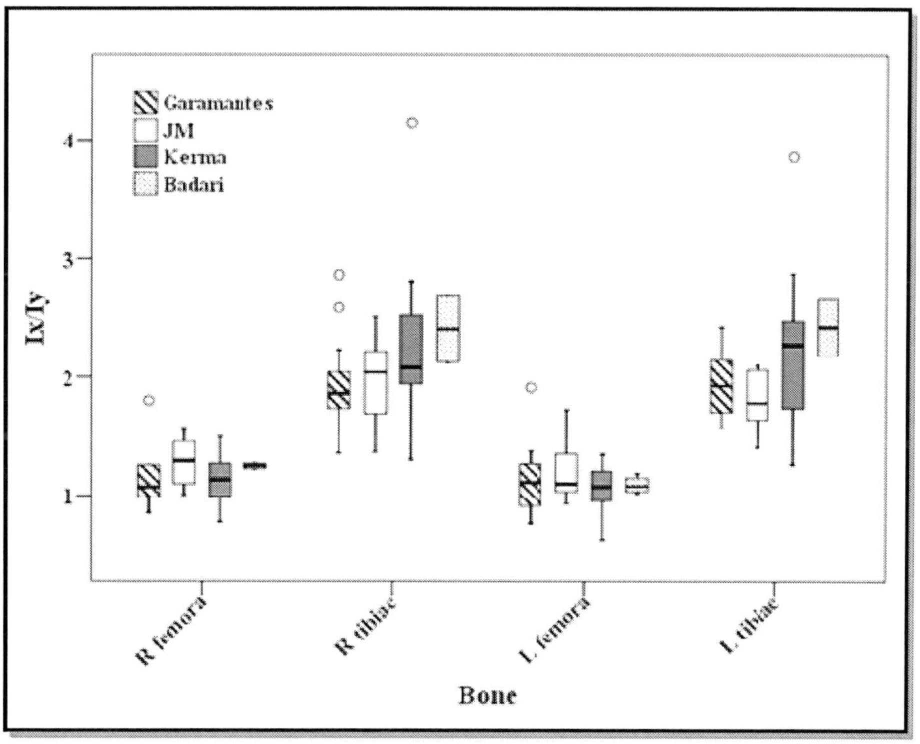

Figure 4. Inter-population comparisons for I_x/I_y (females, lower limbs).

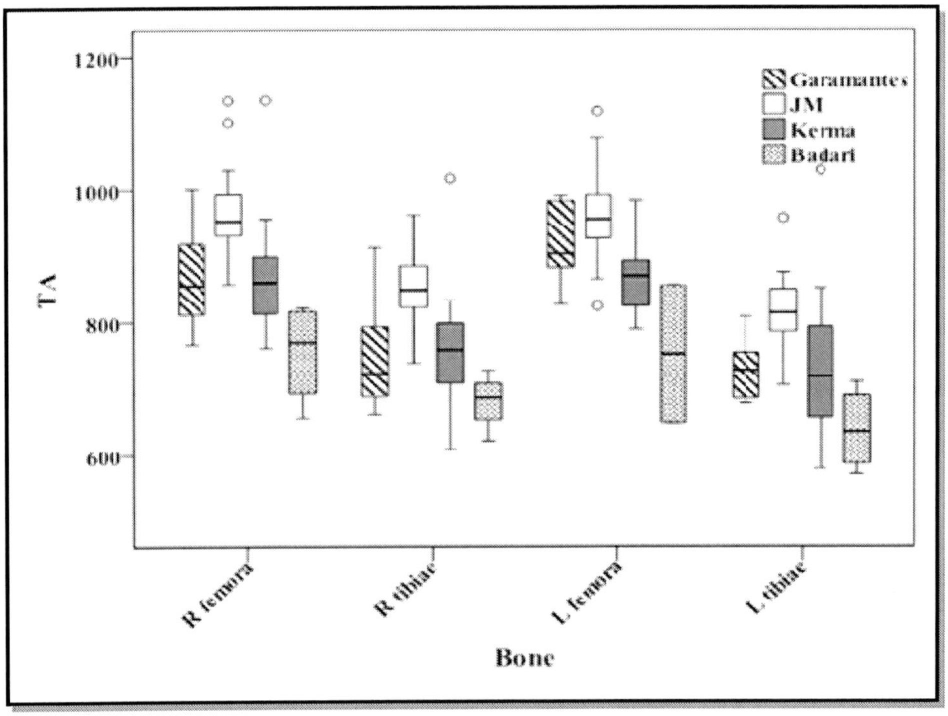

Figure 5. Inter-population comparisons for *TA* (males, lower limbs).

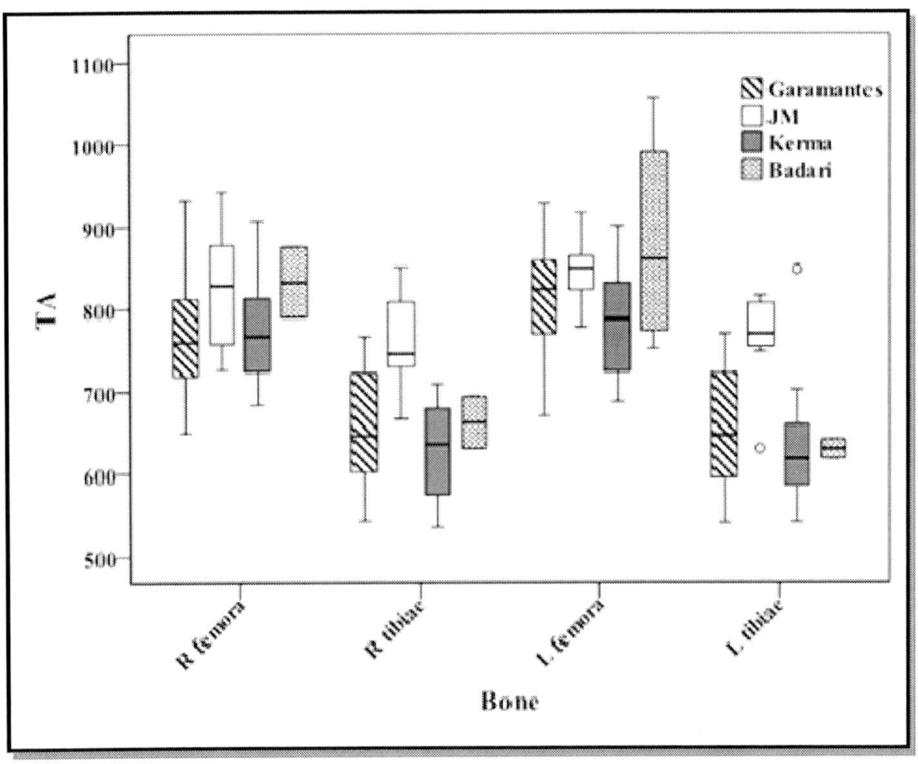

Figure 6. Inter-population comparisons for *TA* (females, lower limbs).

Table 8. Median Percent Bilateral Asymmetry among populations and sexes

		Garamantes		Jebel Moya		Kerma		Badari	
Bone	Property	Males	Females	Males	Females	Males	Females	Males	Females
Humerus	*TA*	2.0	0.03	13.0^{**}	9.5	8.0^{***}	2.8	1.7	0.04
	I_x/I_y	1.3	0.7	5.5	2.6	0.5	1.4	3.1	7.9
	I_{max}/I_{min}	4.1	1.3	0.01	0.4	3.0	1.0	8.2	5.8
Radius	*TA*	2.0	0.1	6.7	1.3	2.2^{*}	4.6^{*}	0.2	3.6
	I_x/I_y	13.2^{*}	4.2	6.6	2.1	14.9^{**}	15.6^{**}	0.7	1.8
	I_{max}/I_{min}	2.5	1.5	3.8	3.7	4.7	9.9^{*}	2.8	3.6
Ulna	*TA*	3.4	0.8	2.1		5.8	1.7		0.4
	I_x/I_y	5.7	16.6	2.7		5.3	18.8^{*}		3.9
	I_{max}/I_{min}	3.9	7.5	7.9		9.3	8.8		18.0
Clavicle	*TA*	6.1^{*}	2.9	11.4	0.3	3.2	5.2	3.1	4.0
	I_x/I_y	3.7	6.6	8.2	8.5	5.4	5.1	12.9	8.9
	I_{max}/I_{min}	9.2	6.3	9.8	4.9	10.7^{*}	11.1	11.3	10.5

* significant at $P < 0.05$, ** significant at $P < 0.01$, *** significant at $P < 0.001$.

Table 9. Mean Percent Sexual Dimorphism among populations and sides

		Garamantes		Jebel Moya		Kerma		Badari	
Bone	Property	Right	Left	Right	Left	Right	Left	Right	Left
Humerus	*TA*	9.0	2.1	20.0^{**}	13.3	22.2^{***}	16.9^{***}	-6.3	-7.8
	I_x/I_y	6.0	8.8	-4.9	-6.8	-6.8^{*}	-8.6^{*}	8.2	13.2
	I_{max}/I_{min}	-6.1	-2.8	0.2	-4.9	-8.3^{**}	-7.5^{*}	7.8	5.4
Radius	*TA*	10.6	9.3	6.6	6.5	15.8^{***}	19.7^{***}	-7.4	-6.4
	I_x/I_y	-10.3	6.7	8.6	-3.1	3.3	3.6	21.4^{*}	24.6^{**}
	I_{max}/I_{min}	2.9	-1.5	-9.4	2.3	-4.4	-2.4	-9.1	-15.6^{**}
Ulna	*TA*	8.0	8.1	24.4	14.7	14.6^{***}	11.5^{**}	-15.8^{*}	-5.2
	I_x/I_y	-3.9	-3.0	7.2	-13.7	5.2	1.6	-7.6	14.7
	I_{max}/I_{min}	5.5	-9.9	17.5	-23.0^{*}	-11.4	-5.0	3.0	-9.6
Clavicle	*TA*	9.7	0.6	7.7	21.2	20.1^{***}	21.7^{***}	-7.3	-6.6
	I_x/I_y	-7.3	-12.8	-15.1	1.7	2.9	10.2	12.7	-3.9
	I_{max}/I_{min}	17.6	2.2	21.9	9.8	2.8	-8.1	-14.2	-0.5
Femur	*TA*	12.6	12.3^{**}	16.2^{***}	13.6^{***}	12.6^{***}	10.7^{***}	-9.4	-14.7
	I_x/I_y	20.7	21.9	2.5	13.5	13.2^{*}	6.7	9.1	33.6
	I_{max}/I_{min}	16.8	7.8	9.3	7.4	6.7	-0.6	12.0	9.1
Tibia	*TA*	14.8^{*}	11.7^{*}	11.8^{**}	7.7	20.6^{***}	15.4^{**}	2.8	1.3
	I_x/I_y	9.6	16.1	-3.9	13.7	-17.2^{*}	-10.2	-10.9	-7.9
	I_{max}/I_{min}	11.4	15.5	-2.7	19.1	-13.6	-9.2	-13.2	-9.8

* significant at $P < 0.05$, ** significant at $P < 0.01$, *** significant at $P < 0.001$.

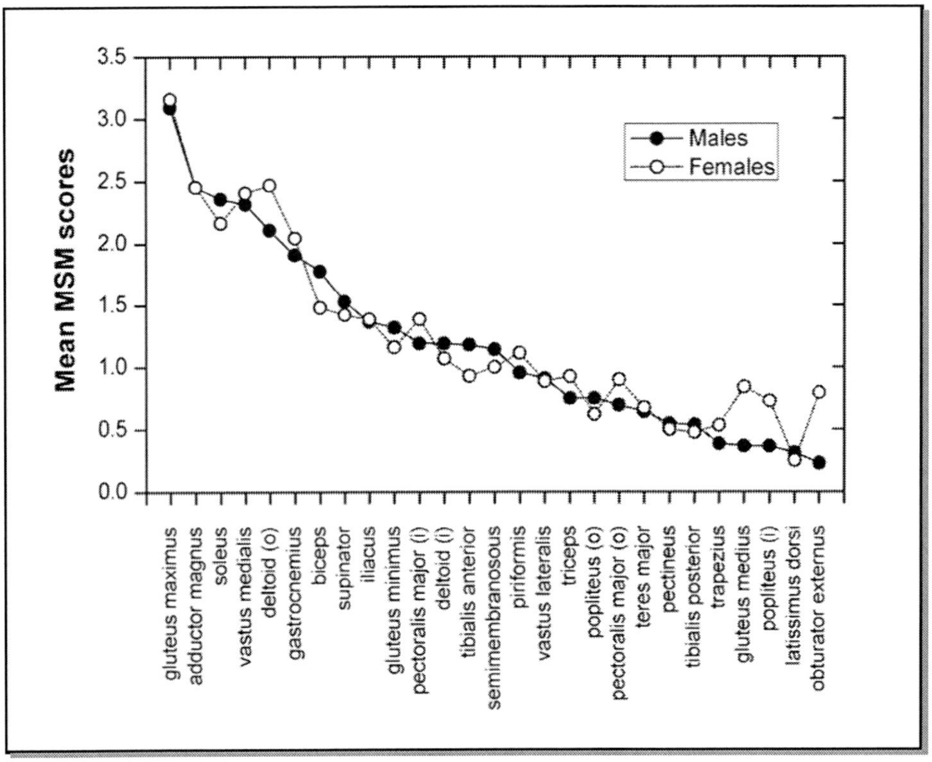

Figure 7. Ranking of the mean MSM scores of males (●) and females (o).

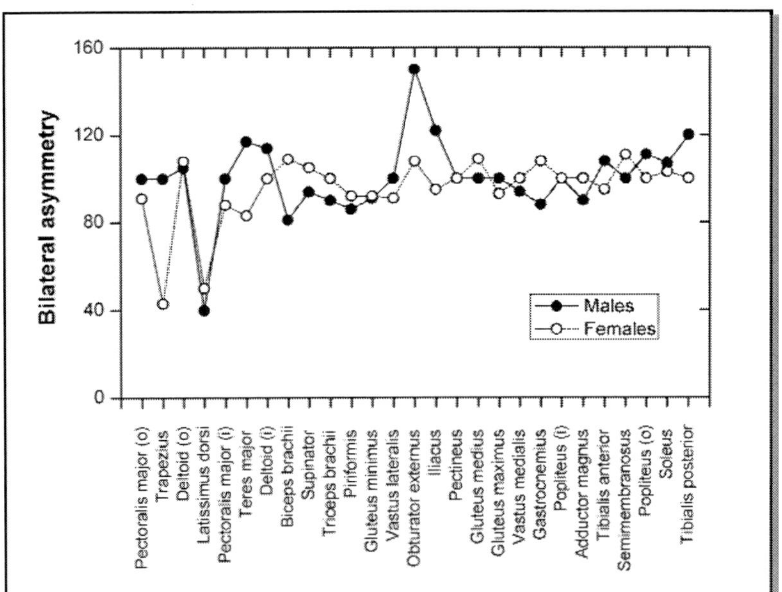

Figure 8. Variation of bilateral asymmetry among the various muscle/ligament attachment sites of males (●) and females (o).

Musculoskeletal Stress Markers

The rank ordering of the MSM scores was very similar between Garamantian males and females, with gluteus maximus, adductor magnus, soleus, vastus medialis and deltoid (o), exhibiting the highest scores (Figure 7). Therefore, the same muscles were used with the same frequency in both sexes. It should be noted that certain exceptions were found, such as the obturator externus, popliteus (i) and gluteus medius, which were used more often among females than males. The male mean MSM values only exceeded those of females in 12 out of the 27 sites, while in 14 out of the 27 sites, female values exceeded the males. Statistically significant differences were traced in the right supinator (p=0.033, M=18, F=21), right latissimus dorsi (p=0.031, M=17, F=23), right gluteus medius (p=0.027, M=10, F=19), and right and left obturator externus (p=0.011, M=10, F=19, and p=0.031, M=9, F=16, respectively). In the former two cases, the male values were higher than female ones, while in the latter three it was the opposite.

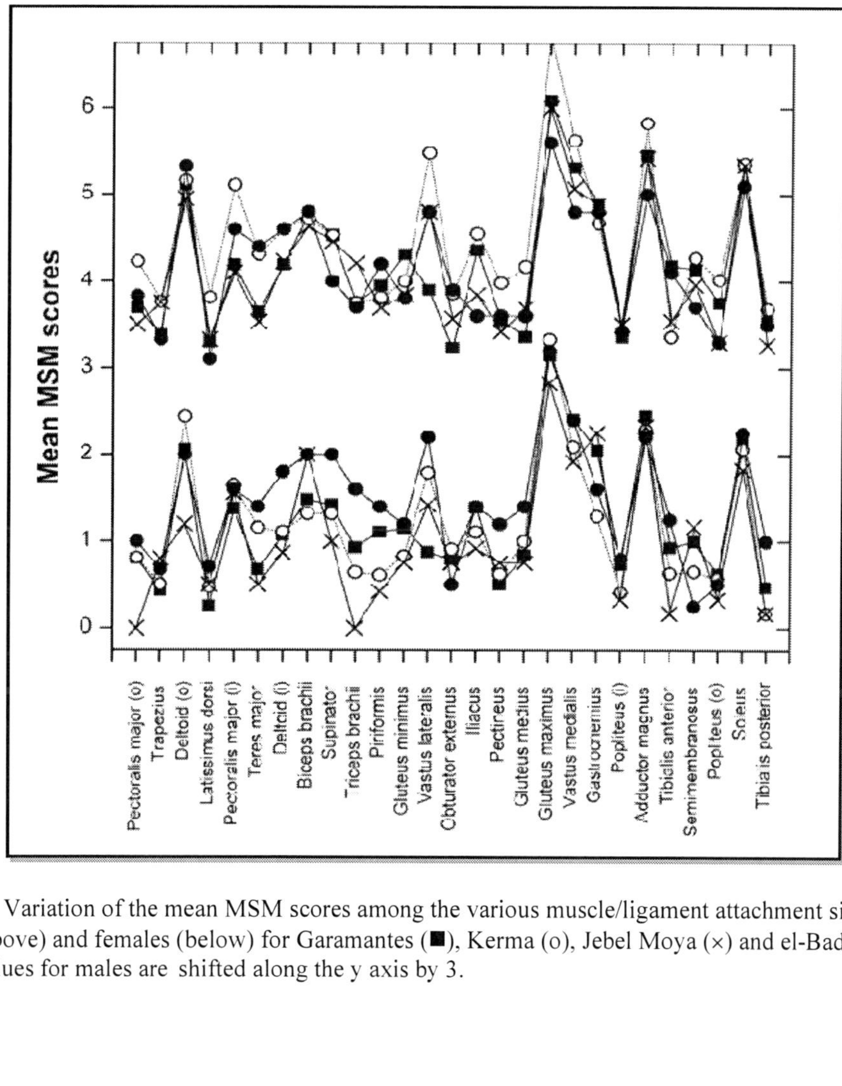

Figure 9. Variation of the mean MSM scores among the various muscle/ligament attachment sites of males (above) and females (below) for Garamantes (■), Kerma (o), Jebel Moya (×) and el-Badari (●). MSM values for males are shifted along the y axis by 3.

The levels of bilateral asymmetry were low and, as expected, higher in the upper limbs than in the lower ones (Figure 8). Among males, 9 out of the 27 sites showed no side preference, 9 showed left-side dominance and 9 showed right-side dominance. Among females, 8 sites showed no side dominance, 8 sites showed left-side dominance and 11 sites showed right-side dominance. It should be noted that in several occasions, the sites that exhibited right-side dominance among males, showed left-side dominance among females, and vice versa. Thus, no clear pattern could be identified within or between the sexes. The statistical testing showed that the only significant bilaterally different MSM sites for males were the deltoid (i) ($p = 0.046$) and the biceps brachii ($p = 0.046$), while for females no significant differences were found.

Given the low levels of bilateral asymmetry, inter-population comparisons were made for pooled sides (Figure 9). The Garamantian males appeared more gracile in muscular activity than the Kerma and more robust than the Jebel Moya, but they also appeared stronger in the lower limbs and more gracile in the upper limbs when compared to the el-Badari. The Garamantian females turned out to be stronger overall than the Kerma and the Jebel Moya but more gracile than the el-Badari.

Discussion

The results of the cross-sectional study and musculoskeletal stress markers show very few significant differences between the Garamantes and the comparative populations. More specifically, *TA* was higher among male Garamantes compared to that in the el-Badari and very similar between the two populations for females. This result is rather surprising, since the el-Badari predate the Garamantes, and their involvement in semi-nomadic pastoralism should have imposed greater loads on their lower limbs. It is possible to attribute this result to the uneven terrain in Fazzān, which must have resulted in increased lower limb robusticity for the Garamantes. However, the small sample size for the el-Badari renders these results inconclusive. The MSM analyses showed Garamantian males and females to be overall less robust than the el-Badari, which partly contradicts the cross-sectional geometry results, but it is to be expected given the date and subsistence economy of the el-Badari.

Only a few significant differences were found in the cross-sectional geometric properties between the Garamantes and Kerma, which suggests similar activity patterns for these populations and agrees with archaeological data for similar societal structure and subsistence economy in these groups. This conclusion is corroborated by the MSM analyses, which identified few differences between the two groups.

Finally, compared to the Jebel Moya, the Garamantian males and females appeared more gracile in cross-sectional geometry, particularly in the upper limbs. However, the MSM analyses showed the Garamantes to be overall more robust than the Jebel Moya, but only in a few muscular attachment sites, which are localized in the lower limbs of males and upper limbs of females.

It appears that the Garamantian activities, focusing primarily on agriculture and livestock keeping, did not impose greater physical loads on the population compared to the daily tasks of other North African groups, who occupied better watered environments. The livestock keeping and other activities that required mobility did not stress the lower limbs of the population with a few exceptions. This indicates that the Garamantes did not have to cover long distances on foot and/or that the uneven terrain in Fazzān was not as stressful as expected. It is also possible that the population occupied mainly the valleys and did not need to traverse the gravel and boulder strewn wastelands and hyper-arid rock plateau on a regular basis. Moreover, the activities that involved the upper limbs were also not more strenuous than in other populations that occupied less extreme environments. Therefore, the life in the Sahara Desert was not particularly more stressful in terms of mechanical loading of the skeleton than in other North African regions.

With respect to intra-population analyses, the cross-sectional geometry and the MSMs showed very low levels of bilateral asymmetry in the Garamantes. This result is in accordance to the pattern obtained in other agricultural populations (Bridges, 1989; Ruff, 2000; Marchi et al., 2006). The low levels of asymmetry indicate bilateral activities, but they may also suggest that unilateral activities may not have been habitual enough. Sexual dimorphism was also limited in the Garamantes, which is again consistent with the pattern expected in an agricultural population (Murdock and Provost, 1973; Ruff, 1987). In cross-sectional geometric properties, the only significant difference was traced in *TA* for the lower limbs, which may relate to the involvement of males in livestock keeping. Furthermore, the ranking of MSM sites was almost identical for both sexes, so males and females were utilizing the same muscles with a similar frequency. The muscles that were used significantly more frequently by females are located in the lower limbs and likely related to females' engagement in grinding the grains. The repetitive nature of this activity during which the individuals crouched over big grinding slabs would have stressed the knees, ankles and feet. Contrary to this, the muscles which were used significantly more by males than by females are those responsible for fine hand movements (Smith et al., 1996), like the ones required in metallurgy and precious stone processing. However, taking into account that only a subset of the population was involved in the above activities and that the differences between the two sexes were very few and localized in specific MSMs, the above interpretation of the underlying factors of sexual dimorphism may be tentative.

SAHARA: BARRIER OR CORRIDOR?

Archaeological Data and Hypotheses Tested

The Sahara has been hyper-arid since the mid-Holocene (Brooks et al., 2003). However, several lines of archaeological evidence suggest that various North African populations were in contact with each other, exchanging material goods and cultural practices. For example, rock art dated to the second half of the Holocene depicts what have been interpreted as multi-ethnic groups (Ottoni, 2007). Furthermore, the Nile River greatly facilitated population movement in North Africa and shaped relations between Egypt and Nubia, either in the form of trade networks and exchange of cultural practices (Gatto, 1997; Williams, 1997; Edwards, 2007) or by hostile encounters (Davies, 2003; Edwards, 2007).

Inter-population contact was further facilitated by the introduction of the horse into North Africa (Mitchell, 2005), as suggested by rock depictions of chariots pulled by horses. The distribution of such depictions shows that the Late Holocene trade networks were focused on the north and northeast, with the Garamantes playing a pivotal role (Mitchell, 2005). In fact, the Garamantes monitored the trade routes that connected Fazzān with the Mediterranean coast, Egypt, Algeria, Tunisia, and sub-Saharan Africa (Liverani, 2000). The existence of these contacts is testified by the presence of Roman glassware, fine ceramics, amphorae, cast bronze statues, and Mediterranean architectural motifs and masonry styles in Garamantian contexts (Mattingly, 2003a). The Garamantes exported salt, gold, semi-precious stones, jewels and natron (Mattingly, 2003a). Similarities in funerary structures (keyhole monuments, antenna tombs) also suggest that the Garamantes were in contact with other Saharan populations (Mattingly, 2003d). Moreover, the knowledge to construct subterranean irrigation channels and medical practices, like trephinations, was most likely introduced from Egypt (Wilson, 2003; Nikita et al., in press b). Finally, according to Roman sources, the Garamantes were trading slaves from sub-Saharan Africa (Mattingly, 2000).

It is therefore reasonable to suggest that North African populations were in contact with each other throughout the Holocene, despite the hyper-arid Saharan environment and the harsh physiography of the desert (Camps, 1982). However, it is unlikely that such an environment would have allowed long-distance population migrations, and it is more feasible that material goods and cultural practices were exchanged via trade networks, probably utilizing specialized traders. Therefore, one would expect the biodistance between the Garamantes who inhabited the core of the desert and other North African populations to be high and high bio-distances in all populations among which the desert had intervened. Finally, given the genetic studies that have found much greater female migration rate in agricultural populations compared to the males (for example, Destro-Bisol et al., 2004; Wooding et al., 2004; Wood et al., 2005), we would expect females to exhibit smaller biodistances than males.

Table 10. Populations and sample sizes used in non-metric and three-dimensional analyses

Population	Symbol	MMD		3D	
		Males	Females	Males	Females
Garamantes	GAR	20	26	11	13
Kerma	KER	45	88	45	88
Soleb	SOL	10	17	10	17
Gizeh	GIZ	85	65	85	63
Alexandria	ALE	11	9	11	9
Naqada	NAQ			42	77
el-Badari	BAD			15	19
Carthage	CAR	12	8	12	8
Algerians	ALG	8	12		

Methods and Material

The North African populations whose biological affinities were analyzed are given in Table 10. The individuals displaying poor preservation or pathologies that could affect cranial shape were excluded from the analyses.

Non-metric Traits

Non-metric traits express genes that affect development (Saunders and Popovich, 1978; Cheverud and Buikstra, 1981a, b, 1982; Hauser and De Stefano, 1989; Tyrrell, 2000; Carson, 2006) and have been used in anthropological studies in order to determine the biological affinities of different groups (for example, Blom et al., 1998; Christensen, 1998; Donlon, 2000; Hanihara et al., 2003). The traits employed in the scoring system used here are taken from Nikita (2010). Each trait was scored as present/absent, and its frequency was calculated based on the 'individual count' method (Sutter and Mertz, 2004). After initial scoring, sexual dimorphism in trait expression was explored with chi-square analysis. Furthermore, non-discriminatory traits across samples and associated traits were removed (Harris and Sjøvold, 2004; Irish, 2010). For the assessment of pairwise biodistances, a modified Smith's Mean Measure of Divergence (MMD) was used (Sjøvold, 1973). The MMD was estimated from Sjøvold (1973) as:

$$MMD = \frac{1}{r}\sum_{i=1}^{r}\{(\theta_{1i} - \theta_{2i})^2 - \frac{1}{n_{1i}+0.5} - \frac{1}{n_{2i}+0.5}\} \tag{10}$$

where r is the number of traits considered; n_{1i} is the number of crania in population 1 examined for trait i; n_{2i} is the number of crania in population 2 examined for trait i; and θ_{1i} and θ_{2i} are the transformed frequencies. The transformed frequencies were estimated using the Freeman-Tukey (1950) transformation. Thus, θ values were calculated from:

$$\theta = \frac{1}{2}\sin^{-1}(1-2\frac{k}{n+1}) + \frac{1}{2}\sin^{-1}(1-2\frac{k+1}{n+1}) \tag{11}$$

where $\sin^{-1}$ is the arcsin function and k is the times that a trait occurs in a sample of size n. Negative MMD values were converted to 0 (Harris and Sjøvold, 2004; Irish, 2006, 2010). The variance of the MMD was calculated from Sjøvold (1973) as:

$$Var_{MMD} = \frac{2}{r^2}\sum_{i=1}^{r}(\frac{1}{n_{1i}+0.5} + \frac{1}{n_{2i}+0.5})^2 \tag{12}$$

The p values for each MMD were estimated by using two methods. The first was based on the fact that the MMD values follow the normal distribution (Sjøvold, 1973; Green et al., 1979) and the second was related to permutations (10,000 in the current study). We have shown that the p values obtained from the above methods converge only if the non-metric traits are properly edited, i.e., when non-discriminatory traits across samples and associated traits are removed (Nikita et al., 2012). The MMD matrices were visualized by dendrograms and Multidimensional Scaling (MDS) plots, performed in SPSS using the Proxscal procedure. Mantel tests were carried out to examine correlations between the biological and geographic distance of the populations under study. Furthermore, partial Mantel tests were used to explore the correlation between biological distance and geographic distance, controlling for the effects of temporal distance among populations. The Mantel tests were performed in the PASSAGE 2 software.

Three-Dimensional Morphometrics

The coordinates of 116 landmarks spanning the entire cranium (Nikita, 2010) were recorded using a Microscribe G2X portable digitizer. The landmarks of all individuals were superimposed with Generalized Procrustes Analysis (GPA), using the PAST software package (Hammer et al., 2001). The Procrustes coordinates were later transformed into Procrustes residuals by using PAST. The Procrustes residuals were analyzed through non-parametric MANOVA (NPM). The NPM was performed in PAST based on the Anderson method using Euclidean distance (Anderson, 2001). Given that the number of Procrustes residuals was greater than the number of cases used for the calculation of the Mahalanobis distance, it was not possible to calculate Mahalanobis distances using the Procrustes residuals as variables. For this reason, the residuals were further processed by Principal Components Analysis (PCA) and the number of PCs that accounted for 99.5% of the total variance was used for estimating the Mahalanobis distance. A home-made macro was written in C++ for the estimation of the Mahalanobis distance and its p value using permutations. The Mahalanobis distance was estimated from the square root of the following equation;

$$D^2 = [(\mu_1 - \mu_2)C^{-1}(\mu_1 - \mu_2)^T] \tag{13}$$

where, μ_1, μ_2 are the matrices of the centroids of the two populations being compared, is the inverse of the pooled variance-covariance matrix of the data, and subscript T denotes the transpose matrix. To visualize the Mahalanobis distances, dendrograms as well as Multidimensional Scaling (MDS) plots were drawn. Finally, in order to examine the correlation between biological, geographical and temporal distance among populations, Mantel tests were run.

Table 11. Significant sex differences in non-metric trait frequencies

Population	Trait
Garamantes	Perietal foramina*
Garamantes	Ethmoidal foramina*
Garamantes	Hypoglossal canal bridging*
Garamantes	Supranasal suture**
Gizeh	Supranasal suture***
Carthagenians	Supranasal suture*
Kerma	Divided mental foramina*
Kerma	Epipteric bone*
Gizeh	Epipteric bone*
Kerma	Apertures at the floor of the acoustic meatus*
Gizeh	Apertures at the floor of the acoustic meatus***
Kerma	Marginal tubercle**
Kerma	Zygomaxillary tubercle*
Gizeh	Zygomaxillary tubercle***
Gizeh	Palatine torus*

*significant at $P < 0.05$, **significant at $P < 0.01$, ***significant at $P < 0.001$.

RESULTS

Non-metric Traits

Pearson's chi-square tests or Fisher's exact tests identified 15 variables that exhibited significant differences in trait frequency between males and females (Table 11). These variables were rarely the same among populations. Because of this, in addition to the uneven representation of the two sexes per population (male/female ratio ranges between 0.4 and 1.5), the sexes were treated separately in all inter-population comparisons.

Pearson's chi-square tests and Fisher's exact tests have shown that 21 traits among males and females did not differ significantly in the pairwise comparisons among populations. The remaining traits included the metopic and supranasal sutures, supraorbital structures, infraorbital and ethmoidal foramina, epipteric bone, coronal and lambdoid ossicles, divided hypoglossal canal, mandibular torus, marginal tubercle, fused ovale and spinous foramina and zygomaxillary tubercle for males. For females, the remaining traits were ossicle at asterion, epipteric bone, mandibular and palatine tori, fused ovale and spinous foramina, infraorbital, parietal and ethmoidal foramina, apertures at the floor of the auditory meatus, marginal and zygomaxillary tubercle, supranasal suture, and supraorbital structures. On the basis of associated traits, the metopic suture, mandibular torus and the fused ovale and spinous foramina in males, and the ossicle at asterion, the mandibular torus and the fused ovale and spinous foramina in females were also removed from the dataset.

Table 12. Biological distances using MMD for males (below the diagonal) and females (above the diagonal)

	GAR	ALG	ALE	GIZ	KER	SOL	CAR
GAR	0	0.079	0.005	0.472***	0.400***	0.085*	0.072
ALG	0.211*	0	0	0.001	0.027	0	0
ALE	0.108	0.117	0	0.084	0.037	0	0
GIZ	0.435***	0.212**	0.113*	0	0.056***	0.209***	0.122*
KER	0.345***	0.221**	0.036	0.090***	0	0.187***	0.067
SOL	0.063	0.029	0	0.170**	0.145*	0	0.002
CAR	0.165*	0	0.070	0.041	0.156**	0	0

MMD values are *significantly different at $P < 0.05$, **significant at $P < 0.01$, ***significant at $P < 0.001$.

Table 12 presents the MMD values for males and females as well as their significance, while Figure 10 shows the variation of these values in pairwise comparisons. The greatest distance of the Garamantes to their neighbors is with the Gizeh and Kerma, for both males and females. The smallest biodistance for male Garamantes is with the Soleb, whereas for females, it is with the Alexandrians. It should also be noted that most distances between the Garamantes and their neighbors are statistically significant. The general pattern of distances in all populations is similar for males and females but females overall exhibit smaller or equal pairwise MMDs to males.

The dendrograms (Figures 11-12) present a similar pattern for both sexes. More particularly, we observe that the Garamantes stand out as a clear outlier, samples from Kerma and Gizeh form a separate cluster, while the Algerians, Carthagenians, Soleb and Alexandrians are grouped together.

The two-dimensional plot of the MDS analyses is given in Figure 13. This clearly shows the separation of the Garamantes on one hand and the Kerma and Gizeh on the other compared to the remaining groups along Dimension 1. This data concurs with the results given in Figures 11-12, although the great distance between the Kerma and Gizeh in the MDS plots is rather surprising. Furthermore, an interesting observation from the MDS plot is that males and females lie close to each other in the samples from Gizeh, Kerma and Garamantes, while this is not the case for the remaining populations.

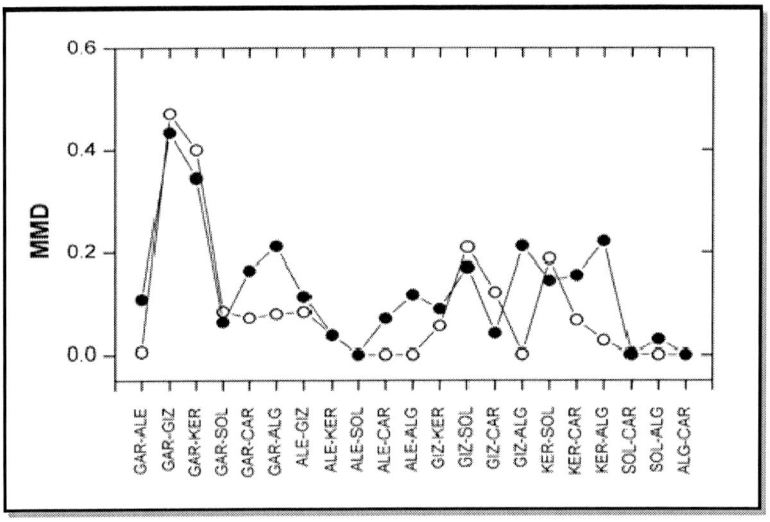

Figure 10. Variation of MMD among the various pairs of populations for males (●) and females (o).

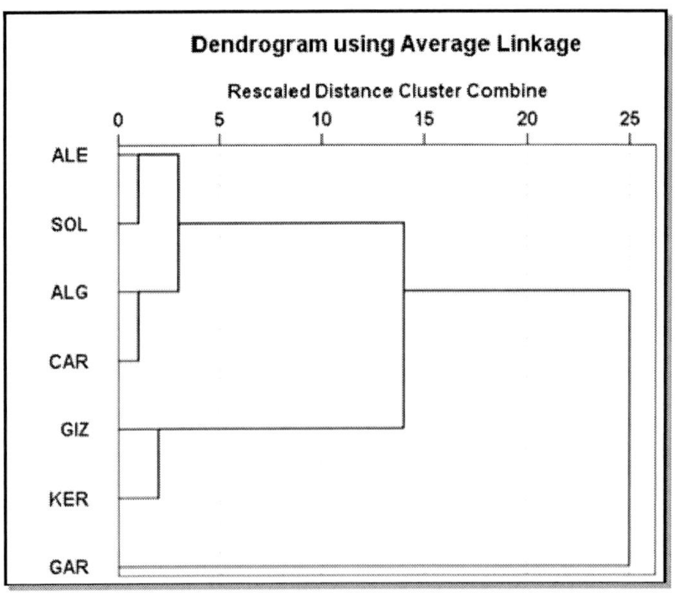

Figure 11. Dendrograms for males based on MMD.

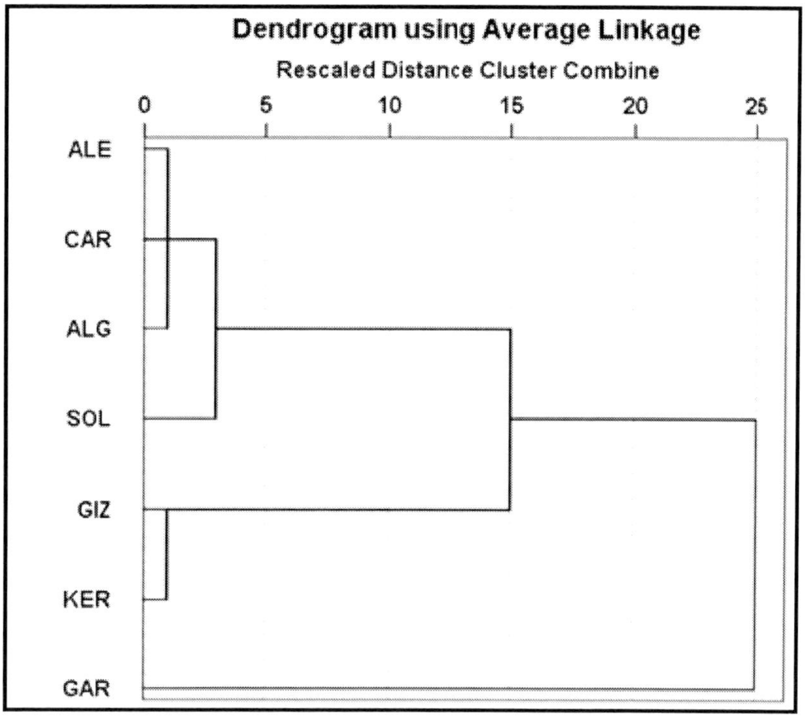

Figure 12. Dendrograms for females based on MMD.

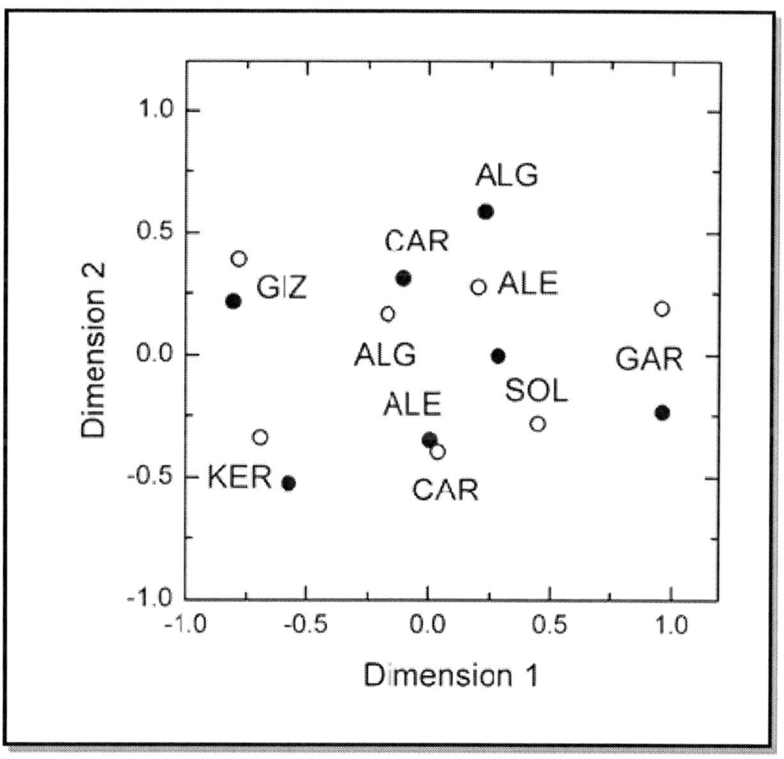

Figure 13. Two-dimensional MDS based on MMD distances for males (●) and females (o).

Table 13. Mantel tests comparing MMD biological, geographic and temporal distances

Sex	Compared matrices	Control matrix	r	P
Males	MMD/geographic	-	0.265	0.33
Males	MMD /geographic	Temporal	0.226	0.39
Males	MMD/ temporal	Geographic	-0.195	0.38
Females	MMD/geographic	-	-0.052	0.85
Females	MMD/geographic	Temporal	-0.081	0.76
Females	MMD/ temporal	Geographic	-0.139	0.57

All correlations between biological distance and linear geographic distance failed to reach statistical significance (Table 13). However, it can be seen that among males, a positive correlation is found between biodistance and geographic distance, whereas this correlation is negligible in the female populations. In all cases, the effect of temporal distance on the above correlations is small.

Table 14. P values after Holm-Bonferroni correction for inter-population comparisons

Populations	Males	Females
GAR-CAR	0.004	0.001
GAR-ALE	0.004	0.001
GAR-GIZ	0	0
GAR-BAD	0.006	0
GAR-NAQ	0.008	0
GAR-SOL	0.004	0
GAR-KER	0	0
CAR-ALE	0.76	0.12
CAR-GIZ	0	0
CAR-BAD	0.001	0.003
CAR-NAQ	0	0
CAR-SOL	0.18	0.015
CAR-KER	0	0
ALE-GIZ	0	0
ALE-BAD	0.007	0
ALE-NAQ	0	0
ALE-SOL	0.43	0.003
ALE-KER	0	0
GIZ-BAD	0	0
GIZ-NAQ	0	0
GIZ-SOL	0	0
GIZ-KER	0	0
BAD-NAQ	0.007	0
BAD-SOL	0.001	0.009
BAD-KER	0	0
NAQ-SOL	0	0
NAQ-KER	0	0
SOL-KER	0	0

THREE-DIMENSIONAL MORPHOMETRICS

According to the results of MANOVA, the Garamantes exhibit a statistically significant distance from all of their neighbors (Table 14). However, this is also the case for most other populations. The exceptions include Carthagenians and Alexandrians, Soleb and Alexandrians, and Soleb and Carthagenians. Nevertheless, these latter results may be biased by the small size of the populations. The results are compatible to those obtained using Mahalanobis distances (Table 15). The pairwise inter-population Mahalanobis distance values for males and females are compared in Figure 14. It can be seen that the biodistances among males are higher than among females. Furthermore, the Garamantian biodistances to their neighbors are higher than among the remaining comparative groups.

Table 15. Mahalanobis distances for males (below the diagonal) and females (above the diagonal)

	GAR	CAR	ALE	GIZ	BAD	NAQ	SOL	KER
GAR	0	21.83***	20.17***	15.94***	19.99***	17.78***	17.63***	16.47***
CAR	25.28***	0	17.22	18.23***	17.91***	15.82***	15.12	17.61***
ALE	27.63***	19.79*	0	16.55***	16.54*	13.85**	14.34	15.80***
GIZ	22.94***	19.18***	21.89***	0	13.64***	10.61***	15.41***	10.40***
BAD	24.30***	22.33***	20.74***	19.20***	0	9.40*	14.48***	11.62***
NAQ	24.24***	18.40*	18.83***	11.99***	13.70	0	13.69***	8.44***
SOL	25.57***	19.57	16.31	20.16***	21.54***	17.80*	0	13.13***
KER	26.57***	20.48***	23.55***	11.42***	23.17***	14.64***	20.71***	0

* values are significantly different at $P < 0.05$, ** significant at $P < 0.01$, *** significant at $P < 0.001$.

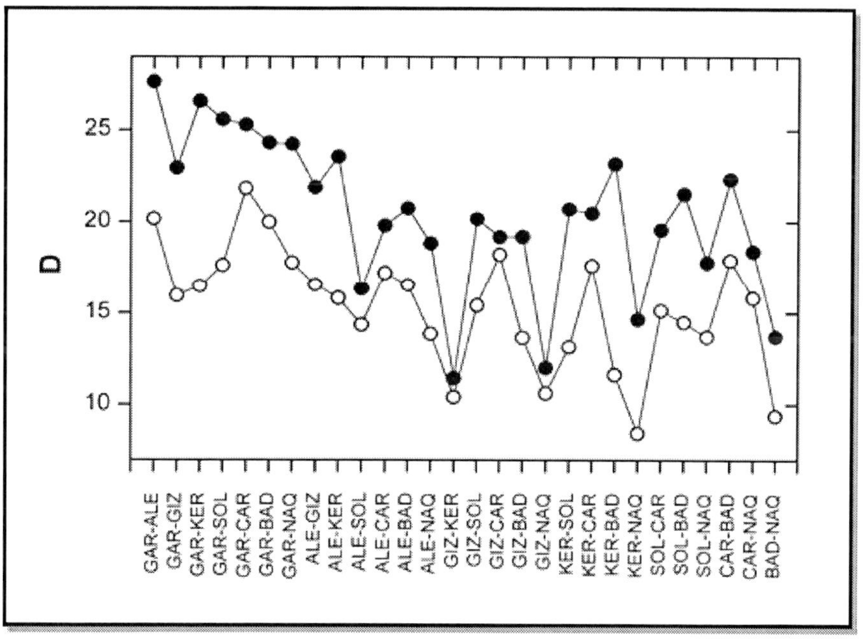

Figure 14. Variation in Mahalanobis distance D among various pairs of populations for males (●) and females (o).

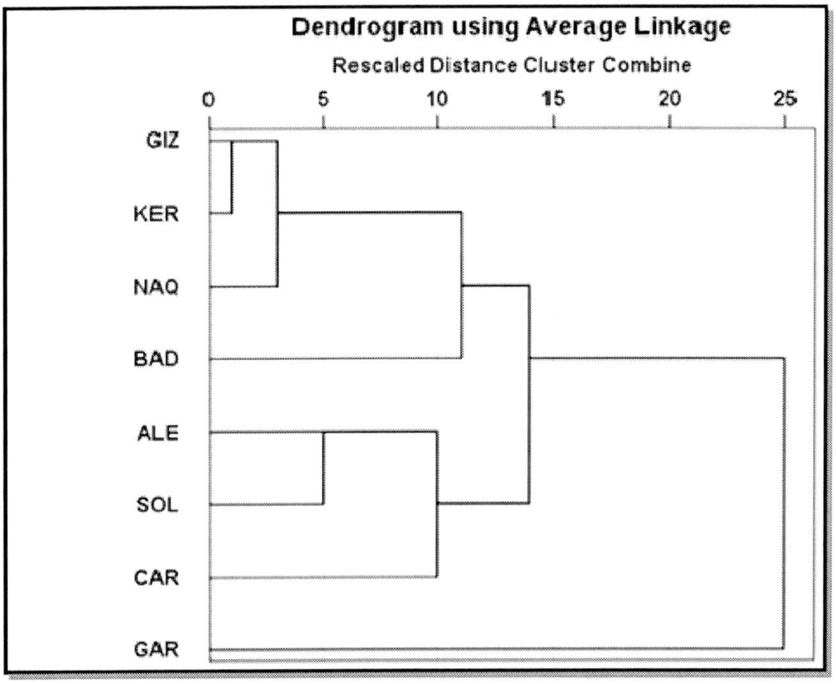

Figure 15. Dendrograms for males based on Mahalanobis distance D.

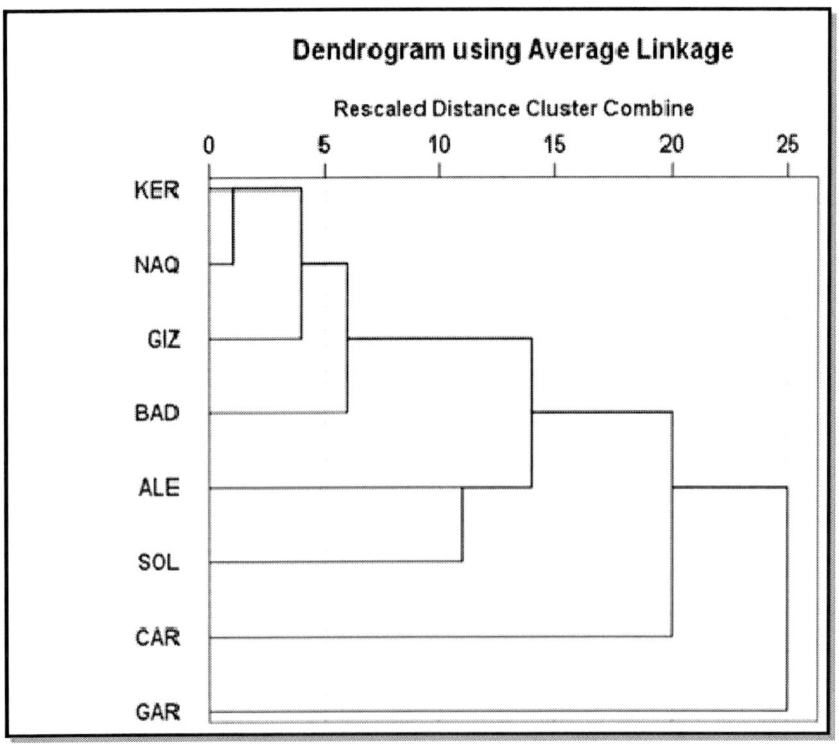

Figure 16. Dendrograms for females based on Mahalanobis distance D.

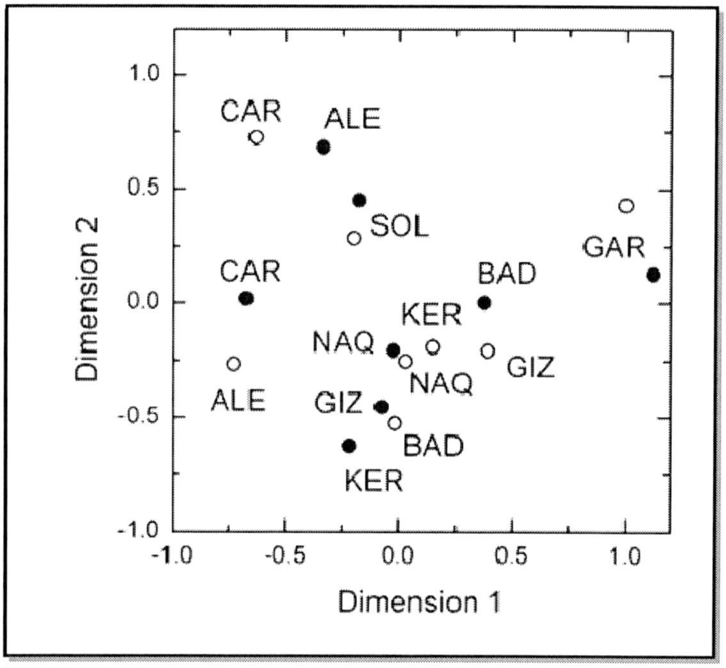

Figure 17. Two-dimensional MDS based on Mahalanobis distances for males (●) and females (o).

The dendrograms show a similar overall pattern for males and females, although some differences are present. In both sexes, the Garamantes appear to be a clear outlier (Figures 15-16). Among the remaining groups, the Gizeh, Kerma, el-Badari and Naqada are close to each other, while the Carthagenians, Soleb and Alexandrians always cluster together. The most notable difference between sexes is that Carthagenian males cluster closely to the Alexandrians and Soleb, whereas, among females they form a separate branch in a wider cluster which includes the Alexandrians and Soleb as one sub-group and the Kerma, Naqada, Gizeh and el-Badari as another sub-cluster.

The two-dimensional MDS visualization of the inter-population biodistances is depicted in Figure 17. The results are in agreement with the dendrograms (Figures 15-16). For males, the Garamantes appear isolated along Dimension 1. Among the remaining populations, three clusters can generally be identified, one includes the Gizeh and Kerma, another consists of the el-Badari and Naqada and the third includes the Carthagenians, Soleb and Alexandrians. It should be noted that the groups including the Kerma, Gizeh, el-Badari and Naqada also cluster close to each other. These results are similar for females, with the Garamantes again standing out along Dimension 1, and the remaining groups clustering in the same way as for males.

It should be stressed that, broadly speaking, the above dendrograms and MDS plots give the same information as those obtained from MMD values (Figures 11-13). The most striking difference is that the Gizeh-Kerma cluster is shifted away from the Garamantes when using MMD data, whereas it is closer to the Garamantes in the three-dimensional data.

The Mantel tests (Table 16) give a positive correlation between biodistance and geographic distance for both sexes. This correlation is statistically significant for females, whether controlling for temporal distance or not. It is interesting to note that using three-dimensional morphometrics, we obtain a positive and significant correlation between biodistance and geographic distance for female populations, whereas the MMD data examined above give practically no correlation between these two distances. In addition, we observe a very small and negative correlation between biodistance and temporal distance, similar to the case of MMD data.

Table 16. Mantel tests comparing Mahalanobis biological, geographic and temporal distances

Sex	Compared matrices	Control matrix	r	P
Males	D/geographic	-	0.351	0.22
Males	D/geographic	Temporal	0.35	0.23
Males	D/ temporal	Geographic	-0.066	0.79
Females	D/geographic	-	0.555	0.03
Females	D/geographic	Temporal	0.555	0.04
Females	D/ temporal	Geographic	-0.08	0.81

Discussion

Our results reveal an interesting pattern of affinities and differences in the Late Holocene North Africa. In particular, the statistical analyses presented above show that the difference between the Garamantes and most comparative populations is significant. In the dendrograms and the MDS plots, the Garamantian males and females appear as an outlier, which must be attributed to their geographic location at the centre of North Africa, separating the bulk of the Garamantian population from all other groups by large expanses of desert. Among the remaining populations, two clusters can be identified, one consisting of the Gizeh and Kerma, and another including the Soleb, Alexandrians, Algerians and Carthagenians. These clusters increase to four when the el-Badari and Naqada are included in the analyses.

The cluster formed by the el-Badari and Naqada is to be expected, since these populations constitute pre-Dynastic Egyptian samples that predate all others, and are located closely to each other along the Nile River, facilitating contacts between them. The cluster of the Carthagenians, Algerians, Soleb and Alexandrians is also reasonable, given that contact between these groups was facilitated by the Nile River and the Mediterranean coast. This is despite the large geographic distance between the Carthagenians and the other three populations, especially as the latter are known to have been highly mobile, maritime people. The case of the Kerma and Gizeh is difficult to explain. These populations were also located along the Nile River and would be expected to cluster with the above populations, instead of what they actually do which is to form a separate group. Given that this cluster is found both in the craniometric and the non-metric trait results, it cannot be an artifact of the analyses, although there is no clear explanation.

In respect to the exploration of correlations between biodistance, geographic and temporal distance, the following hypotheses were formulated; (1) The greater the geographic distance between two populations, the greater the corresponding biodistances are expected to be, (2) the biodistance of two contemporary populations is expected to decrease with time due to mutual gene flow. Therefore, following a so-called isolation-by-distance model

(Konigsberg, 1990), we expect a positive correlation between biodistance and geographic distance and a negative correlation between biodistance and temporal distance. From Tables 13 and 16, we observe that for the populations under study, this model may be assumed for both sexes, although the correlation between MMD and geographic distance for females is negative but very small and therefore practically negligible. However, the major problem is that - with the exception of the correlation between Mahalanobis distance and geographic distance for females - these correlations are not statistically significant. This is due to the great scatter of the experimental points, which may imply that there is no uniform gene flow throughout North Africa. This result may be attributed to the fact that the distances among populations are not equivalent to each other in the sense that travelling across the desert is much more difficult than travelling along the Nile River and the Mediterranean coast. Another possible explanation is massive migration waves. Note that other researchers have found contradictory results. For example, Zakrzewski's (2007) study failed to accept an isolation-by-distance model using the Mahalanobis distance for metric data among populations along the Nile River. The same was verified by Schillaci et al. (2009) but, in contrast to Mahalanobis distance, they found a correlation between MMD distances and temporal and geographic distances.

As far as the relations among men and women are concerned, both sexes exhibited a similar pattern of pairwise biodistance to their neighbors. This pattern supports that both sexes engaged similarly in encounters with the other North African populations. However, females generally exhibited smaller pairwise biodistances compared to males, which may result from the greater mobility of females as part of the various marital networks. This greater mobility agrees with genetic data obtained from other agricultural populations where patrilocality is the general pattern (Seielstad et al., 1998; Destro-Bisol et al., 2004; Wooding et al., 2004; Wood et al., 2005).

CONCLUSION

The study of activity patterns in the Garamantian population revealed that the stresses imposed by a sedentary lifestyle in the Sahara Desert were not substantially heavier than those on populations occupying more favorable environments. This is true for both the upper and lower limbs. With respect to the lower limbs, the results may suggest that the Garamantes did not travel for long distances on foot, that the Fazzanese terrain was not as stressful as expected or that the population lived primarily in the valleys. As for the upper limbs, the manual tasks in which Garamantian males and females were involved were not more demanding than those of their neighbors. Furthermore, activity markers showed low levels of bilateral asymmetry and sexual dimorphism in the Garamantes, which is consistent to an agricultural population (Murdock and Provost, 1973; Ruff, 1987, 2000; Bridges, 1989; Marchi et al., 2006).

The biological affinities of North African Late Holocene populations identified four broad clusters. The first consisted of the Garamantes alone, the second included the Gizeh and Kerma, the third comprised of Soleb, Alexandrians, Algerians and Carthagenians and the fourth included the el-Badari and Naqada. The Garamantes appeared as an outlier despite their involvement in trans-Saharan trade networks. This result can be attributed to the Sahara desert which intervened between them and their neighbors. In contrast, the study of all North African populations suggests that there was no uniform gene flow across North Africa. Therefore, the Sahara desert imposed limitations to gene flow among North African populations and the trade networks, evidenced by the archaeological data, possibly involved a subset of male merchants.

ACKNOWLEDGMENTS

We would like to thank the generous grant of The Society for Libyan Studies and a Philip Leverhulme Prize to Mirazon Lahr towards the fieldwork of the Desert Migrations Project during which the data of this study were collected. E. Nikita would additionally like to thank the Alexander S. Onassis Public Benefit Foundation, the A.G. Leventis Foundation and the Cambridge European Trusts (George and Marie Vergottis Cambridge Bursary) for significant financial support during her Ph.D. project concerning the study of the Garamantian material. Special thanks are to all the archaeologists and experts who participated in the Desert Migrations Project.

REFERENCES

Adams, W.Y., 1977. Nubia. *Corridor to Africa.* Princeton University Press, New Jersey.

Adams, B.J., Konigsberg, L.W., 2004. Estimation of the most likely number of individuals from commingled human skeletal remains. *American Journal of Physical Anthropology* 125, 138-151.

Alston, R., 1997. Philo's 'In Flaccum': Ethnicity and social space in Roman Alexandria. *Greece and Rome* 44, 165-175.

Anderson, M.J., 2001. A new method for non-parametric multivariate analysis of variance. *Austral Ecology* 26, 32-46.

Aubet, M.E., 2001. *The Phoenicians and the West: Politics, Colonies and Trade.* Cambridge University Press, Cambridge.

Auerbach, B.M., Ruff, C.B., 2004. Human body mass estimation: a comparison of 'morphometric' and 'mechanical' methods. *American Journal of Physical Anthropology* 125, 331-342.

Barich, B., 1998. *People, water and grain: The beginnings of domestication in the Sahara and the Nile Valley.* L'Erma di Bretschneider, Roma.

Barich, B., 2002. *Cultural responses to climatic changes in North Africa: beginning and spread of pastoralism in the Sahara.* In: Hassan, F. (Ed.), *Droughts, food and culture. Ecological change and food security in Africa's later prehistory.* Kluwer Academic/Plenum Publishers, New York, pp. 209-223.

Barker, G., 2006. The archaeology and heritage of the Sahara. In: Mattingly, D.J., McLaren, S., Savage, E., al-Fasatwi, Y., Gadgood, K. (Eds.), *The Libyan Desert. Natural resources and cultural heritage. Society for Libyan Studies,* London, pp. 9-26.

Blom, D.E., Hallgrímsson, B., Keng, L., Lozada, M.C., Buikstra, J.E., 1998. Tiwanaku 'colonization': bioarchaeological implications for migration in the Moquegua Valley. *World Archaeology* 30, 238–261.

Bridges, P.S., 1989. Changes in activities with the shift to agriculture in the southeastern United States. *Current Anthropology* 30, 385-394.

Brooks, N., 2006. Cultural responses to aridity in the Middle Holocene and increased social complexity. *Quaternary International* 151, 29-49.

Brooks, S., Suchey, J.M., 1990. Skeletal age determination based on the os pubis: a comparison of the Acsádi-Nemeskéri and Suchey-Brooks methods. *Human Evolution* 5, 227-238.

Brooks, N., White, K., Drake, N., MacLaren, S., 2001. Changes in vegetative cover in the Wadi al-Ajal, southern Libya, over the past half-century. *Proceedings of the First Annual Conference of the Remote Sensing & Photogrammetry Society,* University of London, pp. 247-259.

Brooks, N., Drake, N., McLaren, S., White, K., 2003. Studies in geography, environment and climate. In: Mattingly, D.J., Daniels, C.M., Dore, J.N., Edwards, D., Hawthorne, J. (Eds.), The archaeology of Fazzān. Volume 1, *Synthesis. Society for Libyan Studies,* London, pp. 37-75.

Brooks, N., Chiapello, I., Di Lernia, S., Drake, N., Legrand, M., Moulin, C., Prospero, J., 2005. The climate-environment-society nexus in the Sahara from prehistoric times to the present day. *The Journal of North African Studies* 10, 253-292.

Buikstra, J.E., Ubelaker, D.H., 1994. Standards for data collection from human skeletal remains. *Arkansas Archaeological Survey Report Number 44,* Fayetteville, Arkansas.

Camps, G., 1974. Les civilizations préhistoriques de l'Afrique du Nord et du Sahara. Doin, Paris.

Camps, G., 1982. Beginnings of pastoralism and cultivation in north-west Africa and the Sahara: origins of the Berbers. In: Clark, D. (Eds.), *The Cambridge History of Africa, volume 1: From the earliest times to c. 500BC.* Cambridge University Press, Cambridge, pp. 548-623.

Carlson, D.S., Van Gerven, D.P., 1979. Diffusion, biological determinism, and biocultural adaptation in the Nubian corridor. *American Anthropologist* 81, 561-580.

Carson, E.A., 2006. Maximum-Likelihood Variance Components Analysis of heritabilities of cranial nonmetric traits. *Human Biology* 78, 383-402.

Chang, C., Koster, H.A., 1986. Beyond bones: Toward an archaeology of pastoralism. *Advances in Archaeological Method and Theory* 9, 97–146.

Cheverud, J.M., Buikstra, J.E., 1981a. Quantitative genetics of skeletal nonmetric traits in the rhesus macaques on Cayo Santiago. I. Single trait heritabilities. *American Journal of Physical Anthropology* 54, 43-49.

Cheverud, J.M., Buikstra, J.E., 1981b. Quantitative genetics of skeletal nonmetric traits in the rhesus macaques on Cayo Santiago. II. Phenotypic, genetic, and environmental correlations between traits. *American Journal of Physical Anthropology* 54, 51-58.

Cheverud, J.M., Buikstra, J.E., 1982. Quantitative genetics of skeletal nonmetric traits in the rhesus macaques of Cayo Santiago. III. Relative heritability of skeletal nonmetric and metric traits. *American Journal of Physical Anthropology* 59, 151-155.

Christensen, A.F., 1998. Skeletal evidence for familial interments in the Valley of Oaxaca, Mexico. *Homo* 49, 273-288.

Cremaschi, M., 2002. Late Pleistocene and Holocene climatic changes in the Central Sahara: the case study of the Southwestern Fezzan, Libya. In: Hassan, F. (Eds.), *Droughts, food and culture. Ecological change and food security in Africa's later prehistory*. Kluwer Academic/Plenum Publishers, New York, pp. 65-81.

Cremaschi, M., Di Lernia, S., 1996. Climatic changes and human adaptive strategies in the central Saharan Massifs: the Tadart Acacus and Messak Settafet perspective (Fezzan, Libya). In: Pwiti, G., Soper, R. (Eds.), Aspects of African Archaeology. *Papers from the 10th Congress of the PanAfrican Association for Prehistory and related Studies.* University of Zimbabwe Publications, Harare, pp. 39-52.

Cremaschi, M., Zerboni, A., 2009. Early to Middle Holocene landscape exploitation in a drying environment: Two case studies compared from the central Sahara (SW Fezzan, Libya). *Comptes Rendus Geosciences* 341, 689-702.

Davies, W.V., 2003. Sobeknakht of ElKab and the coming of Kush. *Egyptian Archaeology* 23, 3–6.

Destro-Bisol, G., Donati, F., Coia, V., Boschi, I., Verginelli, F., Caglia, A., Tofanelli, S., Spedini, G., Capelli, C., 2004. Variation of Female and Male Lineages in Sub-Saharan Populations: the Importance of Sociocultural Factors. *Molecular Biology and Evolution* 21, 1673-1682.

Di Lernia, S., Palombini, A., 2002. Desertification, sustainability, and archaeology: indications from the past for an African future. *Origini* 24, 303-328.

Di Lernia, S., Manzi, G., Merighi, F., 2002. Cultural variability and human trajectories in later prehistory of the Wadi Tanezzuft. In: Di Lernia, S., Manzi, G. (Eds.), *Sand, Stones and Bones. The Archaeology of Death in the Wadi Tanezzuft Valley* (5000–2000 BP). Edizioni all'Insegna del Giglio, Firenze, pp. 281–302.

Donlon, D.A., 2000. The value of infracranial nonmetric variation in studies of modern Homo sapiens: An Australian focus. *American Journal of Physical Anthropology* 113, 349-368.

Edwards, D.N., 2007. The archaeology of Sudan and Nubia. *Annual Review of Anthropology* 36, 211-228.

Ehrenberg, M., 1989. *Women in prehistory.* University of Oklahoma Press, Norman.

Eshed, V., Gopher, A., Galili, E., Hershkovitz, I., 2004. Musculoskeletal stress markers in Natufian hunter-gatherers and Neolithic farmers in the Levant: the upper limb. *American Journal of Physical Anthropology* 123, 303-315.

Freeman, M.F., Tukey, J.W., 1950. Transformations related to the angular and square root. *The Annals of Mathematical Statistics* 21, 607–611.

Friedman, R., 1996. The Ceremonial Centre at Hierakonpolis, Locality HK29A. In: Spencer, A.J. (Eds.), *Aspects of Early Egypt.* British Museum Press, London, pp. 16-35.

Garcea, E.A.A., 2006. Semi-permanent foragers in semi-arid environments of North Africa. *World Archaeology* 38, 197-219.

Gatto, M., 1997. Regional variations in the so-called "A-group" culture of lower Nubia. In: Barich, B., Gatto, M. (Eds.), *Dynamics of Populations, Movements and Responses to Climatic Change in Africa.* Bonsignori Editore, Rome, pp. 105–111.

Gerharz, R.G., 1994. *Jebel Moya. Meroitica 14.* Akademie Verlag, Berlin.

Green, R.F., Suchey, J.M., Gokhale, D.V., 1979. The statistical treatment of correlated bilateral traits in the analysis of cranial material. *American Journal of Physical Anthropology* 50, 629-634.

Grine, F.E., Jungers, W.L., Tobias, P.V., Pearson, O.M., 1995. Fossil *Homo* femur from Berg Aukas, northern Namibia. *American Journal of Physical Anthropology* 97, 151-185.

Haas, C., 1997. *Alexandria in Late Antiquity: Topography and social conflict.* The Johns Hopkins University Press, Baltimore.

Hafsaas-Tsakos, H., 2009. The kingdom of Kush: an African centre on the periphery of the Bronze Age world system. *Norwegian Archaeological Review* 42, 50-70.

Hall, R.L., 1982. *Sexual dimorphism in Homo sapiens: a question of size.* Praeger, New York.

Hammer, Ø., Harper, D.A.T., Ryan, P.D., 2001. PAST: Paleontological Statistics software package for education and data analysis. *Paleontologia Electronica* 4, 1-9.

Hanihara, T., Ishida, H., Dodo, Y., 2003. Characterization of biological diversity through analysis of discrete cranial traits. *American Journal of Physical Anthropology* 121, 241-251.

Harris, E.F., Sjøvold, T., 2004. Calculation of Smith's mean measure of divergence for intergroup comparisons using nonmetric data. *Dental Anthropology* 17, 83-93.

Hassan, F.A., 1988. The predynastic of Egypt. *Journal of World Prehistory* 2, 135-185.

Hauser, G., De Stefano, G.F., 1989. *Epigenetic variants of the human skull.* Schweizerbart, Stuttgart.

Hawkey, D.E., Merbs, C.F., 1995. Activity-induced musculoskeletal stress markers (MSM) and subsistence strategy changes among ancient Hudson Bay Eskimos. *International Journal of Osteoarchaeology* 5, 324-338.

Hendrickx, S., Vermeersch, P., 2000. *Prehistory. From the Palaeolithic to the Badarian Culture (c.700000-4000 BC).* In: Shaw, I. (Eds.), The Oxford *History of Ancient Egypt.* Oxford University Press, Oxford, pp. 17-44.

Holl, A.F.C., 2004. *Holocene Saharans: An anthropological perspective.* Continuum, London.

Irish, J.D., 2006. Who were the ancient Egyptians? Dental affinities among Neolithic through Postdynastic peoples. *American Journal of Physical Anthropology* 129, 529-543.

Irish, J.D., 2010. The Mean Measure of Divergence: Its utility in model-free and model-bound analyses relative to the Mahalanobis D^2 distance for nonmetric traits. *American Journal of Human Biology* 22, 378-395.

Keita, S.O.Y., 1988. An analysis of crania from Tell-Duweir using multiple discriminant functions. *American Journal of Physical Anthropology* 75, 375-390.

Keita, S.O.Y., 2005. Early Nile Valley Farmers from El-Badari: Aboriginals or "European" Agro-Nostratic Immigrants? Craniometric Affinities Considered With Other Data. *Journal of Black Studies* 36, 191-208.

Kemp, B.J., 1991. *Ancient Egypt.* Routledge, London.

Knüsel, C., 2000. Bone adaptation and its relationship to physical activity in the past. In: Cox, M., Mays, S. (Eds.), *Human osteology in archaeology and forensic science.* Greenwich Medical Media, Ltd, London, pp. 381-402.

Konigsberg, L., 1990. Analysis of prehistoric biological variation under a model of isolation by geographic and temporal distance. *Human Biology* 62, 49-70.

Kuper, R., 2006. After 5000 BC: The Libyan desert in transition. *Comptes Rendus Palevol* 5, 409-419.

Kuper, R., Kröpelin, S., 2006. Climate-controlled Holocene occupation in the Sahara: motor of Africa's evolution. *Science* 313, 803-807.

Lahr, M.M., Armitage, S., Barton, H., Crivellaro, F., Drake, N., Foley, R., Maher, L., Mattingly, D.J., Stock, J., White, K., 2008. DPM III: Pleistocene and Holocene palaeoenvironments and prehistoric occupation of Fazzan, Libyan Sahara. *Libyan Studies* 39, 263-294.

Lieberman, D.E., Polk, J.D., Demes, B., 2004. Predicting long bone loading from cross-sectional geometry. *American Journal of Physical Anthropology* 123, 156–171.

Liverani, M., 2000. The Garamantes: a fresh approach. *Libyan Studies* 31, 17-28.

Liverani, M., 2004. Rediscovering the Garamantes: archaeology and history. *Libyan Studies* 35, 191-200.

Lloyd, A., 2000. The Late period (664-332BC). In: Shaw, I. (Eds.), *The Oxford history of ancient Egypt*. Oxford University Press, Oxford, pp. 369-394.

Marchi, D., Sparacello, V.S., Hol, B.M., Formicola, V., 2006. Biomechanical approach to the reconstruction of activity patterns in Neolithic Western Liguria, Italy. *American Journal of Physical Anthropology* 131, 447-455.

Mattingly, D.J., 2000. Twelve thousand years of human adaptation in Fezzan (Libyan Sahara). In: Barker, G., Gilbertson, D. (Eds.), *The archaeology of drylands. Living at the margin*. Routledge, London, pp. 160-179.

Mattingly, D.J., 2003a. Background to the archaeology of Fazzān. In: Mattingly, D.J., Daniels, C.M., Dore, J.N., Edwards, D., Hawthorne, J. The archaeology of Fazzān. Volume 1, Synthesis. *Society for Libyan Studies*, London, pp. 1-37.

Mattingly, D.J., 2003b. Fortifications, settlement and domestic architecture. In: Mattingly, D.J., Daniels, C.M., Dore, J.N., Edwards, D., Hawthorne, J. *The archaeology of Fazzān*. Volume 1, Synthesis. Society for Libyan Studies, London, pp. 136-177.

Mattingly, D.J., 2003c. Historical summary. In: Mattingly, D.J., Daniels, C.M., Dore, J.N., Edwards, D., Hawthorne, J. *The archaeology of Fazzān*. Volume 1, Synthesis. Society for Libyan Studies, London, pp. 75-107.

Mattingly, D.J., 2003d. Religious and funerary structures. In: Mattingly, D.J., Daniels, C.M., Dore, J.N., Edwards, D., Hawthorne, J. *The archaeology of Fazzān*. Volume 1, Synthesis. Society for Libyan Studies, London, pp. 177-235.

Mattingly, D.J., 2005. Approaches to the archaeology and environment of the Sahara: the Fezzan Project, 1997-2002 (briefing). *The Journal of North African Studies* 10, 387-396.

Mattingly, D.J., 2006. The Garamantes: The first Libyan state. In: Mattingly, D.J., McLaren, S., Savage, E., al-Fasatwi, Y., Gadgood, K. (Eds.), The Libyan Desert. *Natural resources and cultural heritage*. Society for Libyan Studies, London, pp. 189-204.

McArdle, J., 1982. Preliminary report on the Predynastic fauna of the Hierakonpolis project. In: Hoffman, M.A. (Ed.), *The Predynastic of Hierakonpolis. An Interim Report*. Egyptian Studies Association 1, Cairo, pp. 116-121.

McHenry, H.M., 1992. Body size and proportions in early hominids. *American Journal of Physical Anthropology* 87, 407-431.

Midant-Reynes, B., 2000a. *The Prehistory of Egypt: From the First Egyptians to the First Pharaohs*. Wiley-Blackwell, London.

Midant-Reynes, B., 2000b. The Naqada Period (4000-3200 BC). In: Shaw, I. (Ed.*), The Oxford History of Ancient Egypt*. Oxford University Press, Oxford, pp. 44-60.

Mitchell P., 2005. *African connections: Archaeological perspectives on Africa and the wider world*. AltaMira Press, Walnut Creek.

Molnar, P., 2006. Tracing prehistoric activities: musculoskeletal stress marker analysis of a Stone-Age population on the island of Gotland in the Baltic Sea. *American Journal of Physical Anthropology* 129, 12-23.

Mukherjee, R., Rao, C.R., Trevor, J.C., 1955. *The ancient inhabitants of Jebel Moya (Sudan).* Cambridge University Press, Cambridge.

Murdock, G.P., Provost, C., 1973. Factors in the division of labor by sex: a cross-cultural analysis. *Ethnology* 12, 203–225.

Nikita, E., 2010. *The Garamantes of Fazzān: Bioarchaeological evaluation of desert-induced stress and Late Holocene human migrations through the Sahara.* Ph.D. Thesis, University of Cambridge, UK.

Nikita, E., Lahr, M.M., 2011. Simple algorithms for the estimation of the initial number of individuals in commingled skeletal remains. *American Journal of Physical Anthropology* 146, 629-636.

Nikita, E., Siew, Y.Y., Stock, J., Mattingly, D., Lahr, M.M., 2011. Activity patterns in the Sahara Desert: An interpretation based on cross-sectional geometric properties. *American Journal of Physical Anthropology* 146, 423-434.

Nikita, E., Mattingly, D., Lahr, M.M., 2012. Sahara: Barrier or corridor? Non-metric cranial traits and biological affinities of North African Late Holocene populations. *American Journal of Physical Anthropology* 147, 280-292.

Nikita, E., in press a Estimation of the most likely number of individuals using multiple skeletal elements. *International Journal of Osteoarchaeology,* in review. DOI: 10.1002/ajpa.21624

Nikita, E., Lahr, M.M., Mattingly, D., in press b. Evidence of trephinations among the Garamantes, a Late Holocene Saharan population. *International Journal of Osteoarchaeology.* DOI: 10.1002/oa.1265

Oberlander, T.M., 1994. Global deserts: a geomorphic comparison. In: Abrahams, A.D., Parsons, A.J. (Eds.), *Geomorphology of desert environments.* Chapman and Hall, London, pp. 13-36.

Ottoni, C., 2007. Holocene human peopling of Libyan Sahara. *Molecular analysis of maternal lineages in ancient and extant populations of Fezzan.* Ph.D. Thesis, Università degli studi di Roma 'Tor Vergata', Italy.

Parsons, A.J., Abrahams, A.D., 1994. Geomorphology of desert environments. In: Abrahams, A.D., Parsons, A.J. (Eds.), *Geomorphology of desert environments.* Chapman and Hall, London, pp. 1-12.

Pearson, O.M., Cordero, R.M., Busby, A.M., 2006. How different were Neanderthals' habitual activities? A comparative analysis with diverse groups of recent humans. In: Harvati, K., Harrison, T. (Eds.), *Neanderthals revisited: new approaches and perspectives.* Springer, New York, pp. 135-156.

Petrie, W.M.F., 1907. *Gizeh and Rifeh.* Adamant Media Corporation, London.

Phenice, T.W., 1969. A newly developed visual method of sexing in the os pubis. *American Journal of Physical Anthropology* 30, 297-301.

Phillipson, D.W., 2005. *African archaeology.* Cambridge University Press, Cambridge.

Ruff, C.B., 1987. Sexual dimorphism in human lower limb bone structure: relationship to subsistence strategy and sexual division of labour. *Journal of Human Evolution* 16, 391–416.

Ruff, C.B., 2000. Body size, body shape, and long bone strength in modern humans. *Journal of Human Evolution* 38, 269–290.

Ruff, C.B., 2008. Biomechanical analyses of archaeological human skeletons. In: Katzenberg, M.A., Saunders, S.R. (Eds.), *Biological Anthropology of the Human skeleton. Wiley* Liss, New York, pp. 183-206.

Ruff, C.B., Scott, W.W., Liu, A.Y.C., 1991. Articular and diaphyseal remodeling of the proximal femur with changes in body mass in adults. *American Journal of Physical Anthropology* 86, 397-413.

Ruff, C.B., Holt, B., Trinkaus, E., 2006. Who's afraid of the big bad Wolff?: "Wolff's Law" and bone functional adaptation. *American Journal of Physical Anthropology* 129, 484-498.

Saunders, S.R., Popovich, F., 1978. A family study of two skeletal variants: atlas bridging and clinoid bridging. *American Journal of Physical Anthropology* 49, 193-204.

Schillaci, M.A., Irish, J.D., Wood, C.C.E., 2009. Further Analysis of the Population History of Ancient Egyptians. *American Journal of Physical Anthropology* 139, 235-243.

Seielstad, M.T., Minch, E., Cavalli-Sforza, L.L., 1998. Genetic evidence for a higher female migration rate in humans. *Nature Genetics* 20, 278–280

Shaw, C.N., Stock, J.T., 2009a. Intensity, Repetitiveness, and Directionality of Habitual Adolescent Mobility Patterns Influence the Tibial Diaphysis Morphology in Athletes. *American Journal of Physical Anthropology* 140, 149-159.

Shaw, C.N., Stock, J.T., 2009b. Habitual throwing and swimming correspond with upper limb diaphyseal strength and shape in modern human athletes. *American Journal of Physical Anthropology* 140, 160-172.

Sjøvold, T., 1973. The occurrence of minor non-metrical variants in the skeleton and their quantitative treatment for population comparisons. *Homo* 24, 204-233.

Smith, L.K., Weiss, E.L., Lehmkuhl, L.D., 1996. *Brunnstrom's clinical kinesiology*. F.A. Davis Co, Philadelphia.

Sparacello, V., Marchi, D., 2008. Mobility and subsistence economy: a diachronic comparison between two groups settled in the same geographic area (Liguria, Italy). *American Journal of Physical Anthropology* 136, 485–495.

Sparacello, V., Pearson, O.M., 2010. The importance of accounting for the area of the medullary cavity in cross-sectional geometry: a test based on the femoral midshaft. *American Journal of Physical Anthropology* 143, 612-624.

Sparacello, V., Pearson, O.M., Coppa, A., Marchi, D., 2011. Changes in skeletal robusticity in an Iron Age agropastoral group: The Samnites from the Alfedena necropolis (Abruzzo, Central Italy*). American Journal of Physical Anthropology* 144, 119-130.

Stock, J.T., Shaw, C.N., 2007. Which measures of diaphyseal robusticity are robust? A comparison of external methods of quantifying the strength of long bone diaphyses to cross-sectional geometric properties. *American Journal of Physical Anthropology* 134, 412-423.

Sutter, R.C., Mertz, L., 2004. Nonmetric cranial trait variation and prehistoric biocultural change in the Azapa Valley, Chile. *American Journal of Physical Anthropology* 123, 130-145.

Sutter, R.C., Verano, J.W., 2007. Biodistance analysis of the Moche sacrificial victims from Huaca de la Luna plaza 3C: matrix method test of their origins. *American Journal of Physical Anthropology* 132, 193-206.

Trigger, B.G., 1976. Kerma: the rise of an African civilization. *International Journal of African Historical Studies* 9, 1-21.

Turner, C.G. II, Markowitz, M., 1990. Dental discontinuity between late Pleistocene and recent Nubians. I. Peopling of the Eurafrican-South Asian Triangle. *Homo* 41, 42-53.

Tyrrell, A., 2000. Skeletal non-metric traits and the assessment of inter-and intra-population diversity: past problems and future potential. In: Cox, M., Mays, S. (Eds.), *Human osteology in archaeology and forensic science.* Greenwich Medical Media, Ltd, London, pp. 289-306.

Weiss, E., 2003. Understanding muscle markers: aggregation and construct validity. *American Journal of Physical Anthropology* 121, 230-240.

Wengrow, D., 2006. *The archaeology of Early Egypt. Social transformations in North-east Africa,* 10000 to 2650 BC. Cambridge University Press, Cambridge.

White, K., Mattingly, D.J., 2006. Ancient lakes of the Sahara. *American Science* 94, 58-65.

Williams, B., 1997. Egypt and sub-Saharan Africa: their interaction. In: Vogel, J. (Ed.), *Encyclopedia of precolonial Africa. Archaeology, History, Languages, Cultures and Environments.* AltaMira Press, London, pp. 465-472.

Williams, M.A.J., Adamson, D.A., Abdulla, H.H., 1982. Landforms and soils of the Gezira: a Quaternary legacy of the Blue and White Nile rivers. In: Williams, M.A.J., Adamson, D.A. (Eds.), *A land between two Niles: Quaternary, geology and biology of the central Sudan.* A.A. Balkema, Rotterdam, pp. 111-142.

Wilson, A., 2003. Irrigation technologies: forragas, wells and field systems. In: Mattingly, D.J., Daniels, C.M., Dore, J.N., Edwards, D., Hawthorne, J. *The archaeology of Fazzān.* Volume 1, Synthesis. Society for Libyan Studies, London, pp. 235-279.

Wood, E.T., Stover, D.A., Ehret, C., Destro-Bisol, G., Spedini, G., McLeod, H., Louie, L., Bamshad, M., Strassmann, B.I., Soodyall, H., Hammer, M.F., 2005. Contrasting patterns of Y chromosome and mtDNA variation in Africa: evidence for sex-biased demographic processes. *European Journal of Human Genetics* 13, 867–876.

Wooding, S., Ostler, Ch., Prasad, B.V.R., Watkins, W.S., Sung, S., Bamshad, M., Jorde, L.B., 2004. Directional migration in the Hindu castes: inferences from mitochondrial, autosomal and Y-chromosomal data. *Human Genetics* 115, 221-229.

Zakrzewski, S.R., 2007. Population continuity or population change: formation of the ancient Egyptian state. *American Journal of Physical Anthropology* 132, 501-509.

INDEX

A

B

C

D

E

F

G

H

I

J

K

L

M

Q

R

S

T